ROWAN UNIVERSITY
CAMPBELL LIBRARY
201 MULLICA HILL RD
GLASSBORO, NJ 08028-1701

# ABCs of FT-NMR

# ABCs of FT-NMR

John D. Roberts

Institute Professor of Chemistry, Emeritus

California Institute of Technology

University Science Books
Sausalito, California

**University Science Books**
55D Gate Five Road
Sausalito, CA 94965
*www.uscibooks.com*
Order information: Telephone (703) 661-1572
                    Fax (703) 661-1501

This book is printed on acid-free paper.

Copyright © 1990, 1991, 1993, 1994, 1995, 1996, 1997, 1998, 2000 by John D. Roberts

Reproduction or translation of any part of this work beyond that permitted by Section 107 or 108 of the 1976 United States Copyright Act without the permission of the copyright owner is unlawful. Requests for permission or further information should be addressed to the Permissions Department, University Science Books.

The publisher wishes to express his appreciation to Dr. Roberts who, in addition to writing the manuscript, prepared all of the illustrations. The author is responsible for the design, editing, and composition; final camera-ready pages were prepared at the California Institute of Technology by Dr. Roberts.

---

Library of Congress Cataloging-in-Publication Data

Roberts, John D., 1918–
   ABCs of FT-NMR / John D. Roberts.
     p. cm.
   Includes bibliographical refererences and index.
   ISBN 1-891389-18-1 (alk. paper)
   1. Fourier transform nuclear magnetic resonance spectroscopy. I. Title.
QD96.N8 R63 2000
543'.0877--dc21
                                                           00-063434

---

Printed in the United States of America
10  9  8  7  6  5  4  3  2  1

*Dedicated to Harden M. McConnell*

*A pioneer in NMR whose help and influence as a colleague and friend made much of this book possible.*

# Contents

| | |
|---|---|
| PREFACE: WHAT IS THIS BOOK ABOUT ANYWAY? | xi |

## 1

| | | |
|---|---|---|
| SOME NMR BASICS | | 1 |
| 1-1 | What Should You Know Already | 1 |
| 1-2 | A Qualitative Quantum-Mechanical View of NMR | 3 |
| Exercises | | 20 |

## 2

| | | |
|---|---|---|
| ELECTROMAGNETIC BASICS OF NMR | | 14 |
| 2-1 | Precession, Phase and Excitation | 14 |
| 2-2 | Relaxation | 23 |
| 2-3 | Determination of $T_2$. Echo trains | 29 |
| 2-4 | Determination of $T_1$. Inversion Recovery | 37 |
| 2-5 | More on the 90° Pulse | 39 |
| Exercises | | 42 |

## 3

| | | |
|---|---|---|
| HOW DO WE PERFORM THE NMR FOURIER TRANSFORM? | | 45 |
| 3-1 | The Basics of the Fourier Transform | 45 |
| 3-2 | Applying the Fourier Transform | 48 |
| Exercises | | 52 |

## 4

| | | |
|---|---|---|
| THE BLOCH EQUATIONS. CALCULATING WHAT HAPPENS IN NMR EXPERIMENTS | | 54 |
| 4-1 | The Initial Steps. Relaxation | 54 |
| 4-2 | The Next Step. Taking Into Account the $B_1$ Power Input | 55 |
| 4-3 | NMR at Equilibrium. Slow-Passage Conditions | 58 |
| 4-4 | NMR Under Non-Equilibrium Conditions. Initial Conditions | 62 |
| 4-5 | Numerical Integration Procedure for the Bloch Equations | 62 |
| 4-6 | More on CW-NMR | 66 |
| 4-7 | Resolution in CW Spectra | 68 |
| 4-8 | Rapid-Scan (Correlation) Spectroscopy. The Inverse Fourier Transform | 72 |
| 4-9 | CW Adiabatic Rapid-Passage Dispersion-Mode Spectra | 74 |
| Exercises | | 77 |

# 5

## THE NMR FOURIER TRANSFORM AND ITS PROBLEMS — 79

- 5-1 The Rise of $^{13}C$ NMR and the Need for Fourier Methods — 79
- 5-2 What Are Your Choices and the Restrictions on FT-NMR? — 82
  1. The Nyquist Frequency — 82
  2. Pulse Width — 82
  3. Pulse Shape — 82
  4. Resolution — 83
  5. Data Registers — 87
  6. Truncation — 88
  7. Dead Time — 89
  8. Folding Over — 89
  9. Quadrature Detection — 90
  10. Phasing — 92
  11. Pulse Angle and Repetition Time — 94
  12. Spectral Enhancement by Exponential Multiplication — 99
  13. Integration of NMR Spectra — 102
- 5-3 Alternatives to the Fourier Transform — 105
- Exercises — 113

# 6

## RELAXATION AND THE NUCLEAR OVERHAUSER EFFECT — 118

- 6-1 Some General Considerations of $T_2$ Relaxation — 118
- 6-2 General Considerations of $T_1$ Relaxation — 119
- 6-3 Intermolecular Mechanisms for $T_1$ Relaxation — 120
  1. Relaxation Induced by Unpaired Electrons — 120
  2. Relaxation Induced by Extramolecular Magnetic Nuclei — 122
- 6-4 Intramolecular Mechanisms for $T_1$ Relaxation — 122
  1. Quadrupolar Relaxation — 123
  2. Dipole-Dipole Relaxation — 124
  3. Decoupling of Nuclear Spins and the Nuclear Overhauser Effect — 126
  4. Relaxation by Chemical-Shift Anisotropy — 137
  5. Spin-Rotation Relaxation — 139
  6. Scalar-Coupling Relaxation — 139
- Exercises — 139

# 7

## PULSE EXPERIMENTS IN ONE-DIMENSIONAL NMR SPECTRA — 143

- 7-1 Selective Magnetization Transfer — 143

| | | |
|---|---|---|
| 7-2 | An Inept Preparation for INEPT | 147 |
| 7-3 | INEPT | 155 |
| 7-4 | Phase Cycling | 160 |
| 7-5 | Refocused INEPT | 163 |
| 7-6 | An Inept INEPT Summary | 169 |
| 7-7 | The Role of J. Multiple-Quantum Coherences | 170 |
| 7-8 | DEPT and Its Variations | 174 |
| 7-9 | Soft Pulses and Their Uses | 175 |
| Exercises | | 178 |

# 8

## NMR Spectroscopy in Two Dimensions — 182

| | | |
|---|---|---|
| 8-1 | What is a 2D Spectrum? | 182 |
| 8-2 | J-Modulated 2D Spectra | 184 |
| 8-3 | Some Techniques and Considerations for Multidimensional Spectra | 189 |
| 1. | **Phase Cycling** | 189 |
| 2. | **Cross Relaxation** | 190 |
| 3. | **Spin Locking** | 191 |
| 4. | **The Hartmann-Hahn Condition and Cross Polarization** | 193 |
| 5. | **Magnetic Field Gradients** | 195 |
| 6. | **Waltz Decoupling** | 197 |
| 7. | **Suppression of Solvent Resonances** | 199 |
| 8. | **Multiple-Quantum Coherences Revisited** | 201 |
| 9. | **Bloch-Siegert Shifts** | 204 |
| 8-4 | The Homonuclear COSY Experiment | 205 |
| 8-5 | Some Capabilities of 2D Programs | 207 |
| 1. | **HETCOR** | 208 |
| 2. | **NOESY** | 208 |
| 3. | **TOCSY** | 208 |
| 4. | **ROESY** | 208 |
| 5. | **INADEQUATE** | 209 |
| 6. | **HMQC** | 209 |
| 7. | **Proton Spectra without Spin-Spin Splittings** | 209 |
| Exercises | | 210 |

# 9

## SOME THOUGHTS ON SPIN-SPIN SPLITTING — 212

| | | |
|---|---|---|
| 9-1 | What are our Specific Objectives? | 213 |
| 9-2 | The Two-Nucleus System | 213 |
| 9-3 | Classification of NMR Spin Systems | 221 |

| | | |
|---|---|---|
| 9-4 | More Lines than Expected | 223 |
| 9-5 | Positive versus Negative *J* Values | 227 |
| 9-6 | Unexpected Splittings by Ostensibly Equivalent Nuclei | 232 |
| 9-7 | When What You See is Not Necessarily What is There | 233 |
| 9-8 | How Can We Analyze Spin-Spin Splitting Patterns? | 235 |
| | Exercises | 243 |

# 10

## SOME THOUGHTS ABOUT CHEMICAL SHIFTS — 247

| | | |
|---|---|---|
| 10-1 | Why is the Range of Proton Chemical Shifts So Small? | 247 |
| 10-2 | The Elements of the Second-Order Paramagnetic Effect | 249 |
| 10-3 | The Second-Order Paramagnetic Effect and Nitrogen Shifts | 252 |
| 10-4 | The Second-Order Paramagnetic Effect and Proton Shifts | 253 |
| | Exercises | 256 |

# 11

## MEASUREMENT OF RATES BY NMR — 260

| | | |
|---|---|---|
| 11-1 | An Example of the Determination of Rates by NMR | 260 |
| 11-2 | Derivations of Lineshapes in the Absence of Spin-Spin Splitting | 263 |
| 11-3 | Multi-Site Intermolecular Exchange Processes | 269 |
| 11-4 | Lineshape Analysis to Determine Rates of Intramolecular Processes | 271 |
| 11-5 | Lineshape Analysis for *J*-Coupled Systems | 275 |
| 11-6 | Other Methods of Determining Exchange Rates by NMR | 276 |
| **1.** | **The Carr-Purcell Technique** | 277 |
| **2.** | **Saturation Transfer** | 277 |
| **3.** | **Pulse Methods** | 279 |
| | Exercises | 281 |

## APPENDIX 1  NMR Reference Books — 286

## APPENDIX 2  A Simple Program for Calculating the Fourier Transform — 291

## APPENDIX 3  A Program for Simple Numerical Integration of the Bloch Equations — 295

## APPENDIX 4  A Quantitative Approach to Spin-Spin Splitting — 300

Exercises — 308

## INDEX — 311

# Preface: What Is This Book About Anyway?

If you are one of those people who read the first paragraph of a book before you buy it, I suggest you try to negotiate a free trial period for this book or, in effect, some kind of a money-back arrangement if you don't like it. Why? The problem is that this book has a primary purpose which is to help you, if you are indeed actually interested, to understand what is going on in Fourier Transform (FT) Nuclear Magnetic Resonance (NMR) spectroscopy. Not everyone cares, and you may be one of those. If so, don't buy, or get your money back now.

Modern life is now very largely life with "black boxes." Boxes that do wonderful things like recording and playing back video programs but, except for cleaning dust or coffee stains off the exterior, are not within our ken to maintain or repair and most often carry warning labels, "No user-serviceable parts inside." Many people find black boxes to be a quite acceptable part of modern life, at least as long as they work. The problem is how willing should we be to accept black-box analyses critical to our professional lives without some understanding of how these analyses were obtained.

It seems fair to ask if there is any basic difference between a video recorder and an NMR spectrometer. Although many may look on them as being similar, I think they are different, because a video recorder works with tapes in a very standard manner. About the only judgments you have to make relate to the tape speed and the parameters of the programs that turn the recorder on and off so that you can record something like "60 Minutes" for later viewing, while you go to a Sunday-night party. NMR spectrometers also have some standardized procedures, but it can be quite dangerous to the quality of your spectral data if you rely slavishly on such procedures for FT operation. Although relatively low-field continuous-wave (CW) NMR spectrometers are becoming increasingly passé, I will assume here that you are somewhat familiar with them. In the operation of these spectrometers, either the magnetic field or the spectrometer frequency is varied and the spectrum is recorded relatively slowly using a pen plotter that works from an analog or digital input. Usually, if improper settings of the spectrometer variables are chosen, it is apparent from the appearance of the spectra and appropriate adjustments can be made.

An FT spectrometer is different. Instead of accumulating data, feature by feature, as in CW operation, FT NMR achieves high efficiency by gathering all of the spectral information simultaneously in digitized form and then, through computer processing, a plot of signal intensity is generated as a function of frequency. If the signal is very weak, as it often is, the signal-to-noise ratio (S/N) may be too small to allow positive identification of the peaks and their relative intensities. In this situation, FT-NMR shines, provided the spectrometer has the requisite stability, because it can improve S/N greatly by accumulating spectral data repetitively over long periods of time. In effect, FT allows NMR spectrometers to act like a camera taking a picture in dim light by using a long exposure time. The NMR workload in many laboratories is so heavy that it may not be easy to justify taking a single spectrum over many hours, or even days. If this is the case, it will be very important to be able to choose the spectrometer parameters properly because some "standard" settings and/or procedures may give poor or no spectra, even if spectra are there to be had. Also, when the sample is available in sufficient quantity to require only a few scans and the resulting spectrum appears to be of good quality; if the FT parameters are poorly chosen, there can be serious problems with integrating the spectral peaks areas. For reasons such as these, I think a working knowledge of FT procedures is very important. Furthermore, despite the current trend away from CW spectrometers, CW spectra have some sterling advantages and really should be preserved as an optional operating mode on all high-quality spectrometers. Reasons for this assertion will become clearer later.

The scope of our treatment will be broad rather than deep. It is possible to provide a very detailed and precise quantum-mechanical mathematical treatment of NMR, but this provides more precision than it does insight. Much can be understood in a more qualitative way and that will be the approach taken here for the most part. I will cover only a very few NMR areas where a quantitative approach gives better understanding and, for these, simple mathematics will usually suffice.

Quite a few very good books on FT-NMR are already available. Many written by the ranking experts in the field. Why should there be still another, especially one written by a ranking nonexpert? Over many years, whenever I have wanted to obtain even a rudimentary understanding of a difficult subject, I have found it very helpful to read different explanations by different authors. When finally you can rephrase one or more of their explanations in your own words, in your own way, then you have probably achieved some level of understanding. What follows in this book will be some of those explanations I have been developing over 40 years of using NMR for a variety of chemical purposes.

# Preface

This book has been used as a text at the California Institute of Technology for a three-lecture per week course, *Chemistry 143 - Basic FT NMR Spectroscopy*, for the last ten years. It has been revised after all but two of those years to reflect suggestions from students and teaching assistants, as well as to reflect what I hope are clearer explanations of some rather complex phenomena. But one should not become overconfident about explanations, because there are many levels of "understanding" NMR. Unfortunately, at the highest level, almost all of the simple physical concepts that we employ to try to explain NMR phenomena to intelligent lay persons disappear into a fog of mathematics which is not very useful when one is either trying to introduce others to the subject or designing practical NMR experiments.

It should not be surprising that abstruse mathematics should be "the king of the hill" because NMR phenomena really fall in the realm of time-dependent quantum-mechanics and if you have ever looked into such things you will find it is not easy going and not easy to explain in qualitative terms. The interesting thing about NMR is how much that we can do to produce workable rationalizations from qualitative concepts, on the one hand, of time-independent quantum states and, on the other, of motions of classic magnetic vectors. Ultimate "understanding" comes from the mathematics of the difficult intermediate area, but there is a lot we can do without that ultimate understanding, although to be sure as the field develops the intermediate area becomes the source of a great many of the important advances.

I am deeply appreciative of the suggestions made by the students and teaching assistants, especially Louis A. Madsen who offered many helpful suggestions and also did a searching job of proofreading. Jane Ellis and Susanna Tadlock did wonders in finding inconsistencies and errors in the penultimate draft, for which I am most grateful.

August 2000               John D. Roberts

# 1
# Some NMR Basics

Before we get into discussions of the basic operation of Fourier transform (FT) nuclear magnetic spectrometers, it will be well to make clear what it is that I am expecting you to have in the way of previous NMR experience. The reason is that I do not intend to review the kind of exposure to NMR that is commonly covered in elementary organic chemistry courses. So, please read the following section that outlines what I expect that you should know already and, if there is much that you are not familiar with, or if you are not confident of your ability to use NMR for structural analysis, there are many elementary and advanced books on NMR available for study (see Appendix 1). Also, all modern textbooks of organic chemistry will have at least one chapter in which the rudiments of structure determination by NMR are sketched out.

A fine way to become acquainted with many FT-NMR techniques for structural analysis is with a computer-based problem set, such as the one pioneered by Russell and Chapman.[1] New programs of this kind often appear on the World-Wide Web. A very good example is at *http://www.chem.ucla.edu/~webnmr/*. To be sure, the spectra used in many of these self-teaching programs may have unspecified observation parameters and so are "black box" with respect to the fundamentals of FT spectrometer operation. But this is just the niche to which we hope to contribute here.

## 1-1 What Should You Know Already

I expect by the time you have started to read this book you will have had experience with NMR spectroscopy in structural analysis. At the least, I expect that you will know how to use proton chemical shifts to indicate the different varieties of protons in a sample and integrated signal intensities to inform you about the relative numbers of protons with each chemical shift. You should also have used

---

[1] A. Russell and O. L. Chapman, *FT-NMR Problems*, Alpha-Omega, Inc., 3930 Mandeville Cyn. Rd., Los Angeles, CA 90077.

spin-spin splitting patterns to map out the local environments of the protons with the observed chemical shifts. You should surely know that chemical shifts are almost invariably listed in dimensionless units, parts per million (ppm), with reference to some standard and that, to calculate the separation in Hertz (Hz) of two chemically shifted resonance peaks, you multiply the difference in ppm by observation frequency of the transmitter. Thus, a 1.70 ppm difference in chemical shift at 360 MHz corresponds to a peak separation of 612 Hz ($1.70 \times 10^{-6} \times 360 \times 10^{6}$ Hz).

You should surely also know that spin-spin splittings are reported in Hz and should never be reported in ppm, because they are independent of the spectrometer's observing frequency. You probably will also know that the resonance frequency, $\omega_0$ for NMR-capable nuclei (Table 1-1) is given by the simple equation,

$$\omega_0 = \gamma B_0 \qquad \text{Eqn. 1-1}$$

where $\gamma$ is a nuclear constant and $B_0$ is the value of the magnetic field in the immediate vicinity of the nucleus. If $\omega_0$ has the traditional units of rad sec$^{-1}$ and the magnetic field has units of Tesla, where 1 T = $10^4$ G (gauss), then $\gamma$ will have the units of rad sec$^{-1}$ T$^{-1}$ and, for protons, $\omega_0 = 2.6751 \times 10^8 B_0$ rad sec$^{-1}$. Because NMR spectroscopists customarily use Hz in their work in place of rad sec$^{-1}$, it is convenient to define

$$\nu_0 = \gamma\!\!\!/\, B_0 \qquad \text{Eqn. 1-2}$$

where $\gamma\!\!\!/ = \gamma/(2\pi)$ in analogy to the widely used $\hbar$ for $h/(2\pi)$. Now, for protons, $\nu_0 = 42.575 B_0$ MHz, where $B_0$ is in Tesla.

Further, I hope you will know, at least as catechisms, that protons with the same chemical shift do not split one another's resonances, that viscous liquids and solids give very broad lines and that rapid exchange rates can cause substantial spectral simplifications. You may also have discovered that spin-spin splitting patterns don't always follow "finger-counting" rules, in that the intensities of the components of the patterns may deviate substantially from predictions based on the binomial theorem, and that the presence of nuclei with nuclear spins ($I$) greater than 1/2, such as $^2$H and $^{14}$N, can give odd-looking or not easily observable spectra.

The areas in which you are least likely to have experience in using NMR in connection with structural analysis are nuclear magnetic relaxation, analysis of complex spectra to extract chemical shifts and coupling constants and determination of reaction rates by analysis of line shapes. Of these, relaxation is especially important to FT-NMR.

## 1-2 A Qualitative Quantum-Mechanical View of NMR

NMR phenomena can be explained in terms of quantum mechanics and also, but with some limitations, by classical electromagnetic theory. Some things are easier to understand one way, some the other way. We shall use both, with the choice at any given juncture being what is best for the situation at hand. This approach may seem paradoxical but actually works rather well.

We will start with quantum mechanics, because the property of nuclear spin, $I$, with which nuclear magnetism is associated, is a **quantized** property. Nuclear spins have half integral values: 0, 1/2, 1, 3/2, 2, 5/2, . . and so on. Table 1-1 gives the spins of a few nuclei which are important to organic chemistry. For better or worse, depending on the situation, $^{12}C$ and $^{16}O$ have $I=0$ and exhibit no NMR properties. Nuclei with $I = 1/2$: $^1H$, $^3H$, $^{13}C$, $^{15}N$, $^{19}F$ and $^{31}P$ have paramount importance in giving especially simple NMR spectra. The most widely used nuclei with $I > 1/2$ are $^2H$, $^6Li$, $^7Li$, $^{10}B$, $^{11}B$, $^{14}N$, $^{17}O$ and $^{33}S$.

Nuclear magnetism is associated with this quantized property $I$ called spin and we can understand that a charged body that is spinning could be so constituted as to have circulation of the charge around the spin axis. Circulation of charge, as for example of electric current in a coil of wire, generates a magnetic field. Each kind of nucleus to which we ascribe the property of spin has a distinctive magnetic moment depending on its particular nuclear structure and its associated nuclear angular momentum. Electrons have spin = 1/2, but also have a much larger magnetic moments than do nuclei. For each nucleus with spin, $I$, there are $2I+1$ magnetic states that have $m$, the magnetic quantum numbers equal to $I, I-1, I-2, . . -I$. Examples of the possible spin states are shown in Table 1-2.

In a magnetic field, the magnetic states of the nuclei have different energies, the difference in energy being precisely proportional to the magnetic field. In NMR spectroscopy, energy differences are normally expressed in units of Hz. These may be unfamiliar units, but for an energy change involving **one mole of nuclei**, Hz can be converted to kcal/mole by multiplying by $9.54 \times 10^{-14}$.

The energy states that the nuclei have in a magnetic field correspond to different orientations of the nuclear magnetic moments in the magnetic field are crudely analogous to the way that compass needles line up in the earth's magnetic field. The analogy is quite incomplete in that the nuclear magnets operate in three dimensions; while a compass is basically two dimensional. Also, while the magnet

## Table 1-1
Properties of Some Nuclei of Interest to Organic and Biochemists

| Nucleus | Spin | Freq., MHz (23,400 gauss) | Calcd. rel. sensitivity* | Natural abundance,% | Sensitivity corr. for abundance | $eQ \times 10^{-2}$ |
|---|---|---|---|---|---|---|
| $^1$H | 1/2 | 100.0 | (1.0) | 99.985 | (1.0) | - |
| $^2$H | 1 | 15.4 | $9.6 \times 10^{-3}$ | 0.015 | $1.5 \times 10^{-5}$ | 0.28 |
| $^7$Li | 3/2 | 38.9 | 0.294 | 92.6 | 0.272 | -4.2 |
| $^{11}$B | 3/2 | 32.1 | 0.165 | 80.2 | 0.134 | 3.6 |
| $^{12}$C | 0 | 0 | 0 | 98.89 | - | - |
| $^{13}$C | 1/2 | 25.2 | $1.6 \times 10^{-2}$ | 1.1 | $1.7 \times 10^{-4}$ | - |
| $^{14}$N | 1 | 7.2 | $1.0 \times 10^{-3}$ | 99.63 | $1.0 \times 10^{-3}$ | 2 |
| $^{15}$N | 1/2 | 10.0 | $1.9 \times 10^{-3}$ | 0.37 | $7.0 \times 10^{-6}$ | - |
| $^{16}$O | 0 | 0 | 0 | 99.76 | - | - |
| $^{17}$O | 3/2 | 13.6 | $2.9 \times 10^{-2}$ | 0.037 | $1.1 \times 10^{-5}$ | -4 |
| $^{18}$O | 0 | 0 | 0 | 0.20 | - | - |
| $^{19}$F | 1/2 | 94.0 | 0.834 | 100 | 0.834 | - |
| $^{31}$P | 1/2 | 40.1 | $6.4 \times 10^{-2}$ | 100 | $6.6 \times 10^{-2}$ | - |
| $^{32}$S | 0 | 0 | 0 | 95 | - | - |
| $^{33}$S | 3/2 | 7.7 | $2.3 \times 10^{-3}$ | 0.74 | $1.7 \times 10^{-5}$ | -6 |

*For equal numbers of nuclei at constant field. The observed values will depend on relaxation times and hence on spectrometer parameters for multipulse or multiscan spectra. Relative sensitivity, as used here, is derived from the expected integrated peak intensity, because line widths will usually be broad with nuclei having spin > 1/2.

## 1-2 A Qualitative Quantum-Mechanical View of NMR

**Table 1-2**

The Possible Magnetic Quantum States for Nuclei with Different Values of $I$

| $I = 1/2$ | $I = 1$ | $I = 3/2$ | $I = 2$ |
|---|---|---|---|
|  |  |  | $m = -2$ |
|  |  | $m = -3/2$ |  |
|  | $m = -1$ |  | $m = -1$ |
| $m = -1/2$ |  | $m = -1/2$ |  |
|  | $m = 0$ |  | $m = 0$ |
| $m = +1/2$ |  | $m = +1/2$ |  |
|  | $m = +1$ |  | $m = +1$ |
|  |  | $m = +3/2$ |  |
|  |  |  | $m = +2$ |
| $^1H, ^3H, ^{13}C,$ $^{15}N, ^{19}F, ^{31}P$ | $^2H, ^{14}N$ | $^7Li, ^{11}B,$ $^{17}O, ^{33}S$ |  |

in a compass needle can have a wide range of strengths, the strengths, or moments, of nuclear magnets are strictly quantized. Furthermore, the states of alignment of the nuclei in a field are also strictly quantized.

The nuclear magnetic energy states (see Figure 1-1) are not exactly intuitively rational. For $I = 1/2$, the orientation of a nuclear magnet in a magnetic field acts as though it can only be with, or against the field, while with $I = 1$, the possibilities include with, against, and halfway between. Single-quantum transitions between the states always involve a unit change in magnetic quantum number, $+1/2 \rightarrow -1/2$, $-1 \rightarrow 0$, and so on.

An important question is the distribution of the nuclei between the various magnetic states. For a large assemblage of nuclei of spin 1/2 in the **absence** of a magnetic field, there is no preference for $m = +1/2$ over $m = -1/2$. If we put the assemblage in a magnetic field, the state with $m = +1/2$ becomes the more stable.[2] Attainment of equilibrium is not a simple self-evident proposition because it requires that the assemblage of nuclei give up energy to the surroundings and, indeed, the equal distribution between +1/2 and -1/2 states cannot change until that

---

[2]This will only be true if the spin angular momentum vector and the magnetic moment vector point in the same direction. This is the case for most common nuclei, exceptions include $^{15}N$ and $^{17}O$.

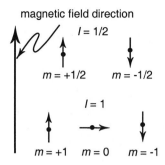

**Figure 1-1** Possible orientations of nuclear magnets in a magnetic field with spins *I* of 1/2 and 1

happens. Consider a compass; in the absence of a magnetic field, it can have any orientation. If we put it in a magnetic field, it will swing around to the field direction and then start to oscillate back and forth across the field direction. It subsequently damps down and comes to equilibrium, the energy of magnetic alignment with the field being converted to heat by friction at the needle's pivot point. This kind of process, where the energy of magnetic alignment is transferred as heat to the surroundings, is called **relaxation**. A more jargony statement applying to nuclei is **spin-lattice** relaxation, where **lattice** refers to the surroundings of the nuclei.

Because atomic nuclei are normally held by frictionless electrostatic levitation in shells of electrons, there is no mechanical connection between the nuclei and the lattice, so the mechanism of energy transfer to the lattice is not simple. Later, we will discuss a number of mechanisms for such energy transfers. For now, let it suffice that relaxation is often quite slow relative to rates of molecular collisions, rotations or vibrations and customarily it occurs over time spans ranging from tenths of seconds to seconds, although even hours are not unusual. Relaxation is related to molecular motions and it has been shown that, near absolute zero where molecular motions are very slow, relaxation may require centuries to attain equilibrium! Rates of relaxation in a given environment are normally first-order processes and can be described by first-order rate constants in units of reciprocal seconds, sec$^{-1}$. Attainment of magnetic equilibrium between different magnetic quantum states, as for the +1/2 and -1/2 states when $I = 1/2$, has a rate constant, $k_1$, and we can write Equation 1-3 for the change in $M_z$ with time.

$$\frac{dM_z}{dt} = k_1(M_o - M_z)$$

Eqn. 1-3

## 1-2 A Qualitative Quantum-Mechanical View of NMR

Here, $M_z$ is the magnetization in the magnetic field direction, that is arbitrarily taken along the Z axis of a three-dimensional Cartesian-coordinate system and $M_0$ is the equilibrium value of the $M_z$ magnetization.

Physicists apparently do not like to think in terms of rate constants, probably because they have unpleasant chemical connotations, and so they describe relaxation rates in terms of "characteristic times," which are reciprocals of the rate constants. Thus, for attainment of equilibrium between magnetic states, the **characteristic time** is $T_1$, usually in units of seconds.

$$\frac{dM_z}{dt} = \frac{M_0 - M_z}{T_1} \qquad \text{Eqn. 1-4}$$

As might be expected, many chemists defer to the physicists' superior wisdom and use $T_1$ rather than $k_1$ although, in some applications, we will see that it is less convenient to do so.

It is important to recognize that $T_1$ does not represent the half-life, the time in which any given nonequilibrium magnetization $M_z$ increases to halfway between a starting value of $M_z$ and the final equilibrium magnetization $M_0$. Actually, the half-life is 0.693 $T_1$. A very useful way to think of $T_1$ is that a starting nonequilibrium magnetization $M_z$ will increase to 0.98 $(M_0 - M_z)$ over a time period $4T_1$.

Another important NMR consideration is the value of the equilibrium magnetization. It is a fact that nuclear magnetic moments are small and, at room temperature, even in very powerful magnetic fields, only a tiny excess of the nuclei are in the more stable magnetic state. To illustrate, protons $^1H$ have the largest moment of any nucleus, except for radioactive tritium $^3H$, and, in a relatively commonly available high field for commercial NMR spectrometers (17.6 T, 750 MHz), the excess of protons in the more stable magnetic state +1/2 can be calculated for a given temperature by Equation 1-5, where $h$ is Planck's constant, $k$ is Boltzmann's constant, $B$ is the field strength and T is the temperature.

$$\frac{N(+\tfrac{1}{2})}{N(-\tfrac{1}{2})} = exp\left(\frac{h\gamma B}{2\pi kT}\right) \qquad \text{Eqn. 1-5}$$

At 25°, the % of excess in the + 1/2 state at 17.6 T is given by Equation 1-6.

$$\% \, excess \, of \, N(+\tfrac{1}{2}) = \left(\frac{N(+\tfrac{1}{2})}{N(-\tfrac{1}{2})} - 1\right) \cdot 100 = 0.012\% \qquad \text{Eqn. 1-6}$$

What keeps the percentage so small? The important factor is thermal motion of atoms and molecules within the sample, because this limits the proportion of the total population in the more stable state at equilibrium, which is a function of the energy difference between the states ε and the temperature T in kelvin (Boltzmann Principle), as in Eqn 1-7, where $K$ is the equilibrium constant and again $k$ is Boltzmann's constant.

$$\ln K = -\frac{\varepsilon}{kT} \qquad \text{Eqn. 1-7}$$

Clearly, the smaller ε and the larger T, the smaller will be the equilibrium constant. The connection between the relative populations of the magnetic states and thermal motions may become clearer through thinking about a wiggly table covered with compasses lined up in the earth's magnetic fields. If we shake the table, the alignments become disturbed, as expected for increased thermal motion. Clearly, if we could increase the earth's magnetic field or the strengths of the compass-needle magnets, it would take stronger shaking to produce the same degree of misalignment. This is the way that the ε effect operates.

Now, with the knowledge that at least a few more nuclei, say protons, occupy the state with the magnetic quantum number $m = +1/2$ than with $m = -1/2$ in a magnetic field, then we can envision an energy input that will perturb the equilibrium distribution by exciting the nuclei in the more favorable $+1/2$ magnetic state to the less favorable $-1/2$ state. The change in energy for each nucleus is ε and ε $= h\nu$, Planck's constant times the frequency, see Figure 1-2. This formulation is the analog of energy absorption in infrared and optical spectroscopy where the changes in state are associated with vibration-rotation and electronic energies, respectively. The units of wavelength used for these other spectroscopies are, of course, usually different, being microns, nm, Å and so on.

With some means for direct measurement of the absorption of radio waves as a function of frequency corresponding to the way that optical and infrared spectra are determined, the quantum-mechanical formulation would probably be an adequate way to understand the NMR phenomenon. The problem is not one of quantization, but one of measurement.

However, you might be curious as to why NMR absorptions often have quite different line widths. What a broad line is telling you is that there is a **spread** of ε values associated with the change in magnetic state from $m = +1/2$ to $m = -1/2$ shown in Figure 1-2. Let us call $\varepsilon_0$ the most probable value for the position of a peak

## 1-2 A Qualitative Quantum-Mechanical View of NMR

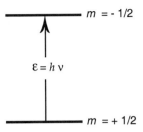

**Figure 1-2** Energy change in the magnetic state of a nucleus with spin $I$ from $m = +1/2$ to $-1/2$ on absorption of a quantum of radio frequency energy

maximum of a typical NMR absorption as shown in Figure 1-3. But you may say, we just read above that the energies of the states are strictly quantized, so how can there be a spread of energies, $\delta\varepsilon$, corresponding to a given transition? Quantum mechanics gets around this dilemma by invoking the **Heisenberg uncertainty principle** which, in this connection, tells us that the problem is not one of quantization, but one of measurement. The uncertainty principle applies to all spectroscopy, but with special cogency to NMR. The point is that if you want to make a very precise measurement of $\varepsilon$, you can only do it if the lifetimes of the states are long. In the NMR arena, it is common to investigate absorption lines which differ in energy by only 1 Hz. The uncertainty principle suggests that, to measure such differences, the lifetimes of the states involved have to be on the order of 1/6 of a second. The relationship is linear and to see ± 0.1 Hz differences requires lifetimes of 1.6 sec, which are very long times indeed compared to those for most electronic or vibration-rotation states. In contrast, if we wish to investigate

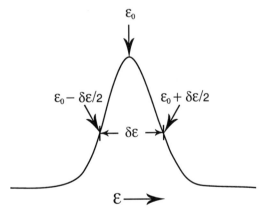

**Figure 1-3** A NMR energy absorption peak showing the uncertainty in the energy that is reflected in the line width. The most probable energy is the peak maximum, $\varepsilon_0$. By convention, the line width here in energy units $\delta\varepsilon$, is taken to be width of the absorption peak at half of its maximum height

electronic absorptions differing by only one Ångstrom in wavelength ($10^{-10}$ meters, 10 nm), the lifetime of the states can be very much shorter, only about $5 \times 10^{-20}$ sec. You can see from this that NMR deals with very small energy changes and rather long lifetimes if we wish to make accurate measurements.

As mentioned above, if we could operate with NMR as we do in optical spectroscopy by measuring the change in intensity of a beam of radio waves as it passes through a sample in a magnetic field, NMR would be experimentally much more like other forms of spectroscopy. In fact, however, the coefficients which measure absorption of radio waves are very small and, consequently, when we measure NMR absorptions, we do it by investigating the magnetic properties of the excited magnetic states. Fortunately, the lifetimes of these states are usually rather long so that the precision of the measurements can be very high. Quantum-mechanical calculations of the behavior of excited magnetic states can be made, but the mathematics involved is complex and little physical insight is gained thereby. Much better qualitative understanding comes from the classical electromagnetic theory and that's what we will turn to next.

**Exercises**

**Exercise 1-1** (This exercise requires a review of a number of basic principles of NMR as applied to the determination of organic structures.) Figure 1-4 shows two proton NMR spectra of a compound of formula $C_2H_6O$. The lower spectrum is of the pure liquid with insets showing expansions of each of the groups of lines by a factor of four in the frequency direction. The upper spectrum differs only in that a trace amount of hydrochloric acid has been added to the sample.
**a.** Account for the presence of three well-separated groups of resonance lines in these spectra and explain how we can rationalize their positions relative to one another. Then display your knowledge of spin-spin splitting to account for the multiplicities and intensities of the lines within each of the three groups and, in detail, why the spectra differ in the observed multiplicities of the left-hand and center groups, but not the right-hand groups.
**b.** What further information would you need about the resonance line positions to be able to calculate the frequency of the spectrometer used to take these spectra?
**c.** Figure 1-5 shows another spectrum of the acidified $C_2H_6O$ compound taken with a spectrometer at a much lower field strength. Here the insets represent expansion of

# Exercises

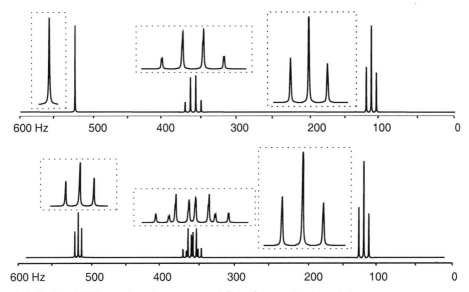

**Figure 1-4** Proton NMR spectra of a compound $C_2H_6O$, see Exercise 1-1

the frequency scale by a factor of two. Explain in as much detail as you can why this spectrum differs from those of Figure 1-4.

**d.** Figure 1-6 shows the proton NMR spectrum of the pure liquid of a different isomer of $C_2H_6O$. Explain why this spectrum shows only one strong line with no splittings. The insets on each side of the principal peak show both vertical and horizontal expansions of two quite small peaks that are multiplets, with a spacing of about 0.3 Hz between the two highest peaks. The multiplets are separated from each other by 134 Hz. Explain in knowing detail how these smaller peaks arise and why they are multiplets with very small spacings.

**e.** Figure 1-7 shows the complete proton NMR spectrum of diethyl sulfite, $(C_2H_5O)_2S=O$, calculated as it would look if it were run at 500 MHz from data obtained

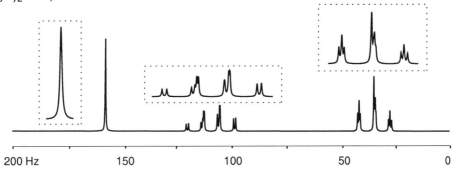

**Figure 1-5** Proton NMR spectra of a compound $C_2H_6O$, see Exercise 1-1c

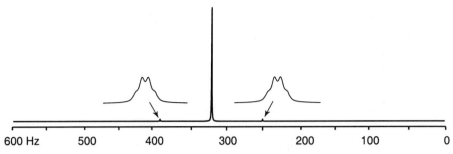

**Figure 1-6** Proton NMR spectra of a compound $C_2H_6O$, see Exercise 1-1d

at 60 MHz. The intensities of the peaks on the left, centered on 1930 Hz, are amplified by a factor of four relative to the intensities of the peaks centered at 523 Hz on the right. Explain in detail the splittings and why they arise.

**Exercise 1-2  a.** The magnetogyric ratio $\gamma$ for protons is $2.675 \times 10^4$ rad sec$^{-1}$ G$^{-1}$ where G is in units of gauss. Boltzmann's constant $k$ is $1.3806 \times 10^{-16}$ erg °K$^{-1}$ and Planck's constant $h$ is $6.626 \times 10^{-27}$ erg sec. For Equation 1-8, derive a value of the constant $C$ to enable one to calculate $N(+1/2)/N(-1/2)$ for protons from the field strength $B_0$ and the temperature, T (°K)

$$\frac{N(+1/2)}{N(-1/2)} = \exp\frac{CB_0}{T} \qquad \text{Eqn. 1-8}$$

**b.** Derive an equation for $N(+1/2)/N(-1/2)$ of the form:

$$\frac{N(+1/2)}{N(-1/2)} = \exp\frac{C\nu}{T} \qquad \text{Eqn. 1-9}$$

where $\nu$ is the resonance frequency in Hz of any nucleus in a specified magnetic field, T is the temperature (°K) and C is a constant.

**Exercise 1-3** Show how to calculate the equivalent of energy of 1 Hz in calories.

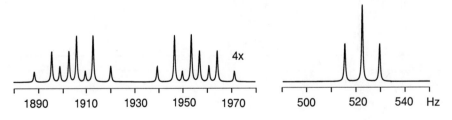

**Figure 1-7** Proton NMR spectrum of diethyl sulfite as it would appear at 500 MHz

# Exercises

**Exercise 1-4** For protons in a magnetic field of 17.6 T, toward the higher end of NMR fields now commercially available, calculate the temperature (K) at which the excess of the nuclei in the + 1/2 magnetic quantum state would be 50%.

**Exercise 1-5** The earth's magnetic field is variable with time and geography, but is on the order of one gauss ($10^{-4}$ Tesla). What would be the resonance frequency for protons in such a field? If the resonance frequency could be measured to an accuracy of 0.01 Hz, what would be the accuracy of measurement by proton NMR of changes in the earth's magnetic field with time or with location?

# 2
# Electromagnetic Basics of NMR

The reason that classical electromagnetic theory is so useful for explaining how NMR spectrometers operate is that the radio-frequency transmitter induces a coherent macroscopic magnetization from an assemblage of nuclei and, except, for some special situations to be discussed later, the behavior of the assemblage can quantitatively be described by classical methods. The key word in all of this is **phase.** The concerted action of nuclei acting in phase is of paramount importance and the concept of phase is much easier to understand in classical than in quantum-mechanical terms.

## 2-1 Precession, Phase and Excitation

As we have seen in Section 1-2, an assemblage of spin 1/2 nuclei at equilibrium in a magnetic field will have a small excess of nuclei with $m = +1/2$ and produce a small net magnetization along the axis of the magnetic field. To illustrate this we will use a cartesian coordinate system with the field axis taken to be along the Z axis. The resultant of the nuclear magnetization produced by the excess of the nuclei with $m = +1/2$ will also be directed along the Z axis. But this is not the whole story. You will remember that a spinning gyroscope when perched on its pedestal tends to precess around the vertical (the gravitational field direction) rather than simply falling over. The higher the rate of spin and the larger the resultant angular momentum, the faster the angular velocity of precession. Also the stronger the gravitational field, the greater the rate of precession. Nuclear magnets also have the property of precession in a magnetic field with the precession being around the magnetic field axis (Figure 2-1). The rate of precession $\omega_0$ is precisely given by:

$$\omega_0 = \gamma B_0 \qquad \text{Eqn. 2-1}$$

where $B_0$ is the magnetic field that is experienced by the nucleus.

The distinction between $B_0$ and $B$, which is the magnetic field of the magnet (the applied field), is very important. The distinction arises because the applied field $B$ of

## 2-1 Precession, Phase and Excitation

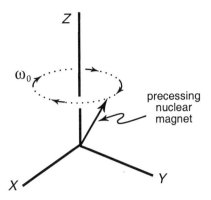

**Figure 2-1** Representation of a nuclear magnetic vector precessing around the magnetic field axis, here taken to lie along the Z axis of the coordinate system

the magnet is "filtered" by the electrons in molecules, some close or directly surrounding the nucleus of interest, some quite far away, so that the field strength $B_0$ at the nucleus can be quite different from the applied field $B$. The most common influence is the **diamagnetic effect**, whereby electron pairs surrounding a nucleus diminish the field at the nucleus by turning out the magnetic lines of force, as shown schematically in Figure 2-2. The field at the nucleus $B_0$ is the algebraic sum of all of the shielding contributions and can be expressed as:

$$B_0 = (1 - \sigma) B \qquad \text{Eqn. 2-2}$$

where $\sigma$ is the dimensionless, overall **shielding constant** for a particular kind of nucleus, such as $^1H$, in a particular chemical location. Thus, with ethanol, $CH_3CH_2OH$, (assuming both rapid tumbling of the molecule and rapid rotation about the C-C bond), there will be different values of $\sigma$ for the $CH_3$, $CH_2$ and OH protons. It should be clear that $\sigma$ is another way of expressing the chemical shift. For the hydrogens of ethanol, the $\sigma$ values are measures of the chemical shift relative to isolated bare protons in the gas phase, for which $\sigma$ is expected to be zero. It should be recognized that the $\sigma$ values for ethanol, if measured for other than gaseous ethanol, will also include contributions from whatever solvent is used. It should also be understood that, besides the diamagnetic influences mentioned above, there can be sizable **paramagnetic** contributions arising from electron circulations which tend to turn the magnetic lines of force inward toward the nucleus. Such contributions will be considered later in connection with other influences on the chemical shift. Rapid tumbling of molecules is important because shieldings are usually anisotropic. What this means is that shieldings change if the orientation of the molecule is changed in the applied magnetic field. Thus, the

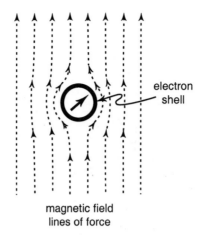

**Figure 2-2** Operation of the diamagnetic effect, where the electrons surrounding a nucleus turn the lines of force of the magnetic field outward, thus shielding the nucleus

hydrogens of ethyne (acetylene) have very different shielding constants if the molecular axis is aligned with, or perpendicular to, the magnetic field. Rapid tumbling will usually average differences in shieldings that correspond to the molecule being oriented one way or the other in a magnetic field to a single value. The direct relation of shielding constants to chemical shifts has led to common usage of "shielding" as terminology for chemical shifts. Thus, we may say "this proton is more shielded than that one" which means that, at constant frequency, resonance will occur at a **higher** magnetic field for the first proton than the second. There is a correlation problem between the current and very widely used chemical-shift scales and shielding parameters. For example, the popular TMS (tetramethylsilane) shift scale for protons has TMS at zero and all of the resonances that correspond to **less-shielded** protons are given positive shift values. This is opposite to the accepted definition of $\sigma$ and can cause serious confusion.

The nuclear magnetic vector $M$ of each nucleus precesses around the field axis in a magnetic field at the rate $\omega_0 = \gamma B_0$, but this might be taken to create a new problem because NMR spectrometers are constructed with a radio-frequency receiver coil, usually taken to have its axis along the $Y$ axis of the coordinate system, as shown in Figure 2-3. To be useful in the NMR experiment, the radio-frequency receiver must be tuned to include the frequency $\omega_0$ and this means that it will respond to an alternating magnetic field along the $Y$ axis with frequencies of alternation near to, and including, $\omega_0$. What "near" means in this connection is determined by the bandwidth of frequencies to which the receiver is set to respond. Be that as it may, it will be seen from the

## 2-1 Precession, Phase and Excitation

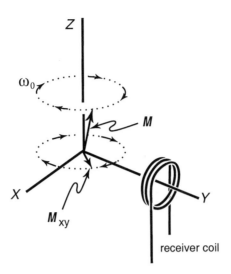

**Figure 2-3** Precession of a nuclear magnet around the magnetic-field axis and the generation of a magnetic moment in the X, Y plane

Figure 2-3 that $M$ has a component $M_{xy}$ in the X, Y plane which precesses at the angular velocity $\omega_0$. Now, in its turn, $M_{xy}$ has a component $M_y$ which will alternate back and forth along the Y axis with the frequency $\omega_0$ (see Figure 2-4). This alternating field could be expected to generate a signal in the receiver coil. But no signal will be observed from a sample placed in the spectrometer unless there is **some perturbing influence**. Why not? It is concerned with what we mentioned earlier, **phase**. A 0.25 ml

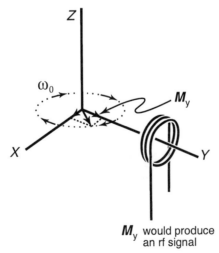

$M_y$ would produce an rf signal

**Figure 2-4** Precession of a nuclear magnet around the magnetic-field axis and the generation of an alternating magnetic moment along the Y axis to which the receiver coil will be sensitive if tuned to the proper frequency

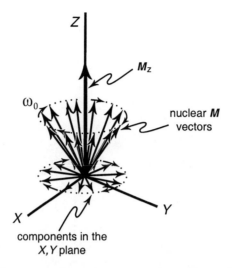

**Figure 2-5** A group of nuclear magnetic vectors precessing with no phase coherence around the magnetic field axis

sample of water, approximately the required volume for a 5-mm NMR sample tube, will contain about $1.6 \times 10^{22}$ hydrogen nuclei precessing at the same frequency, but with wholly **random phases** (Figure 2-5). This means that the $M_y$ of any one nucleus will be canceled by another nucleus which is precessing 180° out-of-phase with it. We can think of the $1.6 \times 10^{22}$ nuclear magnetic vectors as precessing with no **phase coherence** around the Z axis and producing no net rf signal, even though, at equilibrium, there can

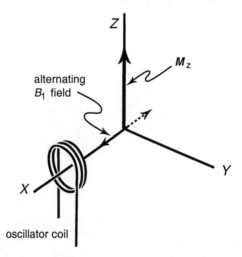

**Figure 2-6** Production of an alternating $B_1$ field by an oscillator coil placed along the X axis of the coordinate system

## 2-1. Precession, Phase and Excitation

be a net magnetization $M_z$ along the Z axis that is the resultant of the components of the individual precessing $M$ vectors.

The trick to getting an NMR signal in the receiver coil is to get the nuclei to precess with at least some degree of phase coherence, so as to have a non-zero $M_y$ component alternating at a frequency within the bandwidth of the receiver. One way to do this is to mount an **oscillator coil** with its axis along the X axis as in Figure 2-6. This coil, when connected with a suitable source of alternating current, can be made to generate an alternating magnetic field along the X axis with a frequency $\omega$ close to $\omega_0$, the nuclear precession frequency. Just how "close" $\omega$ has to be to $\omega_0$ is a very important question that we will address later.

We have seen how the placing of a nuclear magnet in the magnetic field $B_0$ causes it to precess at the frequency $\omega_0 = \gamma B_0$. If we apply a current to the oscillator coil, it will generate a magnetic field along the X axis and this will act to perturb $M_z$. The instantaneous response would be for the perturbed $M_z$ to start to precess around the X axis at the angular frequency $\omega = \gamma B_1$, where $B_1$ is the instantaneous value of the X axis magnetic field. Things get more complex, because $B_1$ will be varying in magnitude and also in direction with time. Perhaps the easiest way to visualize the interaction of $B_1$ with $M_z$ is to formulate $B_1$ in a different way. The idea is to consider $B_1$ to be comprised of two **oppositely rotating** vectors, see Figure 2-7. If we look down along the Z axis toward the X, Y plane, we see $B_1$ clockwise and $B_1$ counterclockwise, as the two components of $B_1$. There will be no detectable Y magnetization, because the two Y components are directed oppositely and cancel. So the net effect will be an alternating magnetization along the X axis. The beauty of the decomposition of $B_1$ into two components

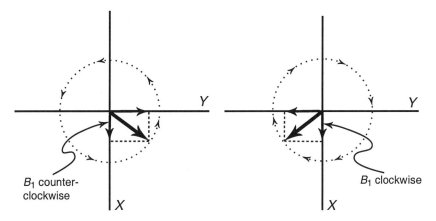

**Figure 2-7** Resolution of the alternating $B_1$ field produced by the oscillator coil along the X axis into two contra-rotating fields in the X, Y plane. The view is down the Z axis from the positive Z direction

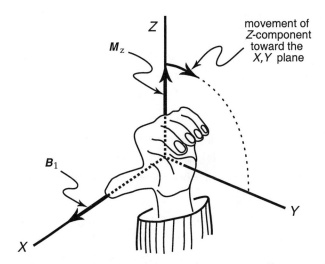

**Figure 2-8** Illustration of the "left-hand rule" for the movement of a magnetic vector under the influence of another magnetic vector at a right angle to it

lies in the influence of the separate components of the constituent nuclear vectors that produce $M_z$. Next, we need to know that a magnetic field applied in the $X$, $Y$ plane, such as $B_1$, will cause the magnetic vector of $M_z$ to tip away from the $Z$ axis in accord with a "left-hand rule". (Incidentally, physicists greatly dislike left-hand rules.) If the thumb of the left hand points in the direction of the applied field, the composite nuclear magnetic moment will tip in the direction of the fingers, see Figure 2-8. So, if $B_1$ is positive, the motion will be toward the $+Y$ direction and, if negative, in the $-Y$ direction.

Now, consider two nuclear magnet moments precessing exactly 180° out-of-phase with one another as in Figure 2-9. Clearly, in this situation, their individual $M_y$ components will cancel one another and they will give no signal in the receiver coil. We next include the clockwise-rotating $B_1$ field and ask about its effect on the nuclear moments. This will be simplest when $\omega$, the angular velocity of the $B_1$ rotating field, is exactly equal to $\omega_0$, the angular velocity of precession. This means that there will be **constant phase angles** between the rotating field and the precessing moments. The applied field will cause the $Z$ components of both of the nuclear moments to tip away from the $Z$ axis. Note that, even though the nuclear moments are precessing 180° out-of-phase, their $Z$ components, by the left-hand rule, tip in the **same** direction with respect to the applied rotating field. As long as $\omega = \omega_0$, the clockwise-rotating $B_1$ will remain just 90° ahead of the tipping components and will cause further tipping as time unfolds.

## 2-1 Precession, Phase and Excitation

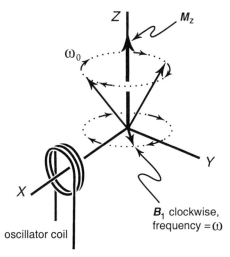

**Figure 2-9** Two nuclear magnetic moments precessing exactly 180° out-of-phase with one another

If at the end of a brief period, we turn off the applied $B_1$ field, the composite moment will stop tipping, but will still precess about the Z axis with the angular velocity $\omega_0$, see Figure 2-10. This moment will have a nonzero $M_y$ component that will oscillate back and forth at the frequency $\omega_0$ and hence give a signal in the rf receiver. Now, $M_z$ the component along the Z axis will be less than $M_{z0}$ and will approach $M_{z0}$ with the time constant $T_1$, (see Section 1-2). The vital element here is that the rotating $B_1$ field will cause **all** of the magnetic nuclei to tip away from the Z axis in unison, even though their individual precessions are not in phase with each other around the Z axis.

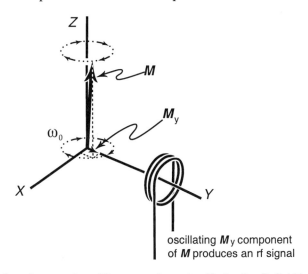

**Figure 2-10** The continued precession of the magnetic vector **M** after the $B_1$ field is shut off

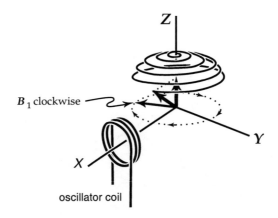

**Figure 2-11** Representation of the precessional motion of a nuclear magnetic moment under the influence of a $B_1$ field with diminution of the component along the Z axis

However, only the excess with $m = +1/2$ will result in a signal. There will be a very large, and equal, number with $m = +1/2$ and $m = -1/2$ that will cancel one another's $M_y$ component, irrespective of how they are tipped by the rotating $B_1$ field.

The counterclockwise rotating field of Figure 2-7 will be ineffective in tipping the nuclear moments toward the Y axis, because the precessions are in the clockwise direction and there is no phase coherence between the nuclear magnets and the counterclockwise-rotating field. However, the counterclockwise rotating $B_1$ field can contribute to the $B_0$ field and this causes the so-called **Bloch-Siegert effect**, to be discussed later.

If $\omega = \omega_0$ and we do not turn off $B_1$, the clockwise rotating $B_1$ field will stay juxtaposed to continuously affect the nuclear moments and what was originally the Z component of the nuclear magnetization will spiral downward toward the X, Y plane as shown in Figure 2-11. If we leave the $B_1$ field on long enough, the magnetization of the nuclei will reach the X, Y plane and make a 90° angle to the Z axis. The time required for this to happen is called the **90°-pulse time** or more often just **90°-pulse**. For FT operation, as we shall see later, oscillators are made powerful enough to have 90° pulse to be on the order of a few **microseconds**. It is interesting to contemplate how many turns there are in the spiraling down to the X, Y plane over such a 90° pulse. Suppose that the 90° pulse is a brief 4 μs and the spectrometer is one with a 600-MHz proton frequency. Then there will be a total of 2400 turns over the 90° pulse time! This could make you dizzy to watch and certainly is hard to depict in a drawing. The way to get around the problem is to use a **rotating frame of reference** with its rotation frequency equal to $\omega_0$. If we sit on a horse in this merry-go-round frame of reference (Figure 2-12) and follow the progress of the composite M vector, it will be seen over the 90° pulse time

## 2-2 Relaxation

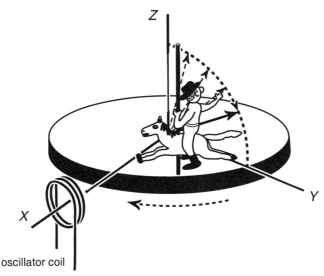

**Figure 2-12** Depiction of the progress of a 90° pulse in a rotating frame of reference. An observer seated on a merry-go-round horse rotating with the frame would see a steady progression of the magnetic moment toward the X, Y plane, but an observer in the laboratory frame would see a spiral precession downward as in Figure 2-11

to move smoothly downward to the X, Y plane. The rotating frame of reference makes it easy to visualize many FT-NMR concepts, but it should not be forgotten that the rotating frame is not the "real world" and that mathematics derived for the rotating frame may require correction to be used in the laboratory frame of reference. Normally, the rotating frame is taken to rotate at the oscillator frequency.

## 2-2 Relaxation

Let us now consider what happens when the oscillator is turned off. In a perfectly homogeneous field with no chemical-shift differences, the nuclei will all have the same precession frequency. This means that their moments will stay in phase and $M$, when viewed in the rotating frame, will relax back to the Z axis, see Figure 2-13. In the absence of other perturbing factors (to be discussed shortly), the rate of relaxation will be consistent with $T_1$ (see Section 1-2). There will be a component of $M$ along the Y axis that will diminish as $M$ becomes coincident with $M_z$. Viewed in the laboratory frame of reference, the Y component of $M$ will oscillate back and forth at the frequency $\omega_0$ and the resultant will induce an alternating current in the receiver coil at the frequency $\omega_0$. This current will decrease with time as $M$ moves back to the Z axis. The decay of the current is not easily displayed on an oscilloscope screen when its period of oscillation is

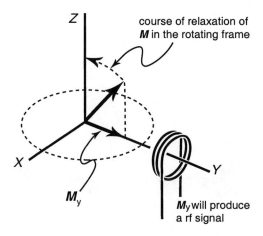

**Figure 2-13** Relaxation of a nuclear magnetic vector **M** in the rotating frame. The $M_y$ component, which oscillates in the laboratory frame of reference, will produce a signal in the receiver coil

as high as 600 MHz. However, if an alternating current is mixed in having a frequency, say 1,000 Hz different from $\omega_0$, then a decaying beat frequency will be produced which is easily viewed. If the oscillator is turned off, the decaying signal is called the **free induction decay** or **FID**, see Figure 2-14.

The free induction decay is a time-dependent phenomenon. It is said to be in the **time domain**. It is a plot of signal intensity vs. time. Of course, NMR spectra are usually thought of as belonging to the **frequency domain**, that is they are plots of signal intensity vs. frequency. This is true of FT-NMR, as we shall see. However, most continuous-wave (CW) spectra do have time-dependent features, such as relaxation wiggles that change with how fast the spectra are run, as will be discussed in Chapter 4.

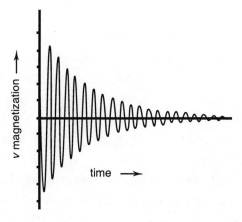

**Figure 2-14** A typical free-induction magnetization decay (FID), the intensity of which for reasons to be discussed later corresponds to what is called *v* magnetization

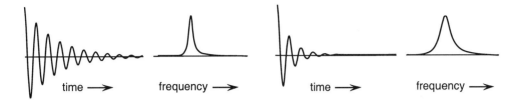

**Figure 2-15** The relationship between rates of decay of FIDs in the time domain and the width of the resulting line in the frequency domain. A long decay gives a sharp line and a short decay a broader line

In the usual FT-NMR, the signal source is almost always a free induction decay (a particular exception will be mentioned later). The magnetization vectors may be massaged in very elaborate ways before taking the FID and, even while taking the FID, such as by radio-frequency irradiation of other magnetic nuclei to either simplify or produce more information in the spectra. Whatever, the result is a **time-domain** signal from which we have to extract frequency information. The most common way of doing that is by the **Fourier transform** (FT, Chapter 3) which is fundamentally a mathematical filter. It searches through the FID for frequency information and allows a plot of signal intensity vs. frequency to be generated. Where the frequency information is imprecise, the frequency peaks will be broad and where it is well defined, the peaks will be sharp.

One reason for broadness or narrowness of the peaks is simple. Suppose a FID is very short, say that it decays so fast that we can observe it for only half of a cycle. Obviously, we cannot know its frequency very accurately, so when we convert its intensity information as a function of time (time domain) to the frequency domain, it will have to be represented by a spread of frequencies, in effect a broad line, with the most probable frequency being the peak of the plot of intensity vs. frequency. Contrariwise, if the FID decays very slowly over many cycles, we can determine its frequency very accurately, and the transform of the FID into the frequency domain will give a sharp peak as shown in Figure 2-15. Just how the Fourier transform does this mathematically will be addressed in Chapter 3.

At this point, we could go in several directions, but it seems desirable to stay with how fast FIDs decay. Just a bit earlier, it was suggested that the FID could decay at the same rate as $M$ moves back to coincide with the Z axis. This can be the case, but is not usually so, because other factors intervene. An especially important effect is associated with the fact that the applied magnetic field is almost never so homogeneous that its lack of homogeneity does not influence the decay of a FID. What this means is best illustrated by a specific example.

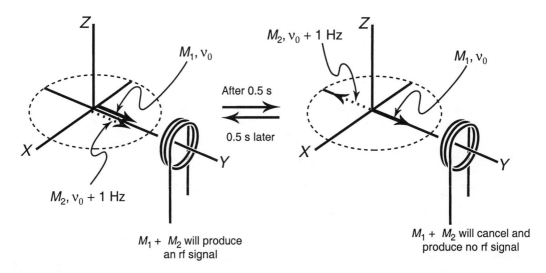

**Figure 2-16** Behavior of two magnetic moments $M_1$ and $M_2$ in the X, Y plane with different precession frequencies that start off in phase and get 180° out-of-phase in 0.5 s

Suppose that in a magnet field, we have a sample divided into two perfectly homogeneous compartments between which the precession frequency differs by exactly 1 Hz. Further, let us suppose that the $T_1$ relaxation time for the protons in both sample compartments is 1 sec. Call the frequency in one compartment $\nu_0$ Hz and the other ($\nu_0$ + 1) Hz. Now, if the 90° pulse takes only 1 µs, the Y components of magnetization in the two compartments will start out essentially completely in phase, see Figure 2-16. If we view the two Y-components in a rotating frame with a rotation rate $\nu_0$ then, in 0.5 second, the Y component of the faster precessing moments will be 180° out-of-phase (**antiphase**) with the slower component. At this point, the signal (which is the net of the two $M_y$ components) will be **zero**. After an additional 0.5 sec, the components will come back into phase again and the overall signal will have just the strength dictated by the diminution required by $T_1$ relaxation. As time goes on further, the signal will wax and wane until $T_1$ relaxation is complete, as shown in Figure 2-17.

Obviously, the sample in a real magnet will not be conveniently divided into two homogeneous compartments. The inhomogeneities of a real magnet are expected to be unevenly distributed in both magnitude and in space. The result is that nuclear moments in different parts of the sample will get out-of-phase with one another more or less randomly. This means that the signal can decrease very much faster than the limit dictated by $T_1$, just because of imperfections in the magnet field. The decay of the signal will look to the observer as relaxation. But it is different from $T_1$ relaxation

## 2-2 Relaxation

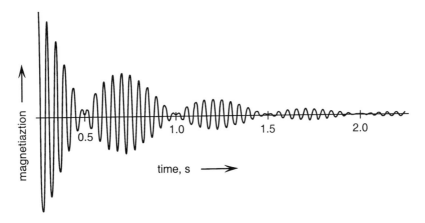

**Figure 2-17** FID corresponding to the signal expected from nuclei with the frequency differences of Figure 2-16

because, if getting out of phase is fast compared to $T_1$ relaxation, the $M_{xy}$ components of the nuclear magnets, shortly after a 90° pulse, could become distributed randomly in the X, Y plane, before any significant $T_1$ relaxation could take place. This means that we need another time constant for the actual decay of the signal. If we simply observe the overall signal decay, this usually turns out to be a first-order process. When the magnetic field is not homogeneous, the characteristic time for decay of the signal is called $T_2'$ and we can write either Equation 2-3 or 2-4.

$$\frac{dM_{xy}}{dt} = -\frac{M_{xy}}{T_2'} \qquad \text{Eqn. 2-3}$$

or

$$\frac{dM_y}{dt} = -\frac{M_y}{T_2'} \qquad \text{Eqn. 2-4}$$

The rate constant for decrease of $M_y$ (the signal-producing part of $M_{xy}$) will be $k_2' = 1/T_2'$. It will be made up of two components; the rate which results from inhomogeneities in the magnetic field, as just described, and the rate intrinsic to the particular sample at the particular temperature and magnetic field strength (Equation 2-5).

$$\frac{1}{T_2'} = \frac{1}{T_2(magnet)} + \frac{1}{T_2(sample)} \qquad \text{Eqn. 2-5}$$

The component associated with the field inhomogeneities will vary from day-to-day, if not minute-to-minute, and is of no particular chemical interest. One way to reduce the

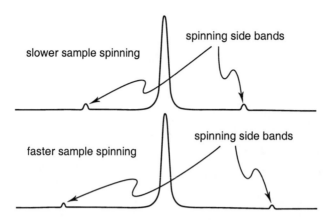

**Figure 2-18** Spinning side bands on NMR spectral peaks are caused by modulation of the signal by the audio frequency of spinning when there are inhomogeneities in the applied magnetic field. They are often mistaken for real peaks and they show up much more clearly on sharp peaks than broad ones. Whenever there is doubt about whether or not a peak is a side band, it is easy to distinguish them by the fact that they change position relative to the center peaks by simply changing the rate of spinning

effect of magnetic-field inhomogeneities is by **spinning** the sample. Spinning can have a sizable averaging influence, but can also make life more difficult by producing **spinning side bands** (see Figure 2-18). The side bands result from audio modulation as parts of the sample experience periodic fluctuating fields at the spinner frequency and the resulting resonances may be large enough to appear to be chemically significant. Spinning side bands may be distinguished from the chemically significant peaks by changing the spinning rate.

The rate of loss of signal inherent in the sample itself is designated by the characteristic time $T_2$ and is very much a chemically interesting property, even if it and $T_1$ are also both functions of the magnetic-field strength. It may seem obvious that $T_2$ is not expected to be longer than $T_1$, because one would not expect that there could be any $X$, $Y$ magnetization left to observe if $M$ were to decay so far as to come to coincide with the $Z$ axis. Thus, $T_2$ seems reasonably restricted to the range of values, $0 < T_2 \leq T_1$. However, while this simple expectation corresponds to what is normally observed, cases are known where there is strong experimental evidence that $T_2 > T_1$.[1]

In the real NMR world, the decay of a FID almost always has some component arising from field inhomogeneity and this has to be separated out if we actually are to determine the **intrinsic** $T_2$ that is a property of the sample itself (here assumed to be a

---

[1] F.A.L. Anet, D.J. O'Leary, C.G. Wade and R.D. Johnson, *Chem. Phys. J.*, **171**, 401 (1990)

## 2-3 Determination of $T_2$. Echo Trains

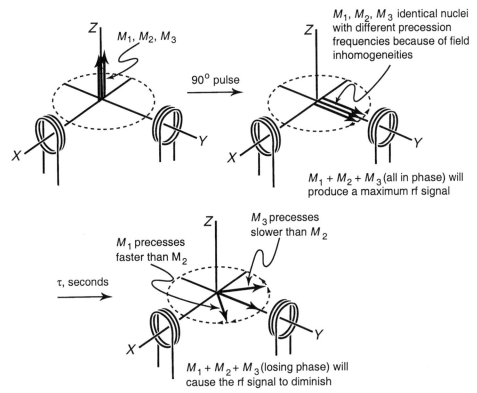

**Figure 2-19** The start of a Carr-Purcell spin-echo sequence in an inhomogeneous magnetic field, as seen in the rotating frame of reference. The example involves three nuclear moments, one that precesses at the frequency $\omega_0$, another that precesses slightly faster and a third that precesses somewhat slower. The initial state is seen at the upper left, a short 90° pulse is given and after a time $\tau$ the moments lose a measure of phase coherence. At this point, the nuclei are given a 180° pulse as described below

material like water with just one NMR frequency). We now turn to seeing how that can be done.

### 2-3 Determination of $T_2$. Echo Trains

How do we disentangle the rate of loss of signal arising from $T_2$ relaxation that is inherent in the sample and the rate which arises from the inhomogeneities of the magnet field? It is experimentally quite easy. For simplicity, let us suppose that $T_2' \ll T_1$ and $T_2 < T_1$. Clearly, with these constraints, signal decay will be fast relative to the reestablishment of magnetic equilibrium along the Z axis. Now, consider a system comprised of three nuclear moments that have slightly different precession frequencies,

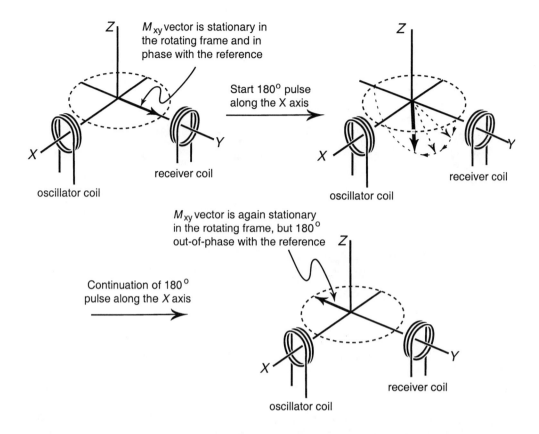

**Figure 2-20** The effect of a 180° pulse applied along the X axis on a nuclear magnetic moment that is precessing in the X, Y plane in phase with the reference as seen in the rotating frame

because of variations in the field over the sample volume. Next, let us enter the rotating frame and give these moments a quick 90° pulse. (If you are a worrier it may occur to you that there could be a problem with getting a clean 90° pulse for each one of an assemblage of nuclei having different precession frequencies. This is a legitimate concern and we will address it farther along. However, there is actually no problem for what follows, provided that the 90° pulse requires only a few µs.) The resulting Y components for the three nuclei just after the 90° pulse will produce a signal in the receiver coil. However, the initially observed Y signal is the resultant of the three individual signals from the nuclear moments, as shown in Figure 2-19. But, we know that these individual moments do not have exactly the same precession frequencies, because of magnetic-field inhomogeneities. Let one of the nuclear moments have the average precession frequency $\omega_0$, let another precess faster and the third precess slower than one at $\omega_0$. Viewed in the rotating frame, the moments will spread out in the X, Y

## 2-3 Determination of $T_2$. Echo Trains

plane; one will go ahead of the moment having the frequency $\omega_0$, while the another will lag behind that moment. The resulting decrease in phase coherence of the moments causes the overall signal intensity to diminish.

Suppose we let the nuclei lose phase over a time period $\tau$. Then, we give the nuclear moments a **180° pulse** along the X axis. The behavior of a nuclear magnet in a 180° pulse is of great importance and let us digress briefly to be sure that you understand what the 180° pulse does. What happens with a 180° pulse to a nuclear magnet that is precessing in phase with the reference (Figure 2-20) is quite straightforward, because it is simply a continuation of the 90° pulse shown in Figure 2-12. Thus, a 180° pulse causes the nuclear magnet (viewed in the rotating frame) to turn around the X axis in the Y, Z plane, go below the X, Y plane, to the -Z direction and, at the completion of the pulse, end up back to the X, Y plane in the -Y direction.

The behavior of a nuclear magnet that is out-of-phase with the reference when pulsed is often found to be confusing. It is most easily understood if the magnetic moment of the nucleus is resolved into its components along the X and Y axes (see Figure 2-21 which uses as an example a moment M that, over the time period $\tau$, has come to be about 45° ahead of the reference). By the "left-hand rule", the component along the X axis will be wholly unaffected by a pulse along the X axis. However, the Y component by definition can only be either in-phase, or 180° out-of-phase, with the reference and, if in-phase, will behave in exactly the same way as the Y component shown in Figure 2-20. So after the 180° pulse, we can reconstitute the moment from its components as shown in the lower part of Figure 2-21.

A very important question is what a 180° pulse along the X axis will mean to the future phase relations of the rotated moments. Clearly, moments that start out being exactly in-phase with the reference will, after a 180° pulse, be 180° out-of-phase as seen in Figure 2-20. Those moments will remain 180° out-of-phase until another 180° pulse is applied. For the magnetic moment depicted in Figure 2-21, you will need to convince yourself that nothing occurring in the 180° pulse will change the precession frequency of the nucleus involved. The reason is that the precession frequency does not depend on the orientation of the nuclear moment in space, rather it is a property of the nuclear moment, the strength of the applied magnetic field and the various external contributions to the electronic shielding. The message then is that a nuclear moment, which is not stationary in the rotating frame, will continue to "speed" or "lag" at the same angular velocity, after the 180° pulse, as before the 180° pulse.

Figure 2-19 shows what happened to our system of three nuclei in the initial 90° pulse and $\tau$ period. Before the 180° pulse, $M_1$ was moving ahead of the reference,

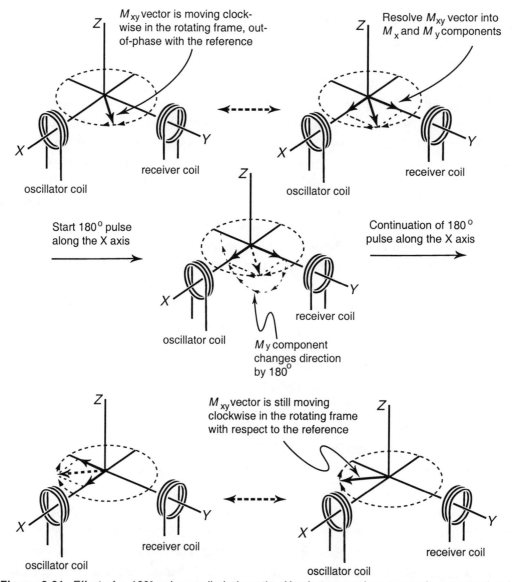

**Figure 2-21** Effect of a 180° pulse applied along the X axis to a nuclear magnetic moment that is precessing in the X, Y plane not in phase with the reference

$M_2$ is in-phase with the reference and $M_3$ is lagging behind. After the 180° pulse (see Figure 2-22), the "lagging" and "speeding" nuclear moments start heading toward one another, rather than spreading farther apart. So, in the same time period $\tau$, they will come back into phase and cause what is called a **spin echo**. In the process, the signal will regain strength, but in the negative sense; because, at the echo, the Y moment will

## 2-3 Determination of $T_2$. Echo Trains            33

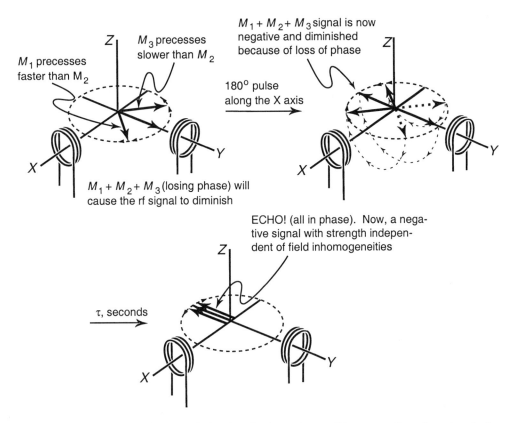

**Figure 2-22** Continuation of Figure 2-19 showing the later stages of the generation of a spin echo in an inhomogeneous magnetic field by the 90°-τ-180°-2τ-180°-2τ-180°- - (Carr-Purcell) pulse sequence. At the upper left, the moments have lost a substantial degree of phase coherence, following the 90° pulse and the time period τ. They are then given a 180° pulse (with the result as in the upper right). Then, after another τ interval, the moments regain phase and give a echo which is a negative signal (below)

be 180° out-of-phase with the rotating-frame reference. The difference between the magnitude of the echo and the original signal (except for possible diffusion effects to be described shortly) represents operation of the inherent $T_2$ decay of the sample nuclei. The reason is that the inhomogeneities in the magnetic field cause the moments to differ from $\omega_0$ and fan out during the time τ. And it is the same inhomogeneities, after the 180° pulse, that cause the moments to catch up with one another and, in so doing, cancel out the effect of those inhomogeneities. If, after the echo, we wait for the time τ the nuclear moments will again lose a measure of phase coherence and, provided $T_1$ is not so short that the magnetization along the Z axis has reached its equilibrium value, we can repeat the 180° pulse and observe an echo again after waiting another increment of time τ. The sequence of waiting and pulses 2τ - 180° - 2τ - 180° - 2τ - 180°- - - - can be re-

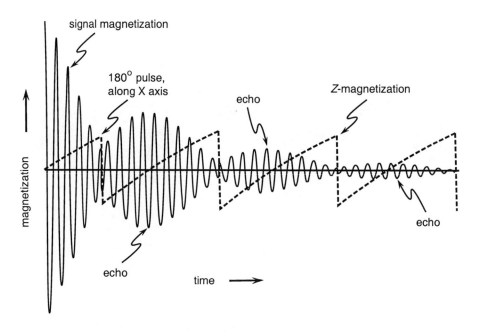

**Figure 2-23** The free-induction decays in the Carr-Purcell 90°-τ-180°-2τ-180°-2τ-- sequence. Note how much faster the initial signal decays than the train of subsequent echo peaks, how the echoes alternate with negative and positive goings and how, after the first period τ there is no overall change of Z magnetization. The signal disappears, because of the $T_2$ decay inherent in the sample. This plot was developed using a computer simulation based on the Bloch equations (to be described later) which involves the sum of the magnetizations of twenty nuclei with a root-mean-square spread in chemical shifts of 0.65 Hz to simulate a non-homogenous magnetic field. The parameters for this calculation were a mean chemical shift of 20 Hz from the reference frequency, $T_1 = 1$ s, $T_2 = 0.5$ s and τ = 0.25 s

peated as long as any component of X, Y magnetization remains. Of course, the echoes will alternate as positive and negative signals. This procedure for separating $T_2$ from $T_2'$ involves what is called the **Carr-Purcell echo train,** and its sequence of pulses, waiting periods and echoes is shown in Figure 2-23. The decay envelope of the absolute values of the peak echo intensities in Figure 2-23 allow calculation of $T_2$ of the sample using Equation 2-6, where $S_0$ is the intensity of the signal after the 90° pulse and $S(t)$ are the absolute values of the echoes at time t.

$$\ln S(t) = \ln S_0 - \frac{t}{T_2} \qquad \text{Eqn. 2-6}$$

The drawbacks in the Carr-Purcell approach are that the echoes alternate in phase with respect to the reference and there is also the possibility of complication by diffusion of the nuclei. How does diffusion complicate matters? Consider the two-

## 2-3 Determination of $T_2$. Echo Trains

compartment sample discussed in Section 2-2 that was assumed to have homogeneous fields in each compartment, but a 1-Hz difference in precession frequency between the nuclei in each compartment. If you followed the discussion there, you will see that, in 0.25 sec, after the 90° pulse, the moments in one compartment will be just 90° out-of-phase with those in the other. If, at this point, a 180° pulse is applied over a few μs, the moments will come back into phase in another 0.25 sec and give a negative echo.

Now, let us suppose that the nuclei in the compartments begin to diffuse from one locale to the other during the 0.25-sec period. As each nucleus crosses the divide, it will lose its phase relationship with the nuclei it left behind and acquire new phase relationships (and a new precession frequency) in its new environment. Depending on how early in the τ period it migrates, it will be out-of-phase to a variable extent with the reference at time τ and so, after the 180° pulse is applied, the echo at time τ will be substantially blurred. As a result, the intensity of the echo signal will depend on $T_2'$ rather than $T_2$. Why $T_2'$? Because diffusion causes migration of a nucleus in one part of an inhomogeneous magnetic field into a different one. We can greatly reduce the influence of the inhomogeneities if we make τ a short time. The reason is that, if τ is short, the molecules will have only a little time to migrate to new volume elements of the field where their nuclei would have different precession frequencies. Just how short τ will have to be is determined both, by how inhomogeneous the field is, and also by how fast diffusion is. The more homogeneous the field, the smaller the problem presented by diffusion, which will usually be expected to be most important for small molecules in nonviscous solvents.

The other problem with the Carr-Purcell sequence is that the signal alternates between positive and negative goings which is inconvenient, particularly with short τ values. A more useful technique is the **Carr-Purcell-Meiboom-Gill sequence** (see Figure 2-24) in which the 90° pulse is applied along the X axis, but the 180° pulse is applied along the Y axis, with the receiver turned off during the pulse. A $180°_y$ pulse will cause the moments to move toward the +Z direction, arc around the Y axis and wind up again in the X, Y plane. Here, the speeding and lagging moments will again be heading toward one another but now, in contrast to the $180°_x$ Carr-Purcell pulse, a **positive** echo will result (see Figure 2-25). Of course, the receiver is turned back on after the $180°_y$ pulse so that the echo can be detected. This procedure can be used with very short τ values, but the $180°_y$ pulse times need to be accurately timed. The advantage of the procedure, besides the all-positive echoes, is that short τ values alleviate the diffusion problem by effectively starting over with a different distribution of nuclei in the magnetic field inhomogeneities for **each** echo.

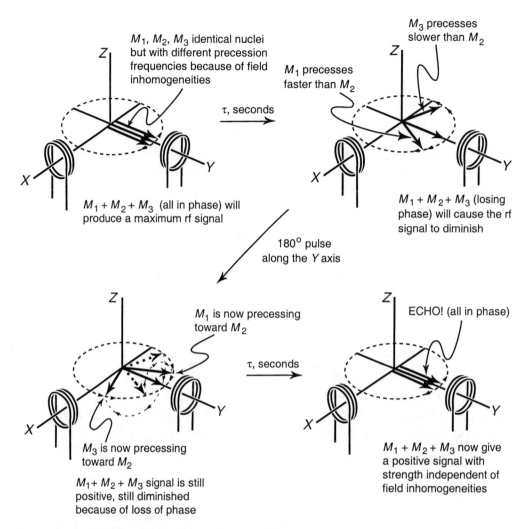

**Figure 2-24** Generation of a Carr-Purcell-Meiboom-Gill spin echo in an inhomogeneous magnetic field by a $90°\text{-}\tau\text{-}180°_y\text{-}2\tau\text{-}180°_y\text{-}2\tau\text{--}$ pulse sequence. The example involves the three nuclear moments of Figure 2-20 to 2-23, one that precesses at the frequency $\omega_0$, another that precesses slightly faster and a third that precesses slightly slower. At the upper left, the 90° pulse has just finished. At the upper right, after a time $\tau$ the moments are given a $180°_y$ pulse with the result shown at the lower left, where the moments have been caused to rotate 180° about the $Y$ axis. Then, after a renewed time $\tau$ the moments come back into phase and give a positive echo

Diffusion of nuclei in a randomly inhomogeneous magnetic field is a pain in the behind when one wants to determine accurate $T_2$ values but, as is often the case, can be turned to advantage when one knows $T_2$ but wants to measure diffusion rates. To do this, one applies a known **uniform field gradient** and compares the $T_2'$ measured with gradient on with $T_2$ measured with gradient off. From this comparison, it is possible to

## 2-4 Determination of $T_1$. Inversion Recovery

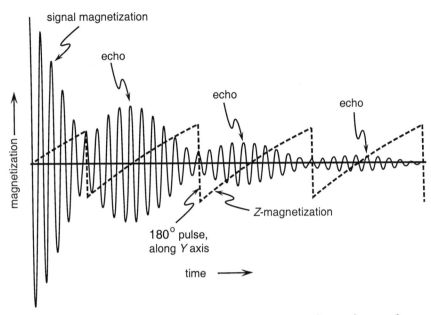

**Figure 2-25** Free induction decays in the Carr-Purcell-Meiboom-Gill $90°$-$\tau$-$180°_y$-$2\tau$-$180°_y$-$2\tau$- - pulse sequence. Note how in comparison with Figure 2-23, the echoes all have positive signs, and again how there is no overall loss of $Z$ magnetization, after the first period $\tau$. This plot was developed by a computer simulation in the same way as for Figure 2-23 with the same parameters

calculate diffusion rates.[2] Because $T_2$ is a measure of the rate of relaxation in the X, Y plane, it is called the **transverse relaxation time**. This contrasts to $T_1$ which is a measure of the rate of relaxation along the Z axis and is called the **longitudinal relaxation time**. Some of the factors that influence $T_1$ and $T_2$ are discussed in Chapter 6.

## 2-4 Determination of $T_1$. Inversion Recovery

Values of $T_1$ can be determined in several ways. The simplest and easiest to understand is called **inversion recovery**. The first step is to give an assemblage of nuclear magnets at equilibrium along the Z axis a 180° pulse. If the pulse is accurately 180°, there will be no X, Y magnetization at all and there will be no signal in the receiver coil. The subsequent decay of the -Z component upward along the Z axis will proceed exactly as expected on the basis of $T_1$. There will be no phase problems, because there will be no X, Y magnetization. But, of course, there is no signal to use to quantify $T_1$. The solution to that problem is to give a 90° pulse at particular intervals after the

---

[2] E.D. Stejskal and J.E. Tanner, *J. Chem. Phys.* **42**, 288-292 (1965).

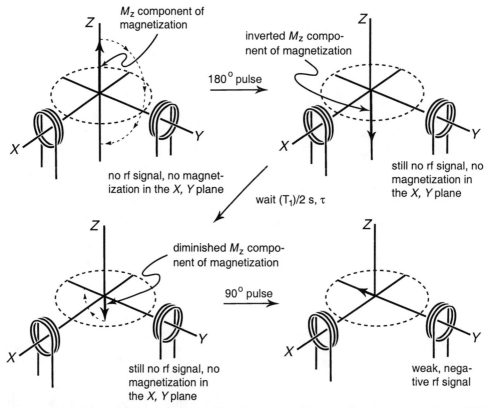

**Figure 2-26** Inversion-recovery sequence for determination of $T_1$. The 180° pulse (upper left) inverts the Z magnetization without generating any X, Y magnetization (upper right). During some desired time period τ here taken to be $(T_1)/2$, the Z magnetization returns by $T_1$ relaxation, at least partially, to equilibrium and can be measured by a 90° pulse that turns the magnetization into the X, Y plane. The sequence is therefore 180°-τ-90°. Then one can repeat with a new value of τ to allow construction of a graph of the change of the Z magnetization with time as in Figure 2-28

original 180° pulse. If the magnetization is in the -Z direction, when the 90° pulse is applied, a negative signal will result. As the Z magnetization passes through zero, no signal will be produced and, when it is in the +Z direction, a positive signal will be observed, see Figures 2-26 and 2-27. The inversion-recovery sequence is thus 180° - τ - 90° with measurements of the signal strength after the 90° pulse to be made with different values of τ. The curve (see Figure 2-28) of signal amplitude $S_\tau$ as function of τ is defined by Equation 2-7.

$$\frac{dS_\tau}{d\tau} = k_1(S_0 - S_\tau) = \frac{1}{T_1}(S_0 - S_\tau) \qquad \text{Eqn. 2-7}$$

where $S_0$ is the magnitude of the signal produced when relaxation is complete.

## 2-5 More on the 90° Pulse

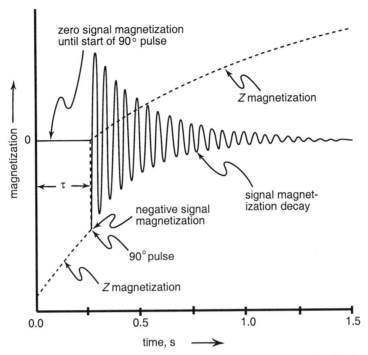

**Figure 2-27** The course of the 180°-τ-90° inversion-recovery pulse sequence for the determination of $T_1$. The 180° pulse takes place over a few microseconds at time 0. Then, the resulting negative Z magnetization starts its climb back to equilibrium. Now, after a chosen interval τ here taken to be 0.25 s, a 90° pulse turns the Z magnetization into the X, Y plane and the magnitude of the resulting Y magnetization is recorded as the measure of the degree of approach to equilibrium. Following the 90° pulse, the Z magnetization resumes its approach to equilibrium, now starting from 0. In this example, $T_1$ was 1 s, $T_2$ was 0.5 s and the magnetic field was taken to be slightly inhomogeneous so that $T_2 > T_2'$

### 2-5 More on the 90° Pulse

Earlier, I alluded to the possible problem of getting a perfect 90° pulse in the rotating frame from nuclear moments that have different precession frequencies because they have different chemical shifts, or because they are associated with different kinds of nuclei with different magnetogyric ratios γ (Section 1-2). Then, I sought to allay your possible worries on this score by suggesting that, if the 90° pulse was very short, this would not be a problem, at least not a problem if the frequency differences are small. I think you can recognize when to expect difficulties by applying the following simple analysis.

Suppose the 90° pulse takes 10 μs, which is not an unusual value in the rather commonly used 500-MHz NMR spectrometers. The precessing nuclear moments would

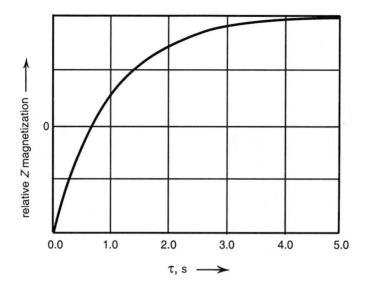

**Figure 2-28** The first-order decay curve exhibited by the Z magnetization as a function of time after a 180° pulse, when $T_1$ is 1 s

then spiral down to the X, Y plane taking some 5000 turns around the Z axis (Figure 2-11). Suppose the sample is composed of material with nuclei having just two precession frequencies, the first at exactly 500 MHz that we will call $\nu_0$ and we will let $\nu_0$ also be the oscillator frequency and, by convention, the rotating-frame frequency. The second frequency we will take to be $\nu_0 + 50{,}000$ Hz. Now, over 10 µs, the second magnetic moment will make 5000.50 turns around the Z axis while the first moment will make 5000.00 turns. If there were no other complications, this means that the second moment

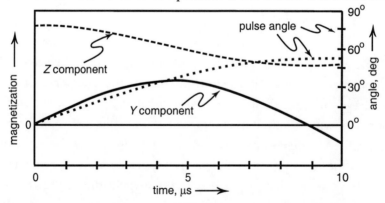

**Figure 2-29** Plots of the pulse angle, the Y and Z components of magnetization as a function of time over a 10-µs pulse, when $\nu - \nu_0$ is 50,000 Hz and $\nu_0 = 500$ MHz. For the situation shown here, a 10-µs pulse is a 90° pulse when $\nu - \nu_0 = 0$

## 2-5 More on the 90° Pulse

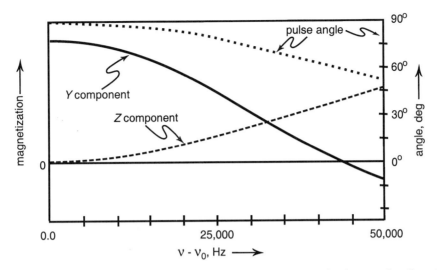

**Figure 2-30** Plots of the pulse angle, the Y and Z components of magnetization as a function of how far the frequency of the nuclei to be excited is from the frequency of the oscillator at the conclusion of a 10 μs pulse, which is a 90° pulse when $v = v_0$ and $v_0 = 500$ MHz. It should be clear that there will be phase and amplitude problems when the difference between $v$ and $v_0$ is large

will land on the X, Y plane exactly 180° out-of-phase with the first moment. Certainly the result would **not** be two perfect 90° pulses. But to make matters worse, there are other complications. You can recognize this, because as the second moment spirals downward, it will be getting out-of-phase with the clockwise rotating component of $B_1$ (Section 2-1). It is possible to calculate how, during a 10-μs (ostensibly 90°) pulse, the resultant pulse angle, Y and Z magnetizations change with time when $v - v_0 = 50,000$ Hz (Figure 2-29). Note how the Z magnetization actually starts to recover when the Y magnetization gets more than 90° out-of-phase with the $B_1$ field. This is because the $B_1$ field begins to turn $M$ back up toward the +Z direction.

Figure 2-30 is a plot of pulse angle, Y and Z magnetizations, at the **conclusion** of a "90°" pulse, as a function of how far the frequency of the nuclei being excited is from $v_0$ of the excitation pulse. When $v - v_0$ is about 45,000 Hz, the Y component goes through zero and, at 50,000 Hz, becomes negative with only 16% of the magnitude of what would be expected after a normal 90° pulse with $v = v_0$.

The point here is that there are problems with getting clean 90° pulses when frequency differences are large. But, remember that, at 500-MHz, a frequency difference of 50,000 Hz corresponds to 100 ppm, far greater than the usual chemical-shift range (~15 ppm) for proton resonances. Furthermore, if the pulse time could be reduced to 1 μs then one should be able to get quite good 90° pulses even with a 50,000-Hz difference in frequency from $v_0$. Nonetheless, if you have taken $^{13}$C or $^{15}$N NMR spectra, you will know that the shift ranges experienced with these nuclei are very much larger than for protons. Thus, carbon shifts range over about 200 ppm and nitrogen over about 900 ppm. This would seem to cause difficulty with the 90° pulse, when as pointed out above, 100 ppm is sure to be troublesome for nuclei that have γ values comparable to those of protons. Fortunately, matters are eased very substantially by the fact that the γ values for $^{15}$N and $^{13}$C (Table 1-1) are such that the 900-ppm shift range for nitrogen in a 500-MHz spectrometer amounts to 45,000 Hz and 200 ppm for carbon amounts to 50,000 Hz. Clearly, both frequency ranges are large enough to cause problems. However, these problems can be diminished by using shorter pulse times or by changing the oscillator frequency. Actually, if the oscillator frequency is set near to the center of the ranges of nitrogen or carbon frequencies, the range of frequencies to be covered is diminished to ± 25,000 Hz. Putting the pulse in the middle of the frequency range can create massive (but soluble) other problems, as we shall see later. The issue now is to be sure you recognize why short pulse times can be important in FT-NMR operation.

If it is not clear, a simple physical analogy may help. Suppose you have two children sitting in two swings side-by-side and each is screaming, "Daddy, get us started swinging together." (Perhaps, you and your brother did this to your father.) But Daddy has a problem because these swings do not have the same frequency. Obviously, if he pushes them in unison, they will soon get out-of-phase and one will want to go up when the other is coming down. However, he can get both swings going at once with a brief hard push. That is just what a short 90° pulse does in FT-NMR.

**Exercises**

**Exercise 2-1** If we were to arbitrarily set the shielding constants of Equation 2-2 for tetramethylsilane (TMS) to be zero, calculate the shielding constant of a proton that comes into resonance at 675 Hz downfield of TMS at 500 MHz.

**Exercise 2-2** Proton chemical shifts measured in the 1970 era of NMR, were often reported on a "τ scale". This scale has TMS at 10 τ where τ has the dimension of ppm. If you found a shift reported as 7.43 τ, what would its shift be on the usual TMS scale?

**Exercise 2-3** The σ value of the protons in $H_2O$ is calculated to be fairly close to 25.97 x $10^{-6}$. The resonance frequency of $H_2O$ protons on the TMS scale is 4.77 ppm. Use this data to estimate the shift of the bare proton on the TMS scale.

# Exercises

**Exercise 2-4 a.** Two identical 0.1-mm diameter tubes of water are placed in a 100-MHz (23,400 gauss) NMR spectrometer at a distance $d$ apart. The magnetic field of the spectrometer varies linearly between the two samples at $1.34 \times 10^{-3}$ gauss/cm. The two samples differ in their resonance frequency by 3.25 Hz. Show how this information can be used to calculate $d$. (This problem illustrates one of the basic principles underlying magnetic resonance imaging, MRI - that is how accurate measurements can be made of small distances with NMR, despite the fact that 100 MHz corresponds to electromagnetic radiation with wavelengths of about 3 meters.)

**b.** Assuming the magnetic-field gradient is precisely linear and its magnitude accurately known, what is likely to be the limiting factor in the measurements of the type described in **a.** above.

**Exercise 2-5** The smallest reported precession frequency for a spin 1/2 stable nucleus is 1.72 MHz at 10,000 gauss for $^{107}$Ag. At this frequency, how long would the 1.72 MHz 90° pulse time, call it $t$, have to be to allow for only **one** precession cycle before the 90° pulse is complete? Sketch out vector diagrams similar to Figure 2-11 (laboratory reference frame) that show the progress of the system under the influence of your calculated one-turn 90° pulse at times, 0, $t/3$, $2t/3$, $t$. Be sure to include both the clockwise and counterclockwise $B_1$ components in your diagrams. Make the corresponding diagrams for the rotating frame of reference.

**Exercise 2-6** With the aid of a vector diagram based on Figure 2-16, explain how $T_2$ can be smaller than $T_1$ because of inhomogeneities in the magnetic field.

**Exercise 2-7** Suppose an NMR spectroscopist mistakenly entered pulse parameters into a Carr-Purcell sequence that called for a 45° pulse instead of a 90° pulse, but entered the correct value for the 180° pulse. Show by vector diagrams similar to Figure 2-22 what the consequences of this 45° - $\tau$ - 180° - $\tau$ sequence would be using the initial conditions of Figure 2-19.

**Exercise 2-8** A Carr-Purcell echo train produces peaks of relative heights, 4.65, -1.85, 0.69 and -0.25 with $\tau = 0.25$ s. Show how you can use these intensities to obtain a value of $T_2$.

**Exercise 2-9** What would be the final result of the pulse sequence $90°_y - \tau - 180°_x - 2\tau - 180°_y - \tau$, starting with the three magnetic moments and initial conditions of Figure 2-19, if the 90° pulse is applied along the Y axis, the first 180° pulse along the X axis and the second 180° pulse along the Y axis?

**Exercise 2-10** An inversion-recovery sequence gives the following relative signal intensities at different $\tau$ values. Use these values to calculate $T_1$ for the experiment.

| $\tau$, sec | Intensity |
|---|---|
| 0 | -7.83 |
| 0.2 | +2.17 |
| 0.4 | +5.75 |
| 0.6 | +7.27 |
| 0.8 | +7.75 |
| 1.0 | +7.93 |

**Exercise 2-11 a.** It is customary in the inversion-recovery sequence for determination of $T_1$ to calibrate the 180° pulse time by varying the pulse time until no signal is observed. Suppose a spectroscopist trying to locate the 180° pulse time accidentally entered twice the correct value? Use appropriate diagrams to show what the consequences would be if this value was used for the 180° pulse and half of it were used for the 90° pulse in the inversion-recovery sequence, in effect $360°_x - \tau - 180°_x$.
**b.** If the operator were to try to locate the best value for the 180° pulse by varying the pulse time ± 15% across the minimum signal-intensity region, explain how he/she could tell the difference between a 180° and a 360° pulse?

**Exercise 2-12** Explain why, in Figure 2-29, the Y-component of magnetization for a nucleus with $v - v_0 = 50,000$ Hz rises to a maximum in about 5 μs of a 10 μs pulse and then decreases, finally becoming negative at the end of the 10 μs.

**Exercise 2-13** Use your knowledge of Faraday's law for induction of an electric current in a coil of wire by changes in magnetic field along the coil axis and Figure 2-10 to determine the phase relationship between the magnitude of the signal observed in the receiver coil and the position of the rotating magnetic vector in the X, Y plane.

# 3
# How Do We Perform the NMR Fourier Transform?

The mathematics of the Fourier Transform are not exciting chemistry and, in the main, you don't need to know a lot about them. However, there are several Fourier concepts with which you will need to be familiar, because they are very important to the practical aspects of taking FT NMR spectra. The basic FT operation, as I have said earlier, is to separate out the frequencies and intensities of the components of the decaying signal of the FID and allow them to be displayed in a plot of intensity vs. frequency. The widths of the resulting spectral peaks depend on how fast the nuclear magnetism decays in the X, Y plane. Fast decays give broad lines and slow decays give sharp lines (Section 2-1). Before entering the domain of the Fourier nitty-gritty, let me offer a caveat to tuck away in your subconscious until you need it; this is that **the Fourier transform technique is not the only way to convert FID to a spectrum**. It is one way and certainly is by far the most commonly used way, but it may not even be the best way. We will return to this point in a later chapter.

### 3-1 The Basics of the Fourier Transform

The Fourier nitty-gritty involves the equation:

$$A(\omega) = \int_{-\infty}^{+\infty} A(t) e^{-i\omega t} dt \qquad \text{Eqn. 3-1}$$

where $A(\omega)$ is the intensity of the signal as a function of frequency, $A(t)$ is signal intensity of the FID as a function of time and $i$ is $(-1)^{1/2}$.

Equation 3-1 is beautifully concise; but, as written, is not very helpful for me. I think it makes more sense when rewritten as Equation 3-2 with the aid of standard substitutions.

$$A(\omega) = \int_{-\infty}^{+\infty} A(t)[\cos(\omega t) - i\sin(\omega t)] dt \qquad \text{Eqn. 3-2}$$

When Equation 3-2 is rewritten in the form of Equation 3-3, there are two important terms, the first representing the **real part** of the frequency spectrum and the

$$A(\omega) = \int_{-\infty}^{+\infty} A(t)\cos(\omega t)dt - i\int_{-\infty}^{+\infty} A(t)\sin(\omega t)dt$$

real         imaginary

Eqn. 3-3

second representing the **imaginary part**. The real part can be taken as the signal component that starts off **in-phase** with the rf oscillator, while the imaginary part is that which starts off 90° **out-of-phase** with the oscillator.

To make further progress, we need to know the form of $A(t)$ and decide how to proceed with the integrations. Equations 3-1 to 3-3 represent continuous functions, but the usual way of taking an FID is to sample the signal strengths at discrete time intervals, digitize and store the information until relaxation reduces the signal strengths to non-useful levels. All electronic signals contain some noise. If the signals from the nuclei are weak and the noise level is high, the FID can be repeated many times and each new set of signal intensities can be digitally added to the earlier values. The principle used here is that noise (static in radio parlance) is expected to be random and have both positive and negative components. If the noise does have those qualities (that is, it is "white noise"), then its summation over a large number of FIDs should cancel to zero, leaving only the pure summed FIDs. Unfortunately, the gain in the signal-to-noise ratio (S/N) is not linear with number of FID scans. The signal-to-noise improves only as the square root of the number of scans, which means that a factor of two improvement requires four times as many scans. The improvement is illustrated by Figure 3-1, where a noisy signal obtained in a single scan is shown to be dramatically improved by signal averaging. How fast we can repeat the scans, the **repetition rate** as it is called, does not depend on $T_2$ (the rate of signal decay), but instead on $T_1$ (the rate of establishment of equilibrium along the Z axis). The reason for this can be seen by supposing that $T_2$ is very short and $T_1$ is long. Then, after a 90° pulse, the signal can quickly decay to zero with little or no significant recovery of magnetization in the Z direction. Now, think about the result of applying a 90° pulse to a system when there is no Z magnetization. For a 90° pulse, it is usual to wait one to four times $T_1$ before pulsing again (the reasons will be discussed in Section 5-2, **11**). If $T_1 \gg T_2$, the FID should be taken only while the signal has a significant intensity, because there is no point in accumulating noise data by itself. Thus, the overall repetition rate will be determined by $T_1$, and the sampling time **(acquisition time)** will be determined by $T_2$.

## 3-1 The Basics of the Fourier Transform

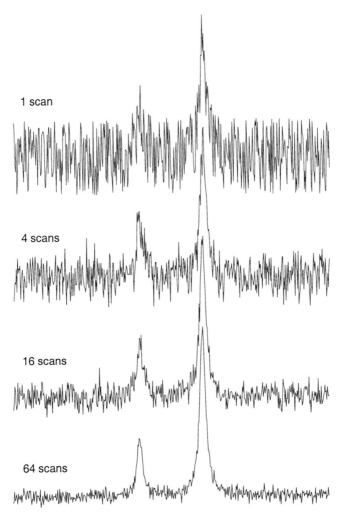

**Figure 3-1** Spectra that illustrate how the S/N ratio of a noisy spectrum can be improved by time averaging. The single-scan spectrum at the top can be seen to be marginally useful, but by increasing the number of scans, the noise can be substantially decreased. The cost is in the extra time needed to accumulate the spectra

When the signal strength has been measured with discrete time-interval sampling, the Fourier integral is replaced by:

$$x(n) = \frac{1}{N} \sum_{n=0}^{N-1} \left[ A(n) \frac{\cos(2\pi kn)}{N} - iA(n) \frac{\sin(2\pi kn)}{N} \right] \qquad \text{Eqn. 3-4}$$

where $N$ is the number of points, $n$ is the $n$th point, and $2\pi k$ is a multiplier that converts point $n$, to the appropriate value corresponding to $\omega t$ (in rad-s). If you have an array, call it $A$, of $N$ signal points as a function of time, calculation of the corresponding arrays of real and imaginary frequency points is quite straightforward and can be easily (if not rapidly) carried out with a desktop computer. The smoothness of plots of the array of the time- and frequency-domain points depends greatly on $n$. The more points, the smoother the plots will be. Calculations with many points are slow, because of the need to compute so many multiplications as well as sine and cosine values. Some speedup can be achieved using lookup arrays of sine and cosine values. Much increased rates of calculating transforms are possible with the Cooley-Tukey fast-Fourier algorithm, which has the special restriction that $N$ be a power of 2.

## 3-2 Applying the Fourier Transform

You should be interested enough in the nitty-gritty of the Fourier transform to the extent of actually trying calculations on your computer, and Appendix 2 contains a simple listing of a 32-point digital Fourier transform program written in True BASIC, an inexpensive and quite powerful language system which uses virtually identical source code on computers from the Macintosh, IBM-PC, Amiga and Sun lines. The program can be easily modified to use more than 32 data points and other programming languages. Many of the curves which I will show you of Fourier transforms were calculated using this program or simple modifications of it.

Our first example is the elementary one shown in Figure 3-2. The topmost plot is of a 32-element array of a decaying cosine function with a frequency of 18.7 Hz that is sampled every 0.01 s from 0 to 0.31 s. Here, $2\pi kn$ is $\omega t$ with $kn$ the frequency in Hz, and $n$ the sampling index. The transform of a cosine function contributes to the real part of $x(n)$ and this is the next plot from the top. This plot looks strange because, when $x(n)$ is calculated, the frequency goes from 0 to 50 and then from -50 to 0. The same frequency range is covered by the imaginary part of the transform, the next plot down. The lowest plot in Figure 3-2 shows the real part of the transform in a more conventional form, with zero in the center, negative frequencies to the left and positive to the right. Normally, one does not show both the positive and negative regimes of the transform, because no new information is obtained thereby. An important exception will be discussed later.

You will note the difference in appearance between the real and imaginary parts of the transform. Here, the real part has the form known as an **absorption** curve while the imaginary part has the form called a **dispersion** curve. These curves are considerably distorted in Figure 3-2 because of the small number of points in the waveform and

## 3-2 Applying the Fourier Transform

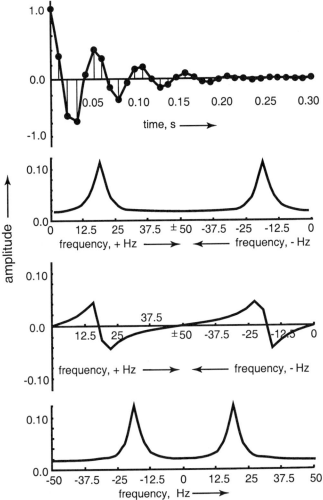

**Figure 3-2** Plot of a 18.7-Hz cosine waveform (upper) with 32 data points in the time domain and a sampling rate of 0.01 s (black dots and vertical lines). The real part of the Fourier transform in the frequency domain is second from the top and the imaginary part is third from the top. The bottom plot is of the real part of the transform with zero in the center, negative frequencies to the left and positive frequencies to the right

the transform. With more points, these two curves take on the shapes shown in the upper two curves of Figure 3-3.

To at least the extent that the NMR FIDs can be regarded as one or more decaying sinusoidal functions, the real and imaginary parts of the transform are closely related. It is matter of phase and, if we use a decaying sine waveform in place of the cosine example of Figure 3-2, which corresponds to introducing a phase shift of 90°, then the shapes of the two curves as absorption or dispersion are interchanged, as can

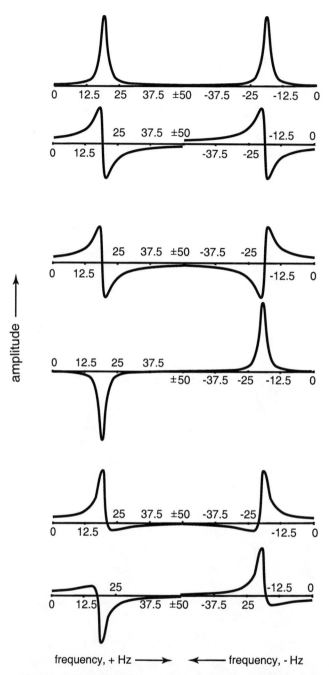

**Figure 3-3** Fourier transforms of decaying cosine and sine waveforms with frequencies of 18.7 Hz. The upper two curves correspond in all respects to the real and imaginary transforms of Figure 3-2 except that they involve 1024 data points. The middle two curves are transforms of a 1024-point array of a decaying sine function, rather than the cosine function of Figure 3-2. The lower two curves are transforms of 1024 data points of an equal mix of decaying 18.7 Hz cosine and sine functions

## 3-2 Applying the Fourier Transform

be seen in the middle part of Figure 3-3. When there is an equal mix of sine and cosine functions, equivalent to a 45° phase shift, then we get the lower two mixed transform curves shown in the lower part of Figure 3-3.

The sets of absorption and dispersion curves in the middle and upper parts of Figure 3.3 are obviously not identical. However, as a practical matter, their information content when viewed as NMR spectra is much the same. Each absorption curve corresponds to a resonance at 18.7 Hz with the same rate of $T_2$-type decay and each has the same integrated intensity. With regard to integrated intensity, it is important to know that the integrals are independent of the rate of decay of the FID. Faster decays give broader resonances with smaller peak heights, but the integrals depend on the signal intensity at the start of the FID. Thus, a pure cosine function with a positive starting value will give an absorption curve with a positive integral, while a sine function will give a dispersion curve with a zero integral. In many instances, the transforms of NMR FIDs will have the appearance of mixed decaying sine and cosine waveforms much like the lower two curves in Figure 3-4. As we will show in Chapter 5, if the spectral parameters are chosen properly, it is possible to use simple mathematical procedures to correct the phases and produce the conventional absorption-curve representations of the NMR resonances.

The range of frequencies covered by the Fourier summation in Figure 3-2 runs from 0 to 50 Hz and -50 to 0 Hz. Why this restricted range? It is a consequence of the **sampling rate**. The 50-Hz figure is called the **Nyquist frequency** and is the highest frequency that can be defined by a sampling rate of 1/100 s. The Nyquist frequency is $1/(2\Delta t)$ where $\Delta t$ is the time interval for each data point. You will see that the Nyquist frequency is that frequency in which **one-half of a cycle occurs in the sampling interval**. It should be obvious that the frequency of a waveform that undergoes more than 0.5 of a cycle in a given sampling interval will not be accurately represented in the transform. If this is not obvious, draw a sine wave with one or more cycles in a given sampling interval and satisfy yourself that its frequency is not defined by the average of the signal measured over that time interval. When you transform a FID that contains a higher frequency than the Nyquist limit, you will usually, but not always, find a peak corresponding to a spurious frequency. This kind of behavior is illustrated by the upper plot in Figure 3-4 that shows a 68.7-Hz waveform, sampled at 0.01 second intervals, while the lower plot shows the frequency domain for the corresponding transform of this waveform. The apparent frequency at 31.3 Hz is a consequence of the **aliasing**, as it is called, of the 68.7-Hz frequency to produce what appears to be a 31.3-Hz frequency.

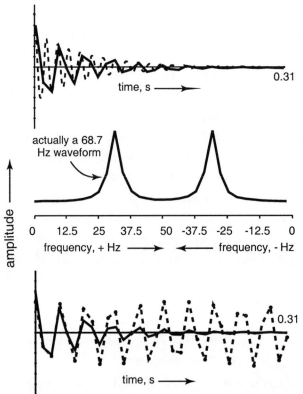

**Figure 3-4** Plots of a 68.7-Hz decaying *cosine* waveform (dashed line) sampled over 0.01 s intervals (top), which is above the limit imposed by the Nyquist frequency. The resulting rather awkward-looking waveform (solid line) transforms to give a real spectrum in the frequency domain (middle) that has both positive and negative 31.3 Hz peaks. The lower plot shows how the sampled waveform is **aliased** by a 31.3-Hz *cosine* waveform, also sampled at 0.01 s intervals

We shall return to the Fourier transform later. Our next task will be to understand better how to calculate simple NMR FIDs and CW spectra.

## Exercises

**Exercise 3-1** Plot a cosine wave with five cycles and explain quantitatively how the Fourier transform procedure works to give a real and imaginary frequency domain spectrum.

**Exercise 3-2** How do we know that the first term of Equation 3-3 corresponds to the **real** part of the Fourier transform and the second term to the **imaginary** part?

# Exercises

**Exercise 3-3** Show qualitatively how the $T_1$ and $T_2$ decays of a cosine-wave FID would influence the Fourier transform in the case where $T_1 \gg T_2$.

**Exercise 3-4** Plot a $\cos(\theta)$ wave with five cycles that starts at $\theta = 45°$. Show qualitatively what the real and imaginary parts of the frequency-domain spectra will look like if the Fourier integral sine and cosine functions each start at $0°$.

**Exercise 3-5 a.** Write a FT-computer program based on the one for which the source code is given in Appendix 2 to determine how the height of the maxima and the integrals in the real part of the transform depend on the waveform decay rate, which is contained in the term, /EXP(I*5/32). Replace the 5 by zero to have infinitely slow decay. Because of the symmetry of the transform; you will only need to sum the first 16 points to obtain the integrals.
**b.** In the program of Appendix 2, determine the integral for the first 16 real points, then replace COS(2*PI*18.7*I/100)/EXP(I*5/32) by the term:

$$(0.5*COS(2*PI*18.7*I/100) + 0.5*SIN(2*PI*18.7*I/100))/EXP(I*5/32)$$

and again calculate the integral. Compare the values with those found in **a**. and comment on the comparison.

**Exercise 3-6** Use a computer with a random-number generator, or a more analytical procedure, to show that the signal-to-noise ratio of a spectrum does not increase linearly with time in the time-averaging mode when the noise is random and has equal probability of positive and negative values. The definition of **signal-to-noise ratio** used for NMR spectrometers is the ratio of the peak absorption-mode signal to twice the **root-mean-square noise**. In practical work, the root-mean-square noise can be approximated as 0.2 times the positive peak-to-negative peak noise amplitude in a spectral section having 100 zero crossings that does not include the peak of interest. For more precise determination of the root-mean-square-noise for $n$ noise points use $\left[\frac{1}{n}\left(\Sigma x_n^2\right)\right]^{1/2}$.

# 4

# The Bloch Equations. Calculating What Happens in NMR Experiments

Although NMR involves quantized changes in nuclear magnetic energy states, much of what happens in the process of taking a spectrum can be calculated quite precisely by classical electromagnetic theory. The differential equations for doing this were derived by Felix Bloch (Nobel Prize in Physics, 1952) and are quite easy to evaluate by numerical integration. Let us see how these equations are set up.

## 4-1 The Initial Steps. Relaxation

We have already encountered (Sections 1-2, 2-2, 2-3 and 2-4) two essential components of the Bloch formulation that describe **relaxation** of a system of magnetic nuclei in a nonequilibrium state:

$$\frac{dM_z}{dt} = \frac{M_{zo} - M_z}{T_1} = \frac{M_o - M_z}{T_1} \qquad \text{Eqn. 4-1}$$

and

$$\frac{dM_{xy}}{dt} = -\frac{M_{xy}}{T_2} \qquad \text{Eqn. 4-2}$$

We will be interested in $M_x$ and $M_y$ which are the components of the magnetization in the X,Y plane $M_{xy}$. The change of $M_x$ and $M_y$ with time in the laboratory frame of reference has two parts. One is governed by $T_2$ and the other arises from precession of $M$ in the magnetic field of the magnet $B_0$. Thus, $M_x$ will **decrease** by relaxation and also **increase** as the $M_y$ component of $M$ (here taken to have a positive sign), moves in the +X direction as the result of precession of $M$ about the Z axis (see Figure 4-1). Thus, we can write:

$$\frac{dM_x}{dt} = -\frac{M_x}{T_2} + M_y \gamma B_o \qquad \text{Eqn. 4-3}$$

where $\gamma B_0$ is the precession frequency. Similarly, we can write:

# 4-2 The Next Step. Taking Into Account the $B_1$ Power Input

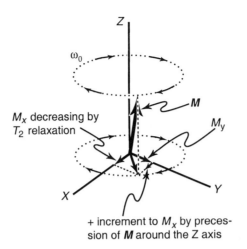

**Figure 4-1** Incremental changes in $M_x$ as the result of precession of the $M$ vector around the field axis, see Equation 4-3 and remember that here we are using the laboratory frame of reference

$$\frac{dM_y}{dt} = -\frac{M_y}{T_2} - M_x \gamma B_o \qquad \text{Eqn. 4-4}$$

The minus sign here, for the second term on the right side of Equation 4-4, reflects the fact that the direction of precession of a positive $M_x$ component of $M$ is such as to **decrease** $M_y$. We now have differential equations for change of each of the three components of $M$ with time as the result of relaxation.

## 4-2 The Next Step. Taking Into Account the $B_1$ Power Input

We now need to take into account the influence of the rf field $B_1$ of the oscillator. The oscillator produces a sinusoidally varying magnetic field along the X axis that, as discussed earlier (Section 2-1), can be usefully resolved into two contra-rotating magnetic vectors, only one of which rotates in the same direction as the precessing nuclei. We can quantify the important clockwise-rotating component of $B_1$ in terms of its X and Y components (don't forget that $\omega$ has the units of rad/s):

$$B_{1x} = B_1 \cos(\omega t)$$
$$B_{1y} = -B_1 \sin(\omega t) \qquad \text{Eqn. 4-5}$$

Now, $B_{1x}$ is a magnetic field that, by the left-hand rule (Section 2-1), when positive, will cause the magnetization equivalent to a positive value of $M_y$ to rotate **downward** in the Y, Z plane toward the -Z axis and will therefore **decrease** $M_z$. On the other hand, $B_{1y}$

will cause that part of $M$ corresponding to positive $M_x$ to rotate **upward** in the X, Z plane in the +Z direction and **increase** $M_z$. If we include relaxation, then we can write:

$$\frac{dM_z}{dt} = -M_y(\gamma B_{1x}) + M_x(\gamma B_{1y}) + \frac{M_o - M_z}{T_1} \qquad \text{Eqn. 4-6}$$

For $M_y$, the component of $M$ which is equal to a positive $M_z$ will rotate away from the Z axis in the +Y direction under the influence of $B_{1x}$ and thus increase $M_y$. As to be expected, $B_{1y}$ does not act to change $M_y$.

$$\frac{dM_y}{dt} = M_z(\gamma B_{1x}) - M_x(\gamma B_o) - \frac{M_y}{T_2} \qquad \text{Eqn. 4-7}$$

Proceeding in the same way (and check my arithmetic) we can write:

$$\frac{dM_x}{dt} = -M_z(\gamma B_{1y}) + M_y(\gamma B_o) - \frac{M_x}{T_2} \qquad \text{Eqn. 4-8}$$

In Equations 4-6, 4-7 and 4-8, you will see that the last term on the right side of the equation is independent of whether the oscillator is on or off.

At this point, we can substitute $B_{1x} = B_1\cos(\omega t)$ and $B_{1y} = -B_1\sin(\omega t)$, that are the terms corresponding to the clockwise-rotating $B_1$ field, and thereby get the somewhat, but not much, simplified equations that follow.

$$\frac{dM_z}{dt} = -\gamma[M_y B_1 \cos(\omega t) + M_x B_1 \sin(\omega t)] + \frac{M_o - M_z}{T_1} \qquad \text{Eqn. 4-9}$$

$$\frac{dM_y}{dt} = \gamma[M_z B_1 \cos(\omega t) - M_x B_o] - \frac{M_y}{T_2} \qquad \text{Eqn. 4-10}$$

$$\frac{dM_x}{dt} = \gamma[M_z B_1 \sin(\omega t) + M_y B_o] - \frac{M_x}{T_2} \qquad \text{Eqn. 4-11}$$

These equations could be evaluated by numerical integration with relative ease, but it is usually considered to be more convenient to recast them in a form that makes it easier to separate the in-phase $M_x$ component from the out-of-phase $M_y$ component of $M$ when the phase reference is the oscillator frequency. With the coordinate system we will use here, the substitutions that achieve this goal are:

$$v = M_x \sin(\omega t) + M_y \cos(\omega t) \qquad \text{out-of-phase} \qquad \text{Eqn. 4-12}$$

$$u = M_x \cos(\omega t) - M_y \sin(\omega t) \qquad \text{in-phase} \qquad \text{Eqn. 4-13}$$

## 4-2 The Next Step. Taking Into Account the $B_1$ Power Input

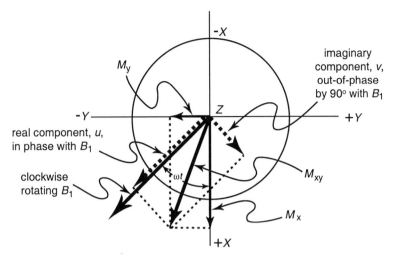

**Figure 4-2** View down the Z axis showing how the $M_{xy}$ vector can be resolved into components that are $u$, in-phase with the clockwise-rotating $B_1$, and $v$, 90° out-of-phase with the clockwise-rotating $B_1$

To see how this works, let us visualize a sizable $M_{xy}$ vector that is rotating clockwise about 20° behind the rotating-field vector as in Figure 4-2. We can resolve $M_{xy}$ into $M_x$ (here large and positive) and $M_y$ (smaller and negative). In order to determine $v$ and $u$, we need to resolve the components of $M_{xy}$ parallel, and perpendicular, to the oscillating field vector. Inspection of the vectors involved shows that $v$ and $u$ are both surely functions of $M_x$ and $M_y$. It is a bit more tedious to show how the $\sin(\omega t)$ and $\cos(\omega t)$ terms fit into the above equations, but it can be done by simple trigonometry.

The next step is to differentiate $u$ and $v$ with respect to $t$ and try to simplify the resulting mess. We will only work out $u$ here before your very eyes (however, it would be instructive for you to check the "arithmetic" for the corresponding treatment of $v$).

$$\frac{du}{dt} = -\omega M_x \sin(\omega t) + \frac{dM_x}{dt}\cos(\omega t) - \omega M_y \cos(\omega t) - \frac{dM_y}{dt}\sin(\omega t)$$

$$= -\omega\left[M_x \sin(\omega t) + M_y \cos(\omega t)\right] + \frac{dM_x}{dt}\cos(\omega t) - \frac{dM_y}{dt}\sin(\omega t) \qquad \text{Eqn. 4-14}$$

Noting that $M_x \sin(\omega t) + M_y \cos(\omega t) = v$ and taking $dM_x/dt$ and $dM_y/dt$ from Equations 4-10 and 4-11, we can write:

$$\frac{du}{dt} = -\omega v + \left\{\gamma\left[M_z B_1 \sin(\omega t) + M_y B_0\right] - \frac{M_x}{T_2}\right\}\cos(\omega t)$$

$$-\left\{\gamma\left[M_z B_1 \cos(\omega t) - M_x B_0\right] - \frac{M_y}{T_2}\right\}\sin(\omega t) \qquad \text{Eqn. 4-15}$$

If we multiply out and combine terms, Equation 4-15 becomes:

$$\frac{du}{dt} = -\omega v + \gamma B_0 \left[ M_x \sin(\omega t) + M_y \cos(\omega t) \right]$$
$$- \left[ M_x \cos(\omega t) - M_y \sin(\omega t) \right] / T_2$$

Eqn. 4-16

Now, using the definitions of $u$ and $v$ (Equations 4-12 and 4-13), as well as remembering that $\omega_0 = \gamma B_0$, we get Equation 4-17.

$$\frac{du}{dt} = -\omega v + \omega_0 v - \frac{u}{T_2}$$
$$= (\omega_0 - \omega) v - \frac{u}{T_2}$$

Eqn. 4-17

Proceeding in the same manner with the differentiation of $v$ gives:

$$\frac{dv}{dt} = -(\omega_0 - \omega) u - \frac{v}{T_2} + \gamma B_1 M_z$$

Eqn. 4-18

Going back to the Equation 4-9 for $dM_z/dt$, we find that:

$$\frac{dM_z}{dt} = \frac{M_0 - M_z}{T_1} - \gamma B_1 v$$

Eqn. 4-19

The differential equations 4-17, 4-18 and 4-19 are the most useful form of the **Bloch equations** and they describe NMR very well, except when quantum-mechanical effects become important, which is most often when spin-spin couplings are comparable to, or larger than, differences in chemical shift between the coupled nuclei.

The Bloch equations give a nice translation to a physical picture in that $M_z$ changes in accord with $v$ when the oscillator is on. Because changes in $M_z$ correspond to energy input into the system, we can take $v$ to be a measure of NMR **absorption** while $u$ is called **dispersion**.

## 4-3 NMR at Equilibrium. Slow-Passage Conditions

Calculation of $v$, $u$ and $M_z$ from the Bloch equations is difficult in closed form for the **general case**. However, it is easy to make such calculations for the antithesis of FT-NMR, where you have a particular combination of values of $M_0$ (determined by $B_0$, $\gamma$, $I$ and temperature, see Section 1-2), $\gamma$, $B_1$, $T_1$, $T_2$ and $(\omega_0 - \omega)$ and the system is allowed to

## 4-3 NMR at Equilibrium. Slow-Passage Conditions

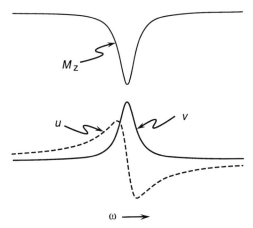

**Figure 4-3** Plots of v, u and $M_z$ calculated for **slow-passage** conditions with the aid of the Bloch equations at a moderate power level. With slow passage, the nuclear spin system is at equilibrium - the inflow of energy from the rf oscillator being equal to the outflow of energy from the system to the surroundings ("lattice"). Note how the dispersion signal u goes through zero when $\omega = \omega_0$

come to **equilibrium**. In FT-NMR, it is usual to be at equilibrium before the initial pulse, in which case $B_1$ is zero. If $B_1$ is not zero and $\omega$ is allowed to change slowly with respect to the rate of establishment of equilibrium, then we have what is called a **slow-passage** spectrum. At equilibrium, energy flows in from $B_1$ and is dissipated to the surroundings by relaxation. In these circumstances, at any given point we will have:

$$\frac{du}{dt} = \frac{dv}{dt} = \frac{dM_z}{dt} = 0 \qquad \text{Eqn. 4-20}$$

Using Equations 4-17 to 4-19 with 4-20 we can now write:

$$u = v(\omega_0 - \omega)T_2$$
$$v = [-(\omega_0 - \omega)u + \gamma B_1 M_z]T_2 \qquad \text{Eqn. 4-21}$$
$$M_z = M_0 - \gamma B_1 T_1 v$$

Manipulation of these equations gives:

$$M_z = \frac{M_0[1 + (\omega_0 - \omega)^2 T_2^2]}{1 + (\omega_0 - \omega)^2 T_2^2 + \gamma^2 B_1^2 T_1 T_2} \qquad \text{Eqn. 4-22}$$

$$v = \frac{\gamma B_1 M_0 T_2}{1 + (\omega_0 - \omega)^2 T_2^2 + \gamma^2 B_1^2 T_1 T_2} \qquad \text{Eqn. 4-23}$$

$$u = \frac{\gamma B_1 (\omega_0 - \omega) T_2^2 M_0}{1 + (\omega_0 - \omega)^2 T_2^2 + \gamma^2 B_1^2 T_1 T_2} \qquad \text{Eqn. 4-24}$$

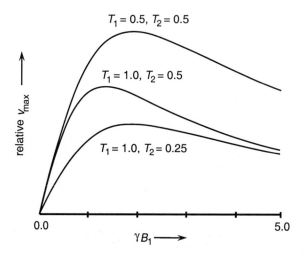

**Figure 4-4** Graphs showing how $v_{max}$, the height of the signal peak when $\omega = \omega_0$, changes with the $B_1$ power level as a function of $T_1$ and $T_2$ under slow-passage conditions. Clearly, a shorter $T_1$ is more effective than a shorter $T_2$ in increasing the signal level, because $T_2$ largely acts to broaden the resonance peak

Plots of $v$ and $u$ as a function of frequency made with the aid of Equations 4-23 and 4-24 (see Figure 4-3) are called **absorption-mode** or **dispersion-mode** spectra, respectively, and depend markedly on the magnitude of $B_1$. If $\gamma^2 B_1^2 T_1 T_2$ is small compared to $1 + (\omega_0 - \omega)^2 (T_2)^2$, then the strength of the absorption-mode signal as a function of $(\omega_0 - \omega)$ will be seen to be essentially linear with $B_1$ the oscillator strength (see Figure 4-4 where $\gamma B_1 < 1$). If the reverse is true, the signal strength will decrease with increasing $B_1$, a condition called **saturation**. We have encountered saturation in a different form in FT-NMR in the situation, where a waiting period determined by $T_1$ is necessary, if a 90° pulse is to be followed by another 90° pulse with good efficiency (see Section 3-1).

The slow-passage equations can be manipulated in various ways to obtain useful information for continuous-wave spectroscopy. Thus, one can differentiate $v$ with respect to $B_1$ to determine the optimum value of $\gamma B_1$ to give the maximum slow-passage absorption signal at $\omega_0 = \omega$. The result is:

$$\gamma B_1(optimum) = (T_1 T_2)^{-1/2} \qquad \text{Eqn. 4-25}$$

Another important use of the equations is to relate $T_2$ to the line width of **slow-passage** spectral peaks. The **line width** is defined as the width of the resonance at half-maximum signal intensity of a given peak (see Figure 4-5). It is easy to show that the value of $v$ at the peak maximum, where $(\omega_0 - \omega) = 0$, is:

## 4-3 NMR at Equilibrium. Slow-Passage Conditions

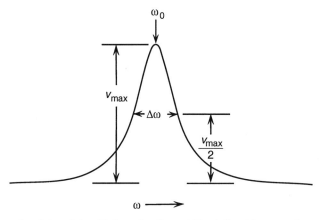

**Figure 4-5** Procedure for determining $T_2$ from the line width, defined here as the width at half height of the peak maximum, $v_{max}$. $T_2$ is equal to $1/(\pi \Delta \omega)$ under slow-passage conditions, if $\Delta \omega$ is in Hz

$$v_{max} = \frac{\gamma B_1 M_0 T_2}{1 + \gamma^2 B_1^2 T_1 T_2} \qquad \text{Eqn. 4-26}$$

Now, we can ask what is the frequency $\omega$ when the intensity is $v_{max}/2$.

$$\frac{1}{2} \cdot \frac{\gamma B_1 M_0 T_2}{1 + \gamma^2 B_1^2 T_1 T_2} = \frac{\gamma B_1 M_0 T_2}{1 + (\omega_0 - \omega)^2 T_2^2 + \gamma^2 B_1^2 T_1 T_2} \qquad \text{Eqn. 4-27}$$

Rearranging Equation 4-27 first leads to:

$$1 + (\omega_0 - \omega)^2 T_2^2 + \gamma^2 B_1^2 T_1 T_2 = 2 + 2\gamma^2 B_1^2 T_1 T_2 \qquad \text{Eqn. 4-28}$$

Further rearrangement gives Equation 4-29:

$$\omega_0 - \omega = \frac{(1 + \gamma^2 B_1^2 T_1 T_2)^{1/2}}{T_2} \qquad \text{Eqn. 4-29}$$

Note here that $\omega_0 - \omega$ is the half line width so that the total line width, call it $\Delta \omega$, will be twice that, as shown in Equation 4-30.

$$\Delta \omega = \frac{2(1 + \gamma^2 B_1^2 T_1 T_2)^{1/2}}{T_2} \qquad \text{Eqn. 4-30}$$

If $\gamma^2 B_1^2 T_1 T_2$ is small compared to 1 then this equation simplifies to Equation 4-31.

$$\Delta \omega = \frac{2}{T_2} (\Delta \omega \text{ in rad/s}) \quad \text{or} \quad \Delta v = \frac{1}{\pi T_2} (\Delta v \text{ in Hz}) \qquad \text{Eqn. 4-31}$$

This calculation for "static" conditions may not seem very germane to either FT-NMR, that operates far from equilibrium, or even ordinary CW-NMR spectroscopy which, as a practical matter, is also a nonequilibrium process and, at best, only approximated by slow-passage equations. However, the line width in FT spectra is also determined by Equation 4-31. Indeed, slow-passage spectra with small $B_1$ (to avoid saturation effects) are fully comparable to FT spectra where full, or essentially full, decay of the FID is involved before the transform. This fact provides a common ground for FT and CW spectral comparisons.

## 4-4 NMR Under Non-Equilibrium Conditions. Initial Conditions

As I have said, the Bloch equations are not easy to solve in closed form but are very well-behaved on numerical integration. The reason they are well-behaved is because of the damping effect of $T_1$ and $T_2$. Setting up the numerical integration is quite straightforward and can easily be done on most hand-held programmable calculators, although plotting can be a problem. A variety of initial conditions can be chosen. For FT-NMR, one would normally assume that the sample has been allowed to come to equilibrium with the magnetic field by $T_1$ relaxation, before the pulse is turned on. For CW-NMR, the usual initial condition has the nuclei in the sample at equilibrium with both the field and the $B_1$ rf power. At the start of a CW sweep, the oscillator frequency $\omega$ is usually reasonably far from $\omega_0$ the resonance frequency.

Another variation is to turn on the oscillator and start the frequency sweep at the same time. This procedure initially produces what are called **Torrey oscillations**. These oscillations are the result of the interaction of the nuclei moments and oscillator, during establishment of an off-resonance quasi–equilibrium along the Z axis with the power input. The result is generation of some off-resonance $v$-mode signal before the main resonance (Figure 4-6). Other conditions can be visualized such as starting a frequency sweep at the instant that the sample is placed in the field.

## 4-5 Numerical Integration Procedure for the Bloch Equations

The way that the numerical integrations are started depends on the initial conditions. In the FT-NMR experiment, the input would be $M_0$, $\gamma B_1$ (a measure of the oscillator power), $\omega_0$ (the chemical shift, here we assume only one shift, but the extension to any desired number is quite easy), $T_1$, $T_2$, the pulse time, and the total desired observation time. A choice also has to be made of the size of the time increment

## 4-5 Numerical Integration Procedure for the Bloch Equations

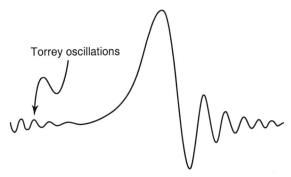

**Figure 4-6** The result of a CW sweep in which the nuclear spin system of the sample is allowed to come to equilibrium with the applied magnetic field, then the oscillator is turned on and the frequency sweep is started. The Torrey oscillations occur as the spin system tries to come to equilibrium with the $B_1$ field of the oscillator. The Torrey oscillations are not seen if the sample is allowed to equilibrate with the field and the rf input, before the sweep is started

suitable for numerical integration $\Delta t$ and also whether or not it is necessary to use some mathematical device, such as Runge-Kutta or predictor-corrector methods to improve the results of the integration. We will return to the latter problem shortly.

At zero time, in the usual pulse FT mode, the sample will be initially at equilibrium with the applied field so that $M_z = M_0$ and $v = u = 0$. Then, the first step in the integration will be derived from Equation 4-18 and is:

$$v = (\gamma B_1 M_0)\Delta t \qquad \text{Eqn. 4-32}$$

If $\gamma B_1$ is large, as for a 10 µs, 90° pulse, then $\Delta t$ should be small, perhaps 0.1 µs. If the total observation time of the pulse and FID is 5 s, then the integration would require a total of $5 \times 10^7$ steps. This number of steps is wholly unnecessary. The value of 0.1 µs can be used in the pulse period and subsequently changed to perhaps 0.001 s throughout the FID, during which period $\gamma B_1 = 0$. The initial $v$ value obtained with Equation 4-32 is then used with Equations 4-17 and 4-19 to calculate initial $u$ and $M_z$ values:

$$u = (\omega_0 - \omega) v \Delta t \qquad \text{Eqn. 4-33}$$

$$M_z = M_0 - (\gamma B_1 v)\Delta t \qquad \text{Eqn. 4-34}$$

Now, to proceed further, we need to increment the time, which at the start was zero.

$$t = 0 + \Delta t \qquad \text{Eqn. 4-35}$$

From here, we set up a loop, in computer terms, with calculations of $v$, $u$, $M_z$ and $t$ by successive repetitions of Equations 4-36 to 4-39.

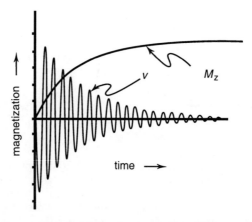

**Figure 4-7** Plot of a 90° pulse and free-induction decay calculated by numerical integration of the Bloch equations. Because the pulse occurs over 10 µs and the decay over 5 s, the change in magnetization during the pulse cannot be seen on the scale used. However, note that $M_z$ appears to start at zero and slowly returns to the equilibrium value. The value of $\omega_0$ for this FID was 5 Hz from the reference frequency, $T_1 = T_2 = 1$ s and $\gamma B_1 = 0.5$ which gives a 90° pulse when $M_z = 5 \times 10^{-6}$. The increment of time in the numerical integration over the FID was 0.001 s

$$v = v + \left[ -(\omega_0 - \omega)u - \frac{v}{T_2} + \gamma B_1 M_z \right] \Delta t \qquad \text{Eqn. 4-36}$$

$$u = u + \left[ (\omega_0 - \omega)v - \frac{u}{T_2} \right] \Delta t \qquad \text{Eqn. 4-37}$$

$$M_z = M_z + \left[ \frac{M_0 - M_z}{T_1} - \gamma B_1 v \right] \Delta t \qquad \text{Eqn. 4-38}$$

$$t = t + \Delta t \qquad \text{Eqn. 4-39}$$

After the pulse period, $B_1$ is set to zero and $\Delta t$ can be changed to 0.001 s. It is important to remember that $\omega$ is in rad/s so you need to divide $\omega$ by $2\pi$ to work with Hz. A simple True BASIC program to calculate $v$, $u$ and $M_z$ as a function of time is given in Appendix 3. The result of the above calculation is shown in Figure 4-7.

In this procedure, as described above, the so-called **Euler method** is used for the numerical integration. Here, each increment falls along the line determined by the current slope at the current value of the time $t$ and this can produce quite considerable error, especially if $\Delta t$ or $\gamma B_1$ are relatively large (see Figure 4-8). This difficulty can be ameliorated by the Runge-Kutta or the predictor-corrector procedures that are useful to

## 4-5 Numerical Integration Procedure for the Bloch Equations

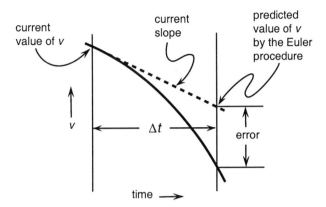

**Figure 4-8** Representation of the error attendant to the use of the Euler procedure for numerical integration of the Bloch equations in which the extrapolated value of v over the period Δt depends on the slope at the previous point

make more accurate predictions of the position of the next point. The Euler procedure works very well with Equations 4-36 to 4-39 for predicting the course of $v$, $u$ and $M_z$ in CW spectra, provided that $\Delta t$ is chosen to be reasonably small and $\gamma B_1$ is not too large. Sample curves for non-equilibrium frequency sweeps illustrate the success of the procedure quite well (Figure 4-9). If you use the Euler method for a particular $\gamma B_1$ and are uncertain as to whether or not it is giving the correct answer, try reducing the size of $\Delta t$ by a factor of two or more. If there is no significant change in the calculated curves, you can be confident that the numerical integration is working satisfactorily.

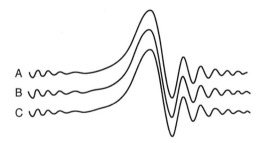

**Figure 4-9.** Results for a CW sweep obtained by numerical integration of the Bloch equations using different mathematical procedures to predict the position of the next point in the integration. For Curve A, the simple Euler procedure was used as described in the text. Curve B was obtained with the third-order Runge-Kutta method and Curve C with the predictor-corrector procedure. All of the methods used are described in detail in D.D. McCracken and W.S. Dorn, *Numerical Methods and FORTRAN Programming*, Wiley, 1964, Chap. 10. The parameters used were $M_0 = 5 \times 10^{-6}$, $\omega_0 = 2.5$ Hz, $B_1 = 1.0$, $T_1 = T_2 = 1$ s, sweep width 5 Hz, sweep time = 10 s and $\Delta t = 0.005$ s

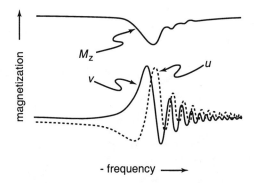

**Figure 4-10** Calculated NMR spectrum for a non-equilibrium frequency-sweep experiment. The spectral parameters $\omega_0$ = 5 Hz, $T_1 = T_2$ = 1 s, sweep width = 10 Hz and $\Delta t$ are as for Figure 4-3, except that here the sweep time is 10 s (not slow passage) and the sweep accords with the usual convention with decreasing frequency to the right. Note how $v$, $u$, and $M_z$ all show the relaxation-wiggle phenomenon

## 4-6 More on CW-NMR

We have seen how we can use numerical integration to calculate the various NMR magnetizations under nonequilibrium conditions as a function of time with relative ease. It is easy to calculate frequency-sweep (or field-sweep) CW spectra, such as the one shown in Figure 4-10 by numerical integration using Equations 4-36 to 4-39. All that is needed is an additional equation which includes the change of $\omega$ with each increment in time, $\Delta t$. For example, if there is to be a **frequency sweep** covering 25 Hz in 15 s, with $\Delta t$ = 0.001 s, then the extra program step will be as given in Equation 4-40.

$$\omega = \omega + \frac{25 \cdot 2\pi}{15} \Delta t \qquad \text{Eqn. 4-40}$$

If a field sweep is used to take the spectrum, $\omega$ will be constant and $\omega_0$ will decrease as the field increases, so the appropriate program step involves changing $\omega_0$ with time.

$$\omega_0 = \omega_0 - \frac{25 \cdot 2\pi}{15} \Delta t \qquad \text{Eqn. 4-41}$$

Such integrations with 1,000-2,000 steps require a few seconds, at most, on a reasonably fast personal computer, such as a PowerMac.

One problem encountered in CW-NMR is that the receiver coil detects the $v$ and $u$ magnetizations simultaneously and the total signal is not the usual absorption-mode signal, see Figure 4-11. The simple way to effectively separate out the absorption-mode signal is to leak a steady rf current from the oscillator to the receiver circuit (see Figure 4-12. This current is sent phase-shifted 90°, so that it is in phase with $v$. As the leakage

## 4-6 More on CW-NMR

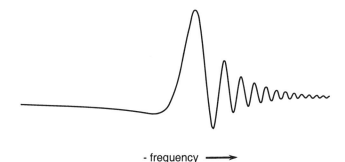

- frequency ⟶

**Figure 4-11** A CW spectrum that shows a mix of *u* and *v* magnetizations. The presence of the *u* mode is signaled by the droop in the spectrum before the principal peak. The spectrum can be corrected to essentially the *v* mode by the leakage technique, as described in the text, or other phase correction

current becomes large with respect to $u$, the resultant detected by the receiver coil becomes (leakage + $v$) and the contribution of $u$ to the signal becomes negligible except for a small phase shift. After detection, the leakage component becomes part of the base-line current and is easily balanced out, so that the nearly pure $v$ mode signal remains. This problem may not seem to be so obvious in FT-NMR, because the transform is expected to separate the signal into real and imaginary parts, but there can be serious phase problems with FT spectra, as discussed in Section 3-2 and as will be seen later in Chapter 5.

The **relaxation wiggles** so clear in Figure 4-9 to 4-11 are very noticeable features of non-slow passage CW-NMR, compare these figures with the slow-passage spectrum of Figure 4-3, which has the same sweep width and $v_0$. The wiggles come after the resonance peaks, and are prominent, especially if the sweep is fast and the resolution

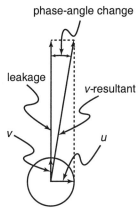

**Figure 4-12** Diagram showing how a rf leakage current that is in-phase with *v* can add to *v* and give a resultant that diminishes the relative importance of the *u* contribution, even when *v* and *u* are equally important, at the cost of a small phase shift

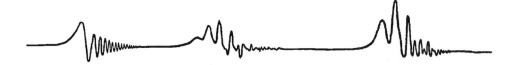

**Figure 4-13** A "high-resolution" 40-MHz proton CW spectrum of ethanol, acidified slightly with hydrochloric acid to cause rapid proton exchange of the hydroxyl protons and thereby eliminate their couplings to the methylene hydrogens. This spectrum was obtained in 1955 on one of the first commercially available spectrometers built by Varian Associates. The stability of the magnet and rf oscillator system was so short that routine spectra were usually taken over 7 s with the aid of a high-speed recorder in which a hot wire traced out the spectrum on a special thermally sensitive strip chart

good and $T_2$ long. In the early days of NMR, the stability of the spectrometer was so poor that fast sweeps were necessary and we used to take spectra at 7 s apiece using a high-speed recorder for perhaps twenty sweeps. After that, we would try to find two or three out of the twenty traces that were similar enough to be called "identical". But these "spectra" were usually plagued by relaxation wiggles that would run over peaks and obscure the spin-spin splittings. An example is a spectrum of ethanol (Figure 4-13) taken at 40 MHz in early 1955.

The form of the relaxation wiggles can be understood as the interaction of the decaying $M_{xy}$ moment, after passage through the resonance, with the changing frequency of the rf oscillator in a frequency-sweep experiment. Unlike an FID, the wiggle spacing decreases with time along the sweep, because of the increase in the absolute value of $\nu - \nu_0$. No relaxation wiggles are seen in FT-NMR after the transform, because all of their information is folded into the frequency domain.

## 4-7 Resolution in CW Spectra

Some of the important factors involved in the question of the **resolution** expected to be observed in NMR are very nicely illustrated by CW spectra. There are three major limitations to being able to detect line separations in NMR spectra. One limitation depends on the lifetimes of the nuclear magnetic states (Section 1-2) as to how the Heisenberg uncertainty principle governs this in the quantum-mechanical formulation of NMR. This limitation is a function of $T_2$, because $T_2$ determines the line width of a NMR transition (Section 4-4). The second limitation is in the way in which we make the observations. As we have stated before (Section 3-2), you cannot observe differences in frequency accurately if you take data over too short a time. The third limitation derives from the apparatus used in our experiments: if the magnetic field is not

## 4-7 Resolution in CW Spectra

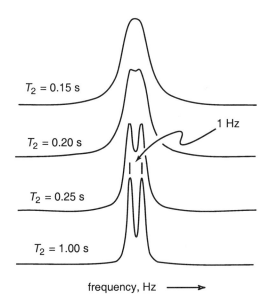

**Figure 4-14** Slow-passage NMR spectra of two resonances separated by 1 Hz, each of which has $T_1 = 1$ s and values of $T_2$ ranging from 1.0 to 0.15 s. The sweep width is 20 Hz

homogenous or not stable enough; if the oscillator or reference frequencies are not stable; if the digitization is inadequate; and so on. In what follows, we will only be concerned with the first two limitations and will assume that the field is perfectly homogenous and the oscillator perfectly stable, that is to say $T_2 = T_2'$.

Suppose that we have two NMR lines separated by 1 Hz. We know that the line widths will depend on $T_2$ and Figure 4-14 shows what we would find in slow-passage spectra as a function of $T_2$. It is interesting that the observable separation between the lines disappears just between $T_2 = 0.15$ s and $T_2 = 0.20$ s. This accords favorably with the 0.16 s suggested by the Heisenberg uncertainty principle (Section 1-2). Clearly, values of $T_2$ provide fundamental limitations to obtainable resolution of closely spaced lines.

We can illustrate how the manner in which we take spectra can limit the obtainable resolution by the curves in Figure 4-15. This figure is similar to Figure 4-14, except that now the $T_2$ value of 1 s is used throughout and, from Figure 4-14, we know that this $T_2$ is long enough to allow the lines to be resolved. However, here we use nonequilibrium sweep conditions and the time taken in passing through two peaks separated by 1 Hz can easily be long enough, or can be too short, to allow a separation to be observed. Thus, when the sweep rate is 1 Hz per 50 s (the lower curve in Figure 4-15), we find that this is essentially slow passage and the curve is identical to that of the

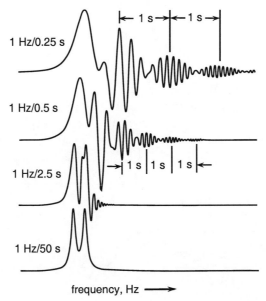

**Figure 4-15** Non-equilibrium-sweep NMR spectra of two resonances separated by 1 Hz, each of which has $T_1 = T_2 = 1$ s and values of the sweep rate ranging from 1 Hz/50 s (essentially slow passage) to 1 Hz/0.25 s. The sweep width is 20 Hz, just as in Figure 4-14. The most rapid-sweep spectrum shows many of the elements of a FID, except that the spacing of the relaxation wiggles changes as the frequency is changed away from resonance. See also Section 2-2 for another example of an echo train in a FID corresponding to a peak separation of 1 Hz

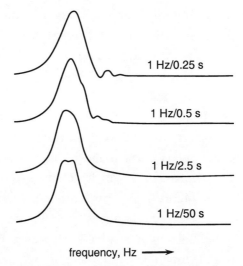

**Figure 4-16** Non-equilibrium sweep NMR spectra of two resonances separated by 1 Hz, each with $T_1 = 1$ s and values of the sweep rate ranging from 1 Hz/50 s (essentially slow passage) to 1 Hz/0.25 s. The sweep width is 20 Hz, as in Figures 4-14 and 4-15. In these spectra, $T_2 = 0.20$ s, close to the limit of observing separate peaks with slow passage, see Figure 4-14. Each spectrum shows evidence of splitting, but except for the slowest sweep experiment, it is not easy to measure the splitting

4-7  Resolution in CW Spectra                                                                 71

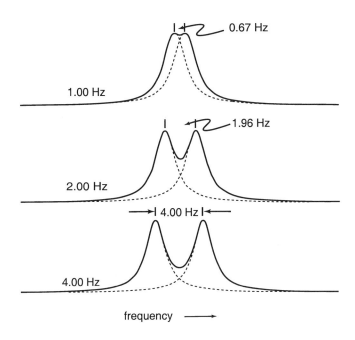

**Figure 4-17** Demonstration of the "pulling-together" effect that results with two closely spaced and partially overlapping resonance peaks. The measured peak separations can be rather far from what would be expected from the actual line positions. Equation 4-23 for *v* under slow-passage conditions can be easily adapted to generate sets of plots of closely spaced lines with the same, or different, line widths and intensities, to allow estimation of how much correction needs to be applied to measured separations to give the true differences in the peak positions

corresponding curve in Figure 4-14. In contrast, the upper curve of Figure 4-15, with a sweep rate of 1 Hz per 0.25 s, provides conditions where we expect that it will be difficult to see the peak separation. It is amusing that, although the peak separation can not be seen directly at this sweep rate, we can detect that there is a 1-Hz line separation by the 1-s intervals between the peaks of the beat frequencies of the two resonances in the relaxation wiggles. It is possible to detect the line separation here simply because of the longer observation period for the relaxation wiggles! We encountered similar behavior earlier, see Section 2-2.

Figure 4-16 shows how difficult it is to observe the peak separation when the $T_2$ value is such that you would expect the separation to be borderline, even in slow passage. Throughout this series of spectra, there is evidence that more than one peak is present, but it is only easy to measure the peak separations under essentially slow-passage conditions. However, you should be aware that measuring the separation of two closely spaced peaks is not as obvious as it might seem. Simple measurement of the separation between the tops of the peaks can give quite the wrong answer. Figure 4-17 illustrates the problem. The overlap of the peaks tends to fill up the center area between them and the result is to make the peaks appear to be closer together than they actually are. Thus, when we have two peaks that are 1.00 Hz apart, with line widths corresponding to $T_2$ values of 0.25 s, the observed separation of the peak tops is about 0.67 Hz, off by more than 30%. Clearly, when these peaks are two or three Hz apart, the error is nearly negligible. When it is important to know the actual line positions of overlapping, closely spaced peaks, it may be necessary to generate sample plots with appropriate line widths, such those in Figure 4-17, with the aid of the slow-passage equations, to see how the measured separations need to be corrected to give the true line separations. Such procedures will be particularly important when the peaks are not of the same height and/or with different line widths.

## 4-8 Rapid-Scan (Correlation) Spectroscopy. The Inverse Fourier Transform

A very useful variation on CW-NMR is to run the sweep quickly and digitize the spectrum, relaxation wiggles and all. The digitized spectrum can be **correlated** with the response of a single resonance line taken under the same conditions (or even generated theoretically) and converted to a normal-appearing FT-NMR spectrum.[1] The top spectrum of Figure 4-15, which is reproduced with more data points in Figure 4-18 (**A**), is just the general kind of spectrum that can be profitably used in this sort of endeavor, because it has characteristics of both a slow-passage CW spectrum and a FID.

The procedure for converting the rapid-scan spectrum to the correlation spectrum has several steps, but these, although computer-intensive, do not necessarily involve operator intervention. The steps do illustrate several important elements in the use of Fourier transforms in dealing with NMR spectra, so they will be described here in some detail. A key operation is the **inverse Fourier transform**. As applied to the top spectrum of Figure 4-18, this provides conversion from the frequency domain to the time domain. The mathematics of the inverse FT are essentially the same as the forward transform (Section 3-2), except that there is a positive sign for the imaginary part of the equation.

$$X(n) = \frac{1}{n}\sum_{n=0}^{N-1}\left[x(n)\frac{\cos(2\pi kn)}{N} + ix(n)\frac{\sin(2\pi kn)}{N}\right] \qquad \text{Eqn. 4-42}$$

---

[1] J. Dadok and R.F. Sprecher, *J. Magn. Resonance* **13**, 243-248 (1974); R.K. Gupta, J.A. Ferretti and E.D. Becker, *ibid.* **13**, 275-290 (1974).

## 4-8 Rapid-Scan (Correlation) Spectroscopy. The Inverse Fourier Transform    73

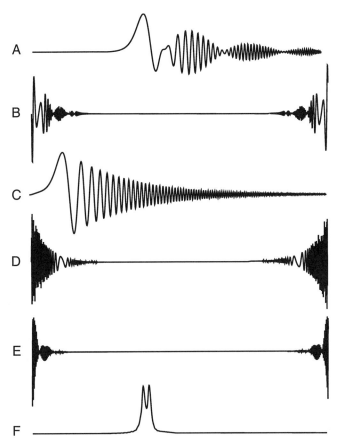

**Figure 4-18** A sequence for the transformation of a rapid-scan CW spectrum **A** to its final form as a correlated spectrum **F**. The original spectrum has two peaks, separated from one another by 1 Hz. Each has $T_1 = T_2 = 1$ s. The sweep time was 5 s and the sweep width 25 Hz. The real part of the inverse transform of the original spectrum is **B**. The reference spectrum **C** is a single resonance with a shift of 5 Hz and was taken to have $T_1 = T_2 = 1$ s. Its inverse transform is **D**. Only the real part of each transform is shown, although both real and imaginary parts were used in the calculations of the correlated FID **E** and in the final forward transform that yields the correlated spectrum **F**. The imaginary counterpart of **F** has a horizontal baseline with no peaks

However, it should be noted that, in calculating the inverse transform, the individual x(*n*) terms, as might be obtained from another Fourier transform, will usually have **both** real and imaginary components, and both must be taken into account in making the summations. It is important to take proper account of sign and remember that $(-i)^2 = -1$. It is simplest to use either a pure real-mode (*u*), or an imaginary-mode (*v*), spectrum as the rapid-scan spectral input into the inverse transform. To save on space, Figure 4-18 shows only the real part (**B**) of this inverse transform. The imaginary part is similar in appearance, but not identical.

The next step in the conversion is to do an inverse transform on the reference spectrum (**C**) and again only the real part of the product (**D**) is plotted in Figure 4-18. The reference spectrum provides the key information content for the range of frequencies to be correlated in the rapid-scan spectral input. (In fact, it is not actually necessary to use the response derived from an experimental resonance in the correlation process. A suitable theoretical spectrum serves quite well.[1])

The correlation can be carried out in more than one way, but the easy way is just to perform a point-by-point multiplication of the two inverse transforms, which operation yields the time-domain transform of the final correlated spectrum.

$$A(n) = X(n) \cdot X'(n) \qquad \text{Eqn. 4-43}$$

In Equation 4-43, $X(n)$ and $X'(n)$ are the two inverse transforms of the input and reference spectra, respectively (both of which will usually have both significant real and imaginary parts). The real part of the correlated transform is shown in Figure 4-18 and it is the equivalent of the real part of a FID of the resonances lines **in the segment of frequencies covered by the sweep**. A forward Fourier transform of this FID into the frequency domain then gives the final correlation spectrum (**E**) of Figure 4-14.

A very obvious question is why go through all of this complex maneuvering when the usual FT procedure for taking spectra seems so much more simple, straightforward and direct. To make the case even weaker for correlation spectroscopy, few modern FT spectrometers are appropriately equipped to offer the possibility of taking rapid-scan CW spectra, or even any CW spectra at all for that matter. The advantage of correlation spectroscopy lies in its ability to examine narrow-frequency ranges in a spectrum, with much more efficient use of observing time than possible with ordinary slow-scan CW spectroscopy. In the usual FT-NMR, the rf pulse excites all of the resonances and, if the spectral range is large, there are potential phase and intensity problems (Section 2-5). Furthermore, trying to observe weak resonances in the presence of very strong ones can cause serious difficulties in FT-NMR, a matter which will be discussed further in Chapter 5. With correlation spectroscopy, interfering resonances can usually be excluded by simply choosing the proper sweep ranges. Hitachi currently makes a spectrometer for routine NMR using the rapid-passage correlation procedure.

## 4-9 CW Adiabatic Rapid-Passage Dispersion-Mode Spectra

Another interesting form of CW spectroscopy which borders on FT produces **adiabatic rapid-passage dispersion-mode** spectra. It is similar to FT-NMR in that it

## 4-9 CW Adiabatic Rapid-Passage Dispersion-Mode Spectra

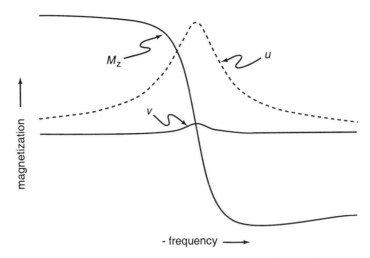

**Figure 4-19** The magnetization components calculated for a CW, rapid-scan, high-power NMR (adiabatic rapid-passage) spectrum. The parameters for this spectrum were: sweep width = 5000 Hz, sweep time = 0.1 s, $\gamma B_1$ (rf power level) = 2000, $T_1$ = 0.5 s, $T_2$ = 0.1 s. Note how, although the v-mode (absorption) signal is essentially saturated, the u-mode (dispersion) signal shows a high fraction of the maximum possible intensity, which would be equal to the equilibrium value of $M_z$

involves a pulse, but a long pulse compared to what we have talked about before. However, it is a short pulse compared to relaxation times (thus the designation **adiabatic**). What is different from FT-NMR is that the signal is observed **during** the pulse. The result of such a procedure calculated for a sweep width of 5000 Hz, a sweep time of 0.1 s, the resonance at 2500 Hz, $T_1$ = 0.5 s, $T_2$ = 0.1 s and the $B_1$ power cranked up sufficiently to give a very substantial turn to the original $M_z$ magnetization is shown in Figure 4-19. This spectrum differs quite a bit from the usual CW spectrum of something like Figure 4-10. Indeed, it might fairly be regarded as downright ugly. What is it good for? Let's look at it a bit.

First, you will note that the $u$ component is quite strong and has a different shape at the resonance frequency as compared to Figure 4-10. Furthermore, the $v$ mode component is quite weak, as the result of being saturated by the strong rf field. Herein lies the advantage of this form of spectroscopy in that the $u$ mode is less subject to saturation than the $v$ mode. However, to be useful, we need to isolate the $u$ signal out of what is actually observed, a composite of the $u$ and $v$ signals, although in this case, the correction will be small, because of the weakness of the $v$-mode signal. The correction is easy to make with the aid of the leakage-signal technique (Section 4-6). Here, where we want $u$, rather than $v$, we leak a signal that is **in-phase** with the oscillator to give $u$ separated from $v$, at the cost of a small phase shift. The outcome, Figure 4-20, is

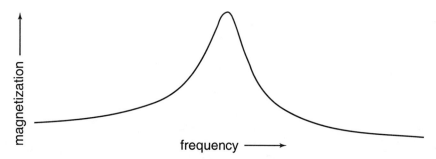

**Figure 4-20** The *u*-part of the CW adiabatic rapid-passage spectrum of Figure 4-19. The *u*-mode component of the signal was separated from the concurrent (weak) *v* component by leakage of a 10-fold stronger *u*-mode leakage signal with the aid of the technique described in Section 4-6. The resultant was adjusted for the baseline level and then inverted to give an upright peak. All of these operations are easy to perform on a CW spectrometer

an essentially pure *u*-mode peak, when the leakage signal corresponds to 10 times that expected from a 90° pulse.

The advantage of the adiabatic rapid-passage technique is that the *u*-mode signal has a lot of intensity, with much less tendency to saturate than a *v*-mode signal. The line widths are not good in adiabatic rapid-passage and, for this reason, it is primarily used when the chemical shifts and couplings are large. A very important early application was to obtain the first NMR signals by Felix Bloch and his coworkers from water in 1946.[2] Another critical early application was for natural-abundance $^{13}C$ spectra.[3] Nowadays, it is most often employed for rapid location of spectral peaks for closer examination by slow-passage or FT-NMR, with the $B_1$ power being adjusted to give the desired compromise between signal strength and line width. You will find that the lineshapes of adiabatic rapid-passage spectra are very strongly influenced by the sweep rate, the $B_1$ power level and both the $T_1$ and $T_2$ relaxation times.

Another use of adiabatic rapid-passage can be seen from Figure 4-19. If we were to stop the sweep when $v \sim v_0$, then we would find $M_z = 0$ and $u \sim M_{z0}$. The result is an essentially 90° pulse produced by a $B_1$ field applied along the X axis, but the magnetization now lies along the +X axis, not along the Y axis! This provides a quite different way of producing magnetization in the X, Y plane. Figure 4-19 also shows that taking the sweep to just beyond $v \sim v_0$ makes $M_z$ strongly negative, although not so much as to give a clean 180° pulse, because *u* is not zero at the maximum negative value of $M_z$.

---

[2] F. Bloch, W.W. Hansen and M. Packard, *Phys. Rev*, **69**, 127 (1946); see also, E.U. Purcell, H.C. Torrey and R.V. Pound, *Phys. Rev.* **69**, 37 (1946). Bloch's formulation of his famous equations and a detailed description of his early experiments is in rather readable form in F. Bloch, *Phys. Rev.* **70**, 460-474 (1946) and F. Bloch, W.W. Hansen and M. Packard, *Phys. Rev.* **70**, 474-485 (1946).
[3] P.C. Lauterbur, *J. Chem. Phys.* **26**, 217-218 (1957).

## Exercises

**Exercise 4-1** Make a diagram similar to Figure 4-1 that illustrates each of the components that contribute to Equation 4-4.

**Exercise 4-2** Show how Equation 4-18 can be derived from Equation 4-12 in a manner similar to the derivation of Equation 4-17 from Equation 4-13.

**Exercise 4-3** Show how Equation 4-19 can be obtained from Equation 4-9.

**Exercise 4-4** Show how the peak height of a slow-passage peak at resonance (*i.e.*, $\omega = \omega_0$) could be used to determine $T_1$ on the basis of Equation 4-23.

**Exercise 4-5** Show how Equation 4-25 can be derived from Equation 4-23.

**Exercise 4-6** Figure 4-3 shows that, in slow-passage conditions, the *u*-mode signal has zero intensity when $\omega = \omega_0$ and positive and negative maxima, respectively before and after the point where $\omega = \omega_0$.
**a.** Use Equation 4-24 to derive an expression for the horizontal separation in Hz between the positive and negative peaks of the *u* mode signal.
**b.** Derive an expression from your result in **a.** for the absolute value of the peak maxima of the *u*-mode signal as a function of the $B_1$ power level.

**Exercise 4-7** Use the numerical integration program listed in Appendix 3 (or write your own program) and the parameters of Figure 4-10 to calculate the course of a CW spectrum when the sample is placed in the field and the frequency sweep is started at the same instant.

**Exercise 4-8** Use the numerical integration program listed in Appendix 3 with the parameters of Figure 4-10 and increase the values of $\Delta t$ until you can see a significant change in the calculated spectrum.

**Exercise 4-9** Use the numerical integration program with parameters of Figure 4-4 to determine how well Equation 4-25 applies to non-slow-passage conditions.

**Exercise 4-10** Use the numerical integration program with the parameters of Figure 4-4 to investigate the relationship between the shape and duration of the relaxation wiggles and $T_2$.

**Exercise 4-11** Figure 4-15 shows how, under conditions where the primary resonances do not show any splitting, that a splitting can be detected and measured from the shape of the relaxation wiggles. Assuming the shift difference and the sweep rate remain the same, what will be the limiting factor that determines whether or not the relaxation wiggles will show the splitting.

**Exercise 4-12** You will note from Equation 4-22 that, at the resonance point ($\omega = \omega_0$), there is no possibility that $M_z$ can be negative in a slow-passage experiment. However, Figure 4-19 shows clearly that this is not so in adiabatic rapid-passage. Explain how and why there is a difference in this regard between these two different ways of running a CW spectrum.

**Exercise 4-13** Use the numerical integration program with the parameters of Figure 4-18 to make a graph of the simultaneous changes in $v$, $u$ and $M_z$ for one of the resonances in this rapid-scan spectrum. Compare your result with what happens in true adiabatic rapid passage as shown by Figure 4-19.

**Exercise 4-14** Show how Equations 4-12 and 4-13 can be derived starting with diagrams like Figure 4-2.

**Exercise 4-15** Make vector diagrams based on Figure 2-6, but using a rotating frame of reference, that show the magnetizations; at the start of the sweep of Figure 4-19, when $v \sim v_0$, and when $v$ is slightly greater than $v_0$.

# 5
# The NMR Fourier Transform and its Problems

$F$ourier lived a long time ago (1768-1830) and NMR got started in 1945, but FT-NMR only became pervasive thirty years later. Why was that? As a historical aside, it is noteworthy in this connection, that Fourier was not able to get his work on the transform published until 15 years after it was first submitted! Possibly that is no more than an omen; because, although FT-NMR did come on rather slowly, today, FT-NMR has come to be used even in elementary organic chemistry laboratories. Some of the reasons for Fourier transforms coming on slowly in NMR are simple. The first requirement was for ways of digitizing spectral data in a straightforward routine fashion; second, relatively inexpensive and quite fast computing equipment was needed to calculate the transforms; and third, a need had to be established.

## 5-1 The Rise of $^{13}$C NMR and the Need for Fourier Methods

The real impetus for FT-NMR came with the demonstration of the utility of natural-abundance $^{13}$C NMR spectra in routine structure determinations. Because $^{13}$C has one-fourth the γ of protons and a natural abundance of only about 1.1%, its NMR in the early conventional CW-mode was a daunting proposition. This was particularly true with compounds where the carbons carried hydrogens or fluorines, because then spin-spin couplings split the resonances into many smaller peaks.

In 1966, a new spectrometer with unprecedented stability and capability for accumulating CW spectra in digitized form became available[1] and demonstrated powerful structural applications of $^{13}$C NMR with a variety of relatively small molecules. However, with larger molecules, such as cholesterol, $^{13}$C spectra (at least at 15 MHz) were almost unintelligible, because of multitudes of resonances arising from $^{13}$C-H spin-spin splittings, involving not only the carbon nuclei being observed and hydrogens directly attached to them, but also the same carbons and hydrogens on adjacent, or even more remote, atoms. The solution to this problem was **proton**

---

[1] F.J. Weigert and J.D. Roberts, *J. Am. Chem. Soc.* **89**, 2967-2969 (1967).

noise (or broad-band) decoupling[2] that removed the proton couplings and caused each carbon resonance to be a single sharp line. Fortunately, nature kept the natural abundance of $^{13}C$ small enough so that the residual splittings arising from carbon-carbon couplings amount to only about 0.02% of the height of the total $^{13}C$ signals and, only in exceptional cases, do these splittings interfere with the interpretation of a $^{13}C$ NMR spectrum in which their absence is assumed. Nonetheless, carbon-carbon couplings are still measurable at the natural-abundance level and are important parameters for determining structures.[3]

Another bonus that comes with proton decoupling is the carbon-proton **Nuclear Overhauser Effect** (NOE) that enhances the signal strength by a substantial degree over and above what would be expected from removal of the proton-carbon splittings. We will discuss the reason for the NOE later. Let it suffice to say for now, that the NOE can increase the heights of the individual carbon resonances substantially more than just removing the couplings would lead one to expect.

The result of the introduction of proton decoupling was to make most $^{13}C$ spectra into a sequence of sharp spikes and because, see Section 2-5, the shifts of $^{13}C$ routinely span 200 ppm, there is often a lot of featureless baseline between the peaks (for example, see Figure 5-1). For this reason, repetitive CW scanning is not very efficient. If there are five peaks, each 6 Hz wide at the base of the peaks, in a 3000-Hz scan, actual signal data will be collected over only about 1% of the total time spent scanning. In contrast, suppose the same peaks had widths corresponding to $T_2$ values of 0.5 s and $T_1$ values of 2.5 s, then, as we will see later, to prevent significant saturation of the resonances in pulse FT-NMR, it is desirable to wait perhaps 10 s between successive 90° pulses. This is a $T_1$ problem and you need to wait; even though, with a $T_2$ of 0.5 s, the signal will effectively disappear in about 2 s. In this circumstance, signal collecting is only fruitful over the first 2 s after a rf pulse (or less, if $T_2' < T_2$) and about a 10-s wait is needed before giving the sample another pulse. This means that you will be collecting data for 2 s out of 10 s, which amounts to a time efficiency for FT-NMR of about 20%. Obviously, while this is not much, it is a vast improvement over the 1% calculated above for the CW-NMR example, where you gather signal for a total of about 30 Hz across a 3000-Hz sweep.

The original answer to the question as to why would we want to use Fourier

---

[2]F.J. Weigert, M. Jautelat and J.D. Roberts, *Proc. Nat. Acad. Sci.* (USA) **60**, 1152-1155 (1968)
[3]F.J. Weigert and J.D. Roberts, *J. Am. Chem. Soc.* **89**, 5962 (1967), **94**, 6021-6025 (1972); for more up-to-date compilations see H.-O. Kalinowski, S. Berger, and S. Braun, *Carbon-13 NMR Spectroscopy*, John Wiley & Sons, New York, 1988 and other references in Appendix 1.

## 5-1 The Rise of $^{13}$C NMR and the Need for Fourier Methods

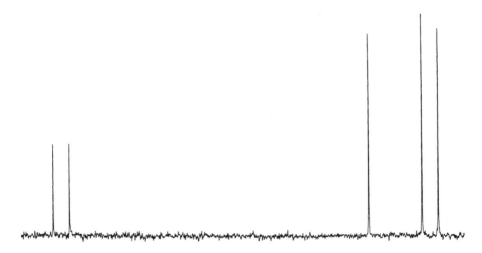

**Figure 5-1** A natural-abundance $^{13}$C spectrum taken with proton noise decoupling at 25 MHz of monosodium glutamate, $Na^+$ $O_2CCH_2CH_2CH(NH_3^+)CO_2^-$. Decoupling greatly simplifies the spectrum by removing the proton-carbon spin-spin splittings and, in addition, enhances the resonances to a variable degree by the proton-carbon nuclear Overhauser effect (NOE). All of the peaks in this spectrum are expected by simple theory to have equal intensities, but the NOEs for the left-hand peaks, that arise from the two carboxyl carbons (which are not directly connected to protons), may be expected to be less than for the peaks on the right-hand side, which arise from the proton-carrying methine and methylene carbons. Another factor that may contribute to the lower intensities of the left-hand peaks is the fact that the relaxation times of the carboxyl carbons are normally much longer than for saturated carbons

transform was because FT provided a way to improve S/N ratios for weak signals and it is still very important for that purpose. However, once one is freed from the necessity of trudging through a CW spectrum with the rf oscillator on at all times, a whole new world of NMR became possible. In this world, the nuclei are pulsed and then, while undergoing what would otherwise be free-induction decay, are massaged with a myriad of carefully timed pulses, delays, magnetic-field gradients, decoupling frequencies, and so on. The result is virtually a new discipline, one that is almost wholly computer driven, because of the complexity of the commands and the vast arrays of accumulated intensity data. The result is that NMR spectra can now be displayed in multidimensional formats and these are revolutionizing analysis of complex NMR spectra by showing relationships between shifts and couplings, which previously had to be dug out by very tedious procedures. Also, multidimensional spectra combined with the aforementioned nuclear Overhauser effects, to be described in detail later, can provide internuclear distance information for peptides, proteins and other complex molecules in solution, and this data can be very useful for determining solution conformations of such molecules.

## 5-2 What Are Your Choices and the Restrictions on FT-NMR?

With the fantastic efficiency of FT-NMR, every pulse potentially working for you, it was not long before $^{13}C$ NMR became really routine. To be sure, as we shall discuss in more detail later, you best should wait something like 5-10 times $T_1$ between 90° pulses to diminish saturation of the resonances if you want accurate integrals, but what else is there to worry about with FT-NMR? The fact is that the programs and hardware of most up-to-date FT-NMR spectrometers offer few problems, and yet there are still quite a few other things you need to be aware of that do not cause comparable concern with the analog output of CW-NMR spectrometers.

**1. The Nyquist Frequency.** In our earlier illustration of the FT procedure (Section 3-2), we found that the frequency range that can be covered by a given digital Fourier summation is determined by the length of the sampling period, $\Delta t$, and the maximum for accurate frequency measurements is $1/(2\Delta t)$, the Nyquist frequency. Clearly, if a wide frequency range is to be covered, the sampling time must be very short. This was once quite restrictive for FT spectroscopy because of the limited number of data registers in the available digitizers, but is not usually a problem with modern digital-sampling equipment. However, remember that if the Nyquist frequency is not large enough to cover the frequencies of the spectral peaks associated with your sample, you will see peaks at spurious frequencies in your spectrum. We will discuss this problem further a little later.

An interesting sidelight on the Nyquist frequency relates to digital sound recording on compact disks (CD's). To have fidelity of sound reproduction at as high as 20,000 Hz (rather well above the auditory range of most humans), the sampling rate is taken to be on the order of 40,000 times per second.

**2. Pulse Width.** In our early discussions (Section 2-5), it was shown how the frequency range that can be covered by FT-NMR is related to the length of the pulse. It takes a very short pulse to produce reasonably uniform 90° pulses over a substantial range of nuclear precession frequencies. Again, with most modern spectrometers this is not a serious restriction, although the difficulties obviously increase with the trend to spectrometers with ever-higher magnetic fields. In any case, you should be aware that the peaks at one or the other (or both) ends of a broad frequency range may not receive a full 90° pulse and also may have serious phase problems.

**3. Pulse Shape.** A 10-µs, 500-MHz 90° pulse will cover 5000 waveform cycles. As a result, one might expect that it would be important for the rf pulse to essentially

## 5-2 What Are Your Choices and the Restrictions on FT-NMR?

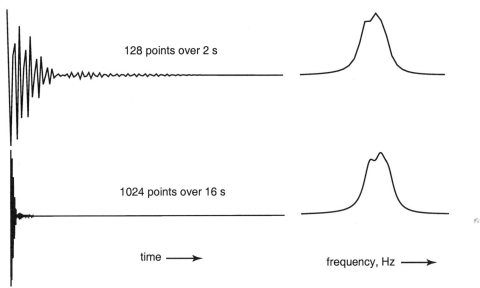

**Figure 5-2** The FIDs resulting from 90° pulses and the corresponding Fourier transforms calculated for a system of three magnetic nuclei, each with $T_1 = 1$ s and $T_2 = 0.25$ s. The resonance frequencies of the nuclei are respectively, 24.5, 25.5 and 26.25 Hz with relative concentrations of 1 : 1 : 0.75. The sampling rate in both FIDs was the same, 0.0156 s, so that, for each, the Nyquist frequency was 32 Hz. The upper curve had 128 data points over 2 s, while the lower one had 1024 data points over 16 s. It is important to recognize that the first 128 points of each of these FIDs are exactly the same. What is different is the duration of the sampling period. Data were taken over a much longer period for the lower curve, even though the signal decay was essentially complete in 1 s. Here, $T_2$ is relatively short; indeed, it corresponds to a linewidth of about 1 Hz. Consequently, we cannot expect to see much separation between the peaks (Section 4-7). Actually, there is not much difference between the resulting transforms, except that the lower one is a bit smoother, as expected with many more data points. Another point worth noting is how the two resonances on the right-hand side in the frequency domain are drawn together, see Section 4-7

be a square wave, with a sharp rise time, a flat top, and a sharp cutoff. However, sample calculations show that, provided the pulse is short, has the prescribed frequency, covers the same time interval and has the same integrated intensity, the shape does not make much difference in producing a 90° pulse, even for well away from the resonance frequency. Some special effects of changing the nature of the pulse will be discussed later.

**4. Resolution.** The resolution obtainable in FT-NMR depends on the $T_2$ or $T_2'$ values and on the manner in which the FID data is acquired. With regard to the latter, one factor is the **sampling period** or **acquisition time**. Another factor is the **number of data points** taken over the acquisition time. First, let us deal with the acquisition time, which, in practice, can be severely limited by $T_2$ (or $T_2'$). Obviously, if the FID has decayed to negligible values in 1 s, an acquisition time of 16 s will not give any better resolution than an acquisition time of 2 s, see Figure 5-2. However, if

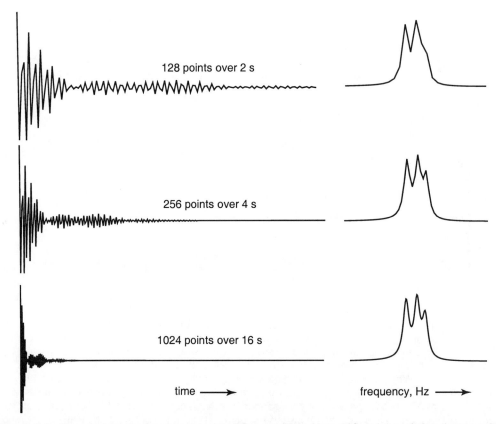

**Figure 5-3** The nuclear parameters for the FIDs resulting from 90° pulses of the three-nucleus system are the same as for Figure 5-2, except that now each nucleus has $T_2$ = 0.5 s. The sampling rate for all of the FIDs was the same, 0.0156 s, so that, for each data set, the Nyquist frequency was 32 Hz. The upper curve had 128 data points over 2 s, the middle one 256 data points over 4 s, while the lower one had 1024 data points over 16 s. Again, it is important to realize that the first 128 points of each of these FIDs are **exactly the same**. The differences in the transforms are in the **sampling periods**. Because $T_2$ is now twice as long as for Figure 5-2, we expect to be able to see separation between peaks separated by even 0.75 Hz. What is interesting here is how, even when the FID information (essentially all of which is contained in the first 2 s) is exactly the same, the transforms with the larger numbers of data points show a much cleaner separation of the lines. Note again, indeed even more clearly than in Figure 5-2, the "drawing-together" effect of closely spaced spectral peaks shown in Figure 4-17

the longer acquisition time involves more points for a given spectral width, the spectral trace can appear to be smoother, even if not actually more informative as to the degree of resolution. In the event that $T_2$ and $T_2'$ decays are long, then the acquisition time is very important to the resolution. If frequencies in a FID differ by say 1 Hz, they will only become 180° out-of-phase in 0.5 s; so to measure their separation one would expect to need an acquisition time longer than 0.5 s (see Section 4-7). This point is well illustrated by Figure 5-3, which shows how a spectrum

## 5-2 What Are Your Choices and the Restrictions on FT-NMR?

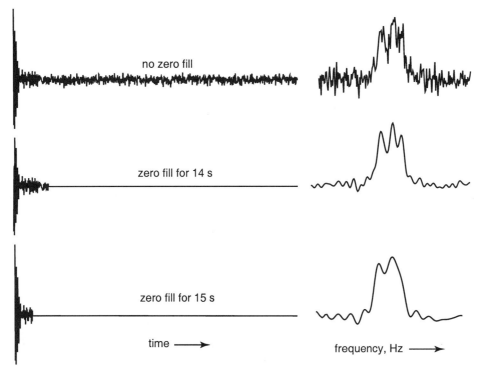

**Figure 5-4** The FIDs and transforms here are for the same set of nuclei and spectrometer parameters as used in the bottom sequence of Figure 5-3, except that a substantial component of real-world electronic noise has been added to the FIDs. From the top-to-bottom sequence, there are different degrees of **zero filling** to smooth out and improve the appearance of the spectra. However, in the bottom sequence, the zero filling is started before all of the key spectral information important to the resolution is in hand

is smoothed and the resolution appears to be much better using a longer acquisition time with more points in the Fourier transform. The surprise is that these benefits are achieved even with **exactly the same** input of actual spectral information into the transform. The only difference between the three frequency-domain spectra of Figure 5-3 is that latter two of the FIDs have many more points of **negligible spectral intensity**. What is happening is, with more points, the transform "grinds" finer, even if it does not actually produce more spectral information.

So, you say to yourself, "Aha, if all of those extra 'non-points' will surely improve the appearance and interpretability of my spectra, I will therefore take advantage of that by increasing my acquisition times." But, you may figure out that a tradeoff will be necessary, because long acquisition times, while accumulating repeated FIDs, but not actually getting significant spectral data, will hardly improve your time efficiency in taking spectra. But wait! Despite the significant gains that

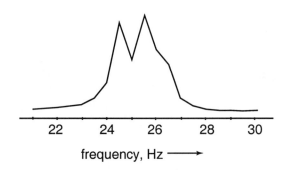

**Figure 5-5** The same spin system and spectrometer parameters as for the lowest spectrum shown in Figure 5-3, except that the digital resolution is only 2 points per Hz. This kind of loss of resolution is very common in FT-NMR spectra taken at strong magnetic fields where the range of chemical shifts to be covered is often very large

appear possible on the basis of Figure 5-3, there is a severe price to pay in the real world by using a strategy which merely involves increasing the acquisition times to get smoother, better-appearing, spectra. This will be seen in the top sequence of FID and FT of Figure 5-4. The problem is that, in the real world, NMR signals contain more or less random noise and if you use a long acquisition time just for the purpose of getting more data points into the transform, rather than actual spectral information from your sample, you will be accumulating noise in those data points. And this noise shows up in your transform to the frequency domain as a hodgepodge of random frequencies.

Now, being clever, you will be quick to realize that you can get those extra data points for **free, with no cost in time,** by simply telling your spectrometer's computer to use a shorter acquisition time, just the one appropriate for your sample; and then, when it reaches the end of that acquisition time, add a string of zeros to fill out a larger spectral array than you might normally have used for your particular acquisition time. This process is called **zero filling**. And indeed, it works, as you can see from the middle sequence of Figure 5-4, in that a much cleaner spectrum is produced, and this spectrum also shows the peaks better resolved. The lower sequence of Figure 5-4 shows that zero filling can be overdone, by starting it before the FID has yielded the maximum of its useful information. Here, the cost is in lessening the spectral resolution.

The number of data points taken over the FID is also important to the resolution, if $T_2$ (or $T_2'$) is not limiting. To see this, suppose that the spectral width is 10 ppm, as would be nominal for a proton spectrum. Then, if the spectrum is taken at 500 MHz, the width will be 5,000 Hz and, if you want five points to define a

## 5-2 What Are Your Choices and the Restrictions on FT-NMR?

**Figure 5-6** The Fourier transform resulting from a FID of a 1:2:1 triplet with a coupling constant of 30 Hz, where overflow of the data registers for the FID has been simulated by having the values in the individual data registers limited to about 50% of the maximum value found for the normal FID. It will be seen that the process causes distortion of the baseline, but does not affect the line positions or the line widths, at least to the degree that the FID intensities are limited in this example

peak 1 Hz wide at half height, you would need about 25,000 points in the FID! With fewer points, one can not be sure that the shape, or width, of the peak is well-defined, particularly if the S/N ratio is low (look at Figure 5-5). One can easily see that with a very high-field spectrometer there will be complex tradeoffs between available memory, spectral width, resolution, $T_2$ (or $T_2'$) and the time of computing the transform.

**5. Data Registers.** Besides the number and spacing of the data registers, their capacity is very important. A simple 16,384 array of 8-bit (1-byte) registers can store 16,384 integers ranging from 0 to 256. A dynamic range of 0 to 256 may seem adequate to store the data for a spectrum, but is actually quite inadequate, especially if there is averaging of many peaks, or some peaks are much stronger than others of interest. The problem is possible **overflow** of the registers, which can be seen in Figure 5-6 to seriously distort the baseline of the transform into the frequency domain. Nowadays, computer memory is quite inexpensive and usually it is possible to provide each data register with enough bits to give a high order of numerical precision so as to cope with the dynamic ranges encountered in most FT-NMR situations. However, this may not be the case when taking the proton spectrum of a dilute solution of a protein in water where the water resonance may be 50,000 times more intense than of some particular proton of interest in the protein. In such circumstances, special techniques are required and, with water, one way is to substitute deuterium for its protons, another is to presaturate the water resonance with a strong rf pulse. This will only be effective if the $T_1$ values of the resonances of interest are shorter than those of water. Efficient water suppression is very

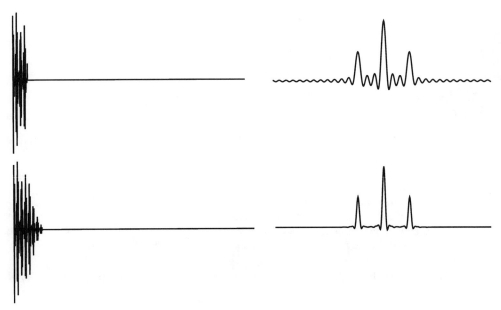

**Figure 5-7** The upper left plot is of a FID with a total accumulation time of 2 s, but truncated after 0.125 s and then zero filled to the end of the accumulation period. The spin system is the same as in Figure 5-6 of a 1:2:1 triplet with a coupling constant of 30 Hz. Fourier transform of the truncated FID into the frequency domain gives a base-line distortion pattern that is typical for an abrupt change in signal strength. Such changes impart some of the frequency characteristics of a square wave to the spectrum. The lower plots are of the same system, except that, between 0.125 s and 0.25 s, the FID signal is apodized by being multiplied by a linear ramp function, such that its value becomes zero at 0.25 s. After 0.25 s, the FID is zero filled to the end. It will be seen that the transform of the apodized FID has a much superior baseline compared to the transform of the simple truncated FID above

important for spectrometers used for study of biological materials by $^1$H NMR.

**6. Truncation.** A different technique from zero filling is **truncation** in which data collection is terminated **before** the signal has decayed to negligible proportions. This procedure enhances S/N by recording the initial stronger part of the FID where the desired signal is proportionately greater than noise. One tradeoff is sacrifice of resolution by shortening the acquisition time. Another tradeoff results because abrupt termination of a FID, whether followed or not by zero filling, can greatly distort the baseline after transform of the FID into the frequency domain, as is shown in the top spectrum of Figure 5-7. It is generally true that a sharp interruption of a time-domain signal has the effect of mixing the multiple frequencies characteristic of square waves into the already-collected spectral information. This problem can be relieved to a more or less satisfactory degree by **apodization**, whereby the data registers following truncation are filled with numbers other than zero. The idea is to rapidly ramp down the FID, but less abruptly than by truncation. Then, the

## 5-2 What Are Your Choices and the Restrictions on FT-NMR?

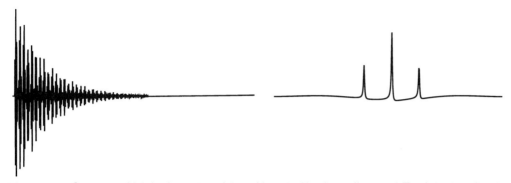

**Figure 5-8** Spectrum (right) of a 1:2:1 triplet with a 30-Hz $J$ coupling and $T_2$ of 0.25 s that is the transform of a FID (on the left) with a 0.01 s delay (dead time) between the end of the pulse and the start of the data acquisition. The dead-time problem is particularly serious when the nuclei giving the signal have very short $T_2$ values; because, then, a significant amount of the signal may be lost during the interval between the pulse and data acquisition. It may not be trivial to reduce dead time to negligible values, because a finite waiting period is usually necessary after a powerful rf pulse to allow "ringing" transients to die away in the spectrometer electronic circuits

balance of the normal decay time can be zero filled to give a smoother transform. A number of decay functions; linear, exponential, trigonometric and so on are suitable for both giving a smoother transform and achieving other ways of improving the spectrum. Some of these will be discussed in Section 5-2-**12**. For the example of Figure 5-7, a simple linear function was used with good results.

  7. **Dead Time.** The short powerful rf pulses used in taking FT spectra cause transients in electronic circuits that can produce spurious signals in the receiver channels and it should not be surprising that there is usually an interval between the time that the pulse ends and data acquisition starts. The result is an incorrect zero time for the FID. This is particularly serious when the $T_2$ values are very short because there can be substantial loss of signal intensity during the **dead time.** In any case, substantial dead time can result in base-line distortion in the transform as shown in Figure 5-8. Dead time can be thought of as **pretruncation of the FID.**

  8. **Folding Over.** We have already noted that the usual Fourier transform has no way of discriminating between positive and negative frequencies relative to the frequency of the pulse. Consequently, if we have a four-peak spectrum and the pulse frequency is chosen so that it falls in the middle of the group of peaks, then, after the transform, the negative-frequency peaks will be **folded over** into the positive frequency domain (see Figure 5-9). The simple solution to this problem is to position the carrier frequency safely to one side, or the other, of the expected resonances. However, this has the potential of creating two problems. The first, that

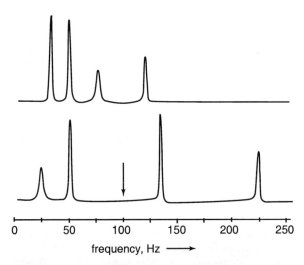

**Figure 5-9** Example of **folding over** in a NMR spectrum. The lower spectrum is a normal spectrum in which the frequencies are 20, 50, 130 and 224 Hz, with different concentrations and $T_2$ values. For this spectrum, the rf pulse frequency (with respect to the reference frequency) was at zero, to the left of all of the resonances. The upper spectrum is of the same nuclear-spin system, but now the rf pulse frequency (with respect to the reference frequency) was set at 100 Hz, the position of the vertical arrow in the lower spectrum. The two left-hand resonances in the lower spectrum, which have negative frequencies with respect to the 100-Hz pulse, are now folded in between the peaks at 30 and 124 Hz, the latter two each being shifted 100 Hz to the left as the result of the change in the pulse frequency

was discussed earlier in Section 2-5, is the problem of achieving a 90° pulse for a nucleus whose resonance frequency is far off the carrier frequency. The other problem is that, because there is no discrimination between positive and negative frequencies, that part of the **electronic noise** which is in the **negative** frequency domain, where there are no spectral peaks of interest, is folded over into the positive domain. The noise problem associated with negative frequencies can be greatly reduced by use of sharply tuned electronic filters that diminish the ability of the negative frequencies to get through to the data registers.

**9. Quadrature Detection.** It should be obvious that, if folding over could be eliminated and the pulse frequency located in the middle of the desired spectral range, the problem of exciting nuclei at widely separated frequencies would be greatly reduced. Also the noise in the spectrum will be more nearly that which is actually associated with the spectrum. The means for doing this is **quadrature detection**, wherein negative frequencies are distinguished from positive frequencies. The basis for doing this is detection and summation of the in-phase and 90° out-of-phase magnetizations. Thus, we use the arrangement of Figure 5-10 in collecting the FID.

## 5-2 What Are Your Choices and the Restrictions on FT-NMR? 91

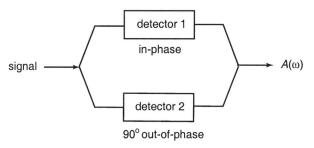

**Figure 5-10** Two-detector scheme used for quadrature detection of NMR signals that allows distinction of signals that have positive or negative frequencies with respect to the reference frequency

The upper two curves in Figure 5-11 show how, if there is but one detector, the FID is the same for a combination of two resonances whether both have positive frequencies, or one has a positive and the other a negative frequency. In contrast, the lower three curves in Figure 5-11 illustrate how the composite FIDs for spectra of two nuclei using two detectors are clearly different when the signals have the same, or opposite, signs for the frequencies in different combinations. These curves show that quadrature detection is a viable means of allowing excitation in the middle of a group of peaks and thereby reducing the phase errors that occur in the 90° pulse and also reducing the noise contributions from outside the spectral region of interest.

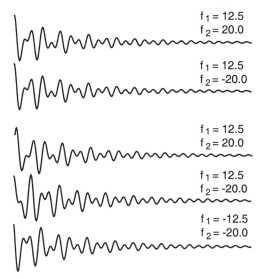

**Figure 5-11** A collection of FIDs with parameters chosen to display (upper two curves) how there is no difference in ordinary FIDs representing two resonances when both resonances have positive frequencies relative to the reference, or one is positive and the other is negative. Here, we restate what has been said before, namely that the transform has no way to distinguish between positive and negative frequencies. In contrast, the lower three FIDs show how quadrature detection with two detectors gathering signals with a 90° phase difference, the composite FIDs are strikingly different

**10. Phasing.** We have occasionally emphasized the importance of phase in NMR, but deliberately chosen conditions and circumstances where our transformed spectra show no, or minimal, phase effects. This hardly reflects the real world of FT-NMR, where it is seldom that the transform of a FID does not require some degree of phase adjustment.

A general phase shift may occur in the transform because of a phase difference between the FID and the reference. Such phase differences have the effect of mixing the $u$ and $v$ magnetizations which differ by 90° in phase. If there is no phase error, then from Equation 3-4 we can write:

$$x_{real}(n) = \frac{1}{N} \sum_{n=0}^{N-1} \left[ A(n) \frac{\cos(2\pi kn)}{N} \right] \qquad \text{Eqn. 5-1}$$

$$x_{imag}(n) = \frac{1}{N} \sum_{n=0}^{N-1} \left[ A(n) \frac{\sin(2\pi kn)}{N} \right] \qquad \text{Eqn. 5-2}$$

If there is a phase error of magnitude $\theta$ then we can correct it by using our "impure" $x_{real}(n)$ and $x_{imag}(n)$ in the equations:

$$y_{real}(n) = x_{real}(n) \cdot \cos(\theta) + x_{imag}(n) \cdot \sin(\theta) \qquad \text{Eqn. 5-3}$$

$$y_{imag}(n) = x_{real}(n) \cdot \sin(\theta) - x_{imag}(n) \cdot \cos(\theta) \qquad \text{Eqn. 5-4}$$

Here, $y_{real}(n)$ and $y_{imag}(n)$ are the real and imaginary data arrays corrected for phase error. Obviously, if $\theta = 0$, there will be no phase error and the $x_{real}(n)$ and $x_{imag}(n)$ will be "pure."

In practice, we adjust $\theta$ until the phase error in the spectrum disappears and we see the simple symmetrical Lorentzian line shape. All well and good, except that the phasing problem is usually rather more complex. Instead of there being just a general phase error, the spectrum will often have a small phase error at one end and a larger phase error at the other. The reason for this is not difficult to understand.

You may remember that we discussed in Section 2-5 how a pulse that produces a 90° rotation of the Z magnetization to the Y axis, when $\omega_0 = \omega$, can give less than 90° rotations of the Z magnetization, as well as phase differences, when $\omega_0 \ll \omega$. This sort of phase problem is expected to be a function of the frequency difference between $\omega$ of the resonance and $\omega_0$ of the reference frequency. Another source of phase error is in dead time (Section 5-2-7) which means that signals from nuclei, which start out in phase, can be somewhat out-of-phase by the time the data collec-

## 5-2 What Are Your Choices and the Restrictions on FT-NMR?

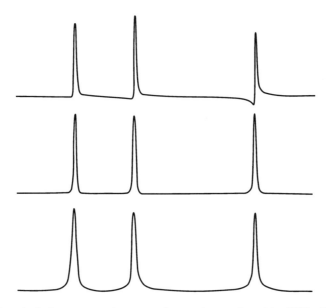

**Figure 5-12** Treatment of phase errors in a v-mode spectra produced by DFT. The upper spectrum shows a typical frequency-dependent phase error with the phase error increasing from left to right. The middle spectrum has the same parameters as the upper spectrum except that both frequency-independent and frequency-dependent phase corrections have been applied to give a good visual fit to the expected Lorentzian line shapes. The lower spectrum is a magnitude spectrum in which $x_{mag}(n) = \{[x_{real}(n)]^2 + [x_{imag}(n)]^2\}^{1/2}$. The admixture of dispersion with absorption removes the need for phase correction, but will be seen to broaden the lower parts of the resonance peaks

tion is started. This will clearly be especially important if the frequency difference is large. What this means is that we have to have both frequency-independent and frequency-dependent factors in our phase-correction routine. The usual way to do this is to let the phase error $\theta$ be the sum of a frequency-independent term and a linear frequency-dependent term.

$$\theta(v) = \rho_1 + \rho_2(v) \quad \text{Eqn. 5-5}$$

The problem then is to determine values of $\rho_1$ and $\rho_2$ that can be used in the phase-correction equations given above with $\theta$ being a function of the frequency $v$.

Making these corrections does not require that the DFT be done over again. The corrections are manipulations of the "impure" $x_{real}(n)$ and $x_{imag}(n)$ values, so the phase correction is actually not very computer intensive. A common procedure is to have $\rho_1$ and $\rho_2$ determined by spectrometer dial settings. The computation is then carried out essentially continuously as the operator adjusts the dials to find values of $\rho_1$ and $\rho_2$ that give the optimum appearance to the spectrum (as a CRT display).

The top spectrum in Figure 5-12 is a spectrum that needs phase correction and corresponds to $\rho_1 = 1.5708$ and $\rho_2 = 0$ ($\theta$ in radians). A trial-and-error variation of the settings gave a good phase correction with $\rho_1 = 0.5708$ and $\rho_2 = -0.00031$. These values are unlikely to be unique; the point is that, at least, they work.

One way to avoid the phase problem is to calculate what is often called the **magnitude spectrum**. This is easy to do by the expression:

$$x_{mag}(n) = \{[x_{real}(n)]^2 + [x_{imag}(n)]^2\}^{1/2} \qquad \text{Eqn. 5-6}$$

The lowest curve in Figure 5-12 is the magnitude spectrum corresponding to the same values of $x_{real}(n)$ and $x_{imag}(n)$ that give the top spectrum (no phase correction). While the magnitude spectrum is simple and avoids some problems in phase correction, the admixture of $u^2$ with $v^2$ causes line broadening, particularly in the lower parts of the peaks. The reason for this can be seen from Figures 4-3 and 4-10, where we see that $u$ has a substantially greater absolute magnitude than $v$ before you reach the point where $\omega_0 = \omega$.

**11. Pulse Angle and Repetition Time.** So far, we have been using 90° pulses to produce magnetization in the $X, Y$ plane. Of course, other pulse angles will do so also, even if producing smaller $X, Y$ magnetizations. Thus, a 60° pulse will produce a $X, Y$ moment equal to 0.87 times $M_0$, compared to the 90° pulse giving an $X, Y$ moment equal to $M_0$. Clearly, for a single pulse and FID, the 90° pulse will give the largest signal. However, if we have weak NMR signals and have to use multiple pulses and signal averaging, you might well ask, "Will a 60° pulse (or some other value) be more efficient of observing time than a 90° pulse?" This is a complicated problem and one of the more difficult decisions to make in carrying out a DFT experiment is to determine the optimal combination of pulse angle and repetition time to use for weak signals in the signal-averaging mode. If you want to completely avoid problems with saturation of the resonances, meaning that you want the Z magnetization to be essentially at its equilibrium value at the beginning of each pulse, then Figure 2-28 shows that you will need to wait 4-5 times $T_1$ seconds, before giving another 90° pulse. Clearly, if $T_1$ is at all long, say > 5 s, a not uncommon value for $^{13}C$ or $^{15}N$ resonances, you will not be able to pulse very often and data collection will be inefficient.

One approach is to find means to shorten $T_1$. And there are indeed substances, such as paramagnetic metal ions, complexes of such ions or even oxygen, that can be dissolved in the sample to shorten $T_1$ and allow you to reduce the repetition times. However, these addends bring problems of their own as we shall see later. For this reason, a natural question to ask is, "Do we really have to wait 4-5 times $T_1$, before initiating another pulse?". The answer is no, but what you should do is wait at least until the $T_2$ process reduces the signal to zero and then turn whatever Z magnetization you have to the X, Y plane to have something to measure. (If you don't wait that long, other problems arise, as we shall see later.) When you think about it a little (but not too much), you might decide that the rate of recovery of +Z magnetization per unit time after a 90° pulse will be greatest when $M_z$ is small. That is of course when $(M_0 - M_z)$ is near 0. Assuming $T_1 \gg T_2$, this could suggest the following scenario: namely, that you first use the 90° pulse to turn the Z magnetization to the X, Y plane so as to give the maximum signal and reduce $M_z$ to 0. After that, you could let the signal decay essentially to zero by $T_2$ (or $T_2'$). Because the +Z magnetization will grow at its fastest possible rate per unit time right after the 90° pulse, you might decide then to turn the relatively small, recovered Z moment into the X, Y plane, accumulate the small FID and repeat.

The problem is that this is not actually the most efficient procedure, because if $T_1 \gg T_2$, a steady-state maximum FID value is set up that depends on $T_1$, the pulse angle and the time between repetitions. If the repetition rate is fast, so that there are short times between pulses, it turns out that low-angle pulses are more efficient than 90° pulses. Indeed, it has been shown by Anderson and Ernst,[4] that there is a quite simple relationship between the optimum pulse angle $\alpha_p$ the time interval between repetitions $t_p$ and the $T_1$ value for a given nucleus. This optimum angle that corresponds to the largest steady-state FID for a given $t_p$ and $T_1$ is known as the **Ernst angle**.

$$\cos \alpha_p = \exp(-t_p / T_1) \qquad \text{Eqn. 5-7}$$

Equation 5-7 can be verified by numerical integration of the Bloch equations using specific $t_p$ values, a desired $T_1$ value and a successive series of five pulses (five is enough to establish a steady-state for a given angle), each series having a different pulse angle.

It may not seem very helpful to have a rather open-ended relationship between $T_1$, $t_p$ and $\alpha_p$, when we can be sure that, in real NMR samples, the various chemically shifted nuclei will have different $T_1$ values. Matters are helped substantially by the fact that an important factor in the choice of $t_p$ will be the

---

[4] R.R. Ernst and W.A. Anderson, *Rev. Sci. Instrum.*, **37**, 93-106 (1966).

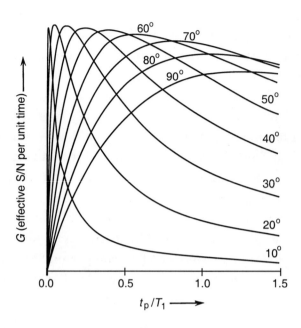

**Figure 5-13** Plots showing how, in the NMR time-averaging mode, $G$ the effective S/N achieved per unit time changes with the pulse angle $\alpha_p$ and $t_p/T_1$, where $t_p$ is the interpulse interval. The plots made in accord with equations developed by Waugh[5] depend on $T_1 \gg T_2$ and assume that the signal has decayed away completely by $T_2$ processes between pulses

resolution required (see Section 5-2-4). But this choice will have to be tempered by consideration of the $T_2$ values. If the $T_2$ values are relatively long, there are potential problems, as mentioned above, with pulsing before the $X, Y$ component of magnetization has essentially completely decayed. Still another relevant problem in the real NMR world is electronic noise, which is present to some degree in every NMR signal. A weak signal mixed in with a relatively large amount of noise will be very hard to detect. In contrast, if $T_2$ is relatively long and you wait for four to five times $T_1$, then the FID amplitude will be larger relative to the noise and much easier to dig out from the noise by time-averaging. But the question still remains how can we determine the optimum pulse angle and repetition rate?

The question has been addressed by Waugh,[5] who has derived equations relating $G$, the effective S/N achieved per unit time, to the ratio of $t_p$ to $T_1$. Informative plots of $G$ of the Waugh equations as a function of $t_p/T_1$ for different values of $\alpha_p$ are shown in Figure 5-13. It is important to remember that the equations used to derive these plots depend on the condition that $T_1 \gg T_2$.

---

[5] J.S. Waugh, *J. Mol. Spect.*, **35**, 298-305 (1970).

## 5-2 What Are Your Choices and the Restrictions on FT-NMR? 97

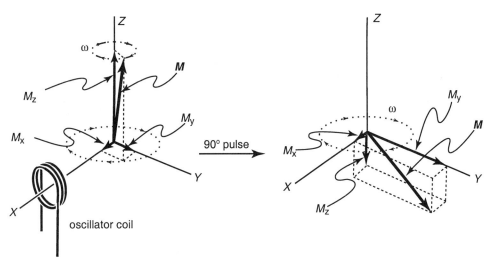

**Figure 5-14** Vector diagrams showing how a 90° pulse applied to a magnetic vector **M** before the magnetization in the X, Y plane has decayed by $T_2$ relaxation can lead to phase problems

It should be clear that rapid pulsing, with $\alpha_p$ of about 20°, is quite efficient in terms of $G$, but requires rather good knowledge of $T_1$, because $G$ falls off rapidly on either side of the maximum, as a function of $t_p/T_1$. Figure 5-13 shows that values of $\alpha_p$ of 50-70° will give rather high efficiencies over a much broader range of $t_p/T_1$ with the optimum values being between 0.5 and 1.0. These angles are clearly more efficient than 90° pulses with long waits between pulses, but it will be seen that, generally speaking, if you choose one of the larger values of $\alpha_p$, the sensitivity to $t_p/T_1$ is smaller than at small pulse angles.

Figure 5-13 helps to illumine why in $^{13}C$ spectra, such as Figure 5-1, there can be substantial differences in peak intensities, because of differences in $T_1$. The peaks to the far left of Figure 5-1 arise from carboxyl carbons which have characteristically longer relaxation times than the carbons of CH, $CH_2$ or $CH_3$ groups. Thus, if $t_p/T_1$ is optimized for the CH and $CH_2$ resonances, the carboxyl resonances are expected to have smaller $G$ values. The plots of Figure 5-13 were derived on an assumption valid for most $^{13}C$ spectra, that $T_1 \gg T_2$ and that $t_p \gg T_2$. These conditions ensure that the signal has essentially decayed completely away, before the next pulse. An example where these conditions are not met is shown in Figure 5-14, where a 90° pulse is applied when there is still significant $X, Y$ magnetization. The pulse, applied along the $X$ axis, will interact with both the $Z$ and $Y$ components of the total magnetization, $M$ (see Section 4-2). For the initial conditions of Figure 5-14, a 90° pulse will result in a negative $Z$ magnetization because, at the time of the pulse,

**Figure 5-15** A 1024-point, 20-repetition FID involving three nuclei, all with $T_1$ equal to 1 s. The chemical shifts are 200, 300 and 375 Hz, with respective $T_2$ values of 0.5, 0.255 and 0.5 s. The repetition rate $t_p$ was 0.5 s

the Y component is positive. However, in the general case, $\omega_0$ will not be equal to $\omega$ and there will be phase changes, so it will be uncertain whether the resulting Y component will be positive, negative or zero. Compounding that uncertainty is the fact, as you can also see from the diagram, that there can be a phase shift resulting from the pulse. Thus, if the Z component, at the time of the pulse, is greater than the Y component and the Y component is positive, as in Figure 5-14, the *M* vector produced by the 90° pulse will be retarded in phase relative to the original *M* vector, as is evident from inspection of the before-and-after sizes of the $M_x$ and $M_y$ vectors.

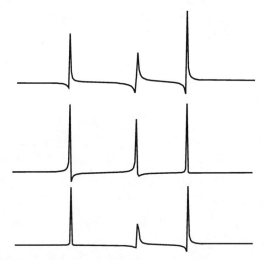

**Figure 5-16** The transform into the frequency domain of the FID of Figure 5-15. The upper spectrum is the unphased transform, the middle one is phased correctly for the peak on the right using the procedure described in Section 5-2-10, but could not be phased correctly for the other peaks. The lower spectrum has been correctly phased for the peak on the left, but could not be phased for those on the right

The situation where a pulse is applied before the magnetization in X, Y plane produced by a previous pulse has decayed away will clearly be complex; the phases of resonances with different frequencies will not necessarily be a simple function of frequency, as envisioned in a nonrepetitive FID (see Section 5-2-10). Figure 5-15 shows a 20-repetition FID in which two of the involved three nuclei have $T_2$ values of 0.5 s and the third has a $T_2$ of 0.25 s. All of the nuclei have $T_1$ equal to 1 s and the repetition rate $t_p$ was 0.5 s. The transform of the FID, Figure 5-16, shows peaks that clearly need to be phased, but if one phases the far left-hand peak correctly, it will be seen that the right-hand peaks are not correctly phased. Phasing the far right-hand peak has the reverse effect.

12. **Spectral Enhancement by Exponential Multiplication.** We have encountered several ways to massage FIDs that produce more or less desirable features on transform into the frequency domain. A very common form of spectral manipulation is by exponential multiplication, which can be used to enhance either resolution or sensitivity but, as we shall see, not both at the same time. The procedure is quite simple, in that each member of the array of points in the FID is multiplied by an exponential function that is itself a function of index number across the FID array. Because the FID is a function of time, the exponential function will also be a function of time. An "all-purpose" function that is widely used for exponential multiplication is:

$$\exp(At - Bt^2) \qquad \text{Eqn. 5-8}$$

in which we have unlimited choices of A and B, including zero. Consider first the situation where B is zero and A is positive. We know that the decay of the FID is an exponential decay in accord with $1/T_2$. So now, if we start with $A > 0$, we cause the FID to decrease more slowly than it would otherwise and, if A is large, the FID will actually grow, rather than decline to zero as shown in the series of Figure 5-17.

When A is large and the FID has a large amplitude at the end of the acquisition period, the transform will show some effects of truncation (Section 5-2-6), but greater difficulties with noise, because the noise is amplified more than the signal. Nonetheless, at smaller values of A, improvement can be seen in the resolution. Transform into the frequency domain of FIDs that decay by a first-order process produces **Lorentzian** line shapes and these typically have rather broad skirts near to the baseline. If you use exponential multiplication with both A and B positive, you can achieve a more or less complete **Lorentz-to-Gauss transformation**.

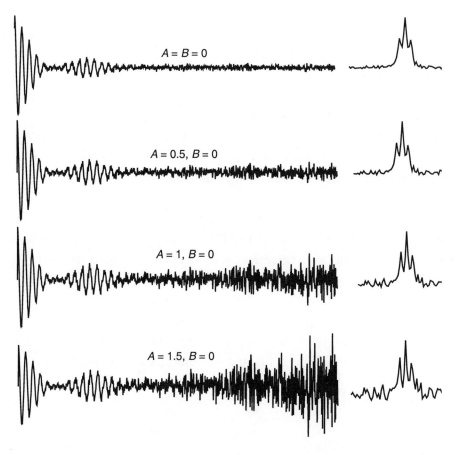

**Figure 5-17** The FIDs and corresponding transforms for a 1:2:1 triplet spin system with $J = 2$ Hz and $T_2$ values of 0.25 s. The FIDs have undergone exponential multiplication with the indicated values of $A$ and with $B = 0$

This is helpful on two counts; first, Gaussian line shapes are narrower than the Lorentzian line shapes as you approach the baseline. Second, while you can use $A$ to increase the magnitude of the FID over the part of the acquisition time where the resolution is largely determined, incursion of the $B$ term causes the FID to decay smoothly and diminishes the effects of noise and truncation compared to the situation shown in the lower plots of Figure 5-17, where $B$ is zero. Some results of this type of exponential multiplication are shown in Figure 5-18 and it will be seen that, with large and positive values of $A$ and $B$, the resolution is markedly improved. However, this improvement does not come for free. Whenever, you increase the resolution by exponential multiplication, you also increase the noise level, because with positive values $A$, you in effect slow down decay of the FID and,

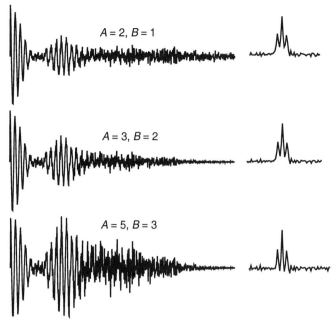

**Figure 5-18** The FIDs and corresponding transforms for a 1:2:1 triplet spin system with $J = 2$ Hz and $T_2$ values of 0.25 s. The FIDs have undergone exponential multiplication with the indicated values of A and B. Note how the latter part of the FID is accentuated with large A and then attenuated with large B. The resolution here is determined mostly by A and the noise level mostly by B

in so doing, make the noise a more important part of the total signal.

With $B = 0$ and negative values of $A$, the reverse process results from exponential multiplication, in that now, the decay of the FID is accelerated, the noise is suppressed relative to signal (which is of course strongest in the early part of the FID) and the signal becomes broadened. The overall process is rather similar to the truncation-apodization sequence described earlier (Section 5-2-**6)**, except that the induced decay now extends over the whole of the FID, and is exponential rather than linear. Starting with the same nuclear-spin system whose FID is used in Figures 5-17 and 5-18, where the noise level is relatively low and the resolution not very good to begin with, exponential multiplication with $A = -1.5$ and $B = 0$, doesn't do much for you, as can be seen in Figure 5-19. However, the same nuclear spin system with a high noise level can be seen to give transforms with a very marked improvement in signal-to-noise ratio (Figure 5-20) by using relatively large negative $A$ values in exponential multiplication.

Exponential filtering is optimal when $A = -1/T_2$ (and $B = 0$), a condition that is known as using a **matched filter**. Because matching involves having the filter function decay with the same time constant as the FID, the simple exponential

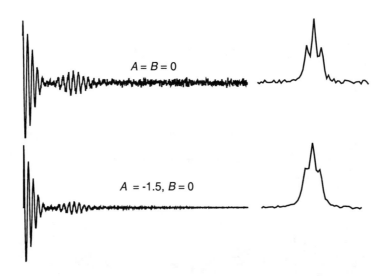

**Figure 5-19** The FIDs and corresponding transforms for a 1:2:1 triplet spin system with $J = 2$ Hz and $T_2$ values of 0.25 s. The FIDs have undergone exponential multiplication with the indicated values of $A$ and $B$. Note how the noise component in the FID has been attenuated with respect to the signal by having the decay occur more rapidly. However, while there is a substantial improvement in signal-to-noise, it is not much needed, because the noise level is not very large to begin with. Furthermore, the improvement is accompanied by a significant loss in resolution. The point is that this is not a very good example of when to use exponential filtering

function decay with the same time constant as the FID, the simple exponential matched filter described here, with $A = -1/T_2$ and $B = 0$, will not be a matched filter when used with a FID that overruns the capacity of its data registers (Section 5-2-5) or otherwise undergoes nonexponential decay. With a matched filter, those line widths for which $A = -1/T_2$ will be doubled.

Clearly, there are a variety of mathematical manipulations for FIDs which may be able to help to clarify their information content. Note that these manipulations do not require that the spectrum be taken again. The data array of the original FID can be archived and a copy of it can be manipulated to yield the desired spectral enhancements.

**13. Integration of FT Spectra.** The digitization of transformed NMR spectra makes integration very easy to perform, but the integrals, even if there is no overlap of the resonances and no impurities to worry about, are quite inaccurate and it behooves us to consider the factors that can detract from accuracy, because quantitative NMR has many applications in structural and chemical analysis. Most of the factors complicating integration derive from topics discussed earlier in this Section.

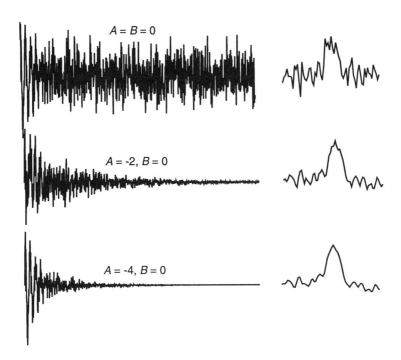

**Figure 5-20** The result of exponential multiplication for a 1:2:1 triplet spin system with $J = 2$ Hz and $T_2$ values of 0.25 where $A < 0$ and $B = 0$ and the noise level of the FID is high. The large negative value of $A$ causes the FID to decay faster and decreases the noise level, although at the cost of sacrificing resolution. You may be surprised how exponential multiplication can extract a more or less intelligible spectrum from a FID with the level of noise in the top unfiltered FID, but the key is to look and see if there is a waveform clearly visible in the earliest part of the FID. If there is, some spectral peak or peaks will come out of the transform

Whenever time-averaged spectra are integrated, the first consideration should be for the **length of time between pulses**. The $T_1$ values of various chemically shifted nuclei in an NMR spectrum are unlikely to be the same and, to the degree that insufficient time is allowed for $T_1$ relaxation, the integrals of the different resonances can be far from the expected values. If accuracy is needed to 1% or less, the time between pulse should be on the order of ten, or even more, times the longest $T_1$ in the sample. Using such long delay times does not of itself ensure accuracy, but little accuracy is possible whenever a range of $T_1$ values is involved, unless the delay times are long.

To gain sensitivity and improve the **signal-to-noise ratio** with $^{13}C$ spectra of most organic compounds, it is customary to use proton decoupling, as described in Section 5-1, to remove the proton-carbon spin-spin splittings and also to take advantage of the proton-carbon **nuclear Overhauser effect** (NOE), which can enhance the $^{13}$C resonance substantially more than expected by simply collapsing the proton-carbon splitting multiplets. Unfortunately, the resulting NOE enhancements can be quite variable and NOE effects must be avoided if integrations are to be useful. Figure 5-1 is an excellent example of how far off the peak intensities can be in $^{13}$C spectra, as the result of differential NOE enhancements and/or too-short delay times. Fortunately, there is a procedure called **inverse-gated decoupling** that allows the proton-carbon splittings to be collapsed, but avoids NOE enhancement of the resonances. This procedure which will be described in greater detail in Chapter 6, is based on the fact that the proton-carbon NOE builds up to its maximum signal strength at a finite rate and, if proton decoupling is begun just following the pulse, then the initial points of the FID are not enhanced but the splittings are removed. Buildup of the NOE in the later stages of the FID does not affect the resonance intensities even if there are strong differential NOE influences (see Section 3-2).

Another consideration is the number of **points/Hz** used in taking the spectrum. Narrow resonance lines may be defined by a very few points, while broad lines are normally defined by many. So, for accuracy in integrations, it is important that the peaks be defined by sufficient numbers of points. If a spectrum has very sharp lines and there are not many points defining these lines, it may be desirable to broaden the lines by decreasing the homogeneity of the magnetic field or by stopping the spinner, which then cannot help to average out the field inhomogeneities (Section 2-2).

Still another consideration is the **position of the carrier frequency** and that will be important whenever the spread of resonance frequencies is large. We have seen in Section 2-5 how even a very short pulse cannot produce a clean 90° flip when the shift of the nucleus to be flipped is far from the carrier frequency. This situation can often be greatly alleviated by quadrature detection (Section 5-2-9). However, in accurate work, it is likely to be important to determine how much relative change in the integrals occurs with different placements of the carrier frequency. The **pulse time** can also be important when the spread of resonance frequencies is large, because with shorter pulses (and smaller pulse angles), nuclei with widely different resonance frequencies will have less time to get out of phase during the pulse period, as we have seen in Section 2-5.

The flatness of the **baseline and phasing** can be major concerns and we have seen earlier in this Section how truncation, dead time, data-register capacity and phasing can distort baselines. Many modern spectrometers have both manual and automatic baseline-correction routines that can be helpful in improving integrations. **Dead time** can also reduce signal intensities, if $T_2$ is short.

If all of the above potential difficulties seem too much, there is still another rather subtle problem, that of **analog-to-digital conversion**. This is expected to be especially important if very weak resonances are to be integrated and compared with integrals of strong resonances in the same spectrum. This is an important consideration in molecular-weight determinations of polymers. Thus, one might wish to determine the molecular weight of a polymer such as $C_6H_5(CH_2CH_2)_nH$ by comparing the integrated intensities of the five phenyl hydrogens relative to those of the aliphatic protons. Clearly, if $n$ is large the aromatic protons will have small integrals relative to the others and, of course for accuracy, enough scans will have to be accumulated to produce good **signal-to-noise ratios** for the weaker signals.

The potential difficulties in analog-to-digital conversion arise in the way that modern NMR spectrometers collect intensity data by counting pulses produced by an analog-to-digital converter. The crux of the matter is the size of the analog magnetization required to produce a pulse in the counter. At some concentration level, the analog signal associated with a minor component may be too small to trigger a contribution to the FID and much of this occurring would be likely to distort the relative integral sizes. Among the ways to check whether this is a problem would be to compare the relative integrals when changes are made in the gain of the analog-signal amplifier, or by changes in the heights of the peaks by changing the spectral resolution. If such changes were found to produce different relative integrals, then one might reasonably conclude that some part of the analog signals are not being converted.

## 5-3 Alternatives to the Fourier Transform

When one sees how powerful the Fourier procedures are, and how the FID can be manipulated, it might seem surprising that anyone would be so dissatisfied with Fourier procedure for the transform into the frequency domain as to try to find alternatives. But the fact is that as we have already discussed, there are problems of resolution, of phase, of dead time and, especially, of sensitivity. Consider sensitivity. If we could somehow improve the transform in such a way as to increase the S/N ratio by a factor of two, this would result in a saving of a factor of four in observation time. Even this minor-sounding improvement would make many experiments using NMR at least feasible, and many others much more feasible. Electronic noise of the so-called "white-noise" variety is incoherent, while the signals we desire are coherent. It seems natural to expect that some form of data processing could reject the unwanted incoherent signals without loss of resolution. The quest for such NMR procedures is sort of a quest like those for the Holy Grail or Ponce de Leòn's Fountain of Youth. It is given impetus by the fantastic success of image-processing techniques that convert a very weak, noisy data stream of picture information from a spacecraft passing a moon of Neptune, 2.7 billion miles away, into a detailed high-resolution photograph. What are the problems with performing the same kind of miracle with NMR data?

If we look at a FID that involves first-order $T_2$ decays, we can expect that each contributing resonance will be some sort of a decaying sinusoidal waveform with a specific frequency ν, a $T_2$, an intensity $c$ and a phase angle θ. $T_1$ will be unimportant unless we do time averaging without waiting for complete relaxation between pulses (see Section 5-2-11). With four unknowns per resonance, it seems reasonable that a FID with the usual large numbers of data points would contain ample information to determine the properties of the contributing resonances, provided that there are not too many of them and their intensities are not negligibly small. After all, the FT procedure does recover the same information.

Suppose that we have the essentially noise-free FID at the top of Figure 5-21 and we need to know the frequencies, $T_2$ values, concentrations and phase angles of the nuclei from which it was generated. We can take a minimalistic approach to trying to replicate the coherent signals of the spin system that we clearly see it contains. The simplest, most gross possible approach would be to start by assuming there is only one nucleus in the spin system and guess at a set of particular values of ν, $T_2$, $c$ and θ (frequency, $T_2$, relative concentration and phase angle). The grossness of the approach with one nucleus is evident from the fact that the FID in question clearly shows interference patterns that indicate that it has to be represented by more than one frequency. No matter - we can calculate the FID that corresponds to any particular one-nucleus system and then we can determine its goodness of fit to the experimental FID with the aid of the least-squares correlation coefficient. Assuming our first guess will not be a very good one, we can then repeat the process with slightly different values of ν, $T_2$, $c$ and θ, until the correlation coefficient is as close to 1.000 as we desire, or else the correlation coefficient reaches some sort of a plateau value incapable of further improvement for a one-nucleus system. At this point, the goodness of fit may be so poor as to, automatically or by visual inspection, trigger

## 5-3 Alternatives to the Fourier Transform

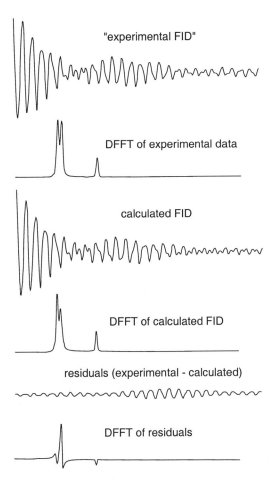

**Figure 5-21** At the top is an "experimental" FID with 256 data points generated for three resonances with chemical shifts 25.0, 27.0 and 47.0 Hz; $T_2$ values of 0.25, 0.5 and 0.65 s; and relative concentrations of 0.59, 0.29 and 0.12, respectively. Low frequencies were chosen to allow easy visual comparison of the "experimental" and calculated FIDs. A three-nucleus analysis with a correlation coefficient of 0.9872 showed chemical shifts of 24.91, 26.87 and 46.98 Hz; $T_2$ values of 0.303, 0.271 and 0.672 s and relative concentrations of 0.50, 0.38 and 0.13, respectively. In the analysis program, the variables for the resonances were changed in a wholly random order, except that the three frequencies were varied first, because of the great sensitivity of the correlation coefficients to frequency changes. When the variation of a particular parameter ceased to improve the correlation coefficient by at least 0.001, then a new parameter was varied. The corresponding "best" fits to the one- and two-nucleus analyses had correlation coefficients of 0.880 and 0.902, respectively

a decision that one nucleus does not suffice and that the treatment should be carried further with two nuclei, each contributing to the FID with its own $\nu$, $T_2$, $c$ and $\theta$. This process can then be repeated, taking into account more and more nuclei, until

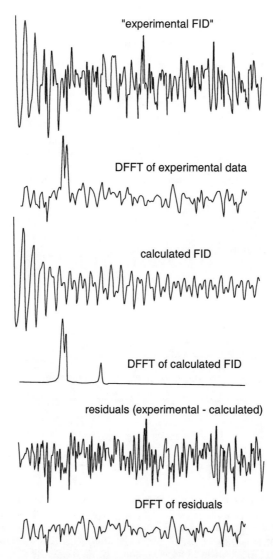

**Figure 5-22** The topmost FID has the same parameters as for Figure 5-21, but with substantial added random noise. The three-nucleus analysis, done in the same way, showed a correlation coefficient of 0.7415 with chemical shifts of 25.03, 27.08 and 47.17 Hz; $T_2$ values of 0.236, 1.455 and 1.291 s and relative concentrations of 0.66, 0.15 and 0.19, respectively. The calculated frequencies are quite accurate, but the $T_2$ values and relative concentrations have substantial errors. As might be expected, errors in the $T_2$ values and relative concentrations tend to offset one another - a low concentration can be in part compensated by a large $T_2$ and vice versa. The corresponding one- and two-nucleus analyses had correlation coefficients of 0.616 and 0.680, respectively

the operator is tired, the computer has expired or some, more or less arbitrary, criterion has been satisfied with respect to the goodness of fit. The third curve of Figure 5-21 shows a FID which is the final result of a three-nucleus analysis of the

## 5-3 Alternatives to the Fourier Transform

uppermost FID of Figure 5-21 by this crude procedure, in which the $c$ and $T_2$ values were constrained during the fitting process to be greater than 0.1 and 0.05, respectively. The reason these constraints were used is because the fitting process is not helped by consideration of resonances that have $c$ or $T_2$ equal to, or near to, zero. To reduce the time required for the calculations, the phase angles were assumed for the "experimental" FID to be 0° throughout, but this is not necessary.

You will note that, after subtraction of the calculated FID from the "experimental" FID, there is a residuum of signal that clearly contains coherent frequency components. The FFT of this residual signal only shows significant frequency components at just the same frequencies that produce a good fit to the data. This FFT shows that the inadequacies of the calculated fit result from imprecise $c$ and $T_2$ values. In this regard, it will be generally true that the correlation coefficient of the least-squares fit is far more sensitive to variation of the frequencies than it is to variation of either $c$ or $T_2$.

Figure 5-21 is hardly impressive, because the FFT of the original FID is more than adequate to define the spectrum. The proof of whatever there is in this kind of pudding is to work with a noisy FID, as at the top of Figure 5-22, that has the same frequency components as Figure 5-21, but with the addition of a heavy component of random noise. Here, the peak expected at 47 Hz is not clearly visible in the corresponding FFT. This "experimental" FID was analyzed by the same procedure of adding resonances and varying their properties one-by-one. Now, although the FFT of the final three-nucleus fit looks rather impressive and is very accurate with respect to frequencies, it is not so satisfactory in its fit to the concentrations and $T_2$ values. However, you can see that the FFT of the residual signal after subtraction of the calculated FID from the starting FID corresponds quite closely to the ups and downs of the noise seen in the FFT of the original spectrum. This would indicate that the coherent spectral information has been quite effectively extracted from the FID.

The important question when we are trying to improve sensitivity by this kind of method is how small can we allow $c$ to be and still have confidence that we are seeing real resonances. Our calculated FID is wholly noiseless and its transform, in Figure 5-22, gives a beautiful noiseless spectrum. Furthermore, because our mathematical analysis gives frequencies and relative concentrations, we can calculate transforms for which, by arbitrarily increasing the $T_2$ values, there can be very sharp spectral lines. Such a procedure can clearly achieve any desired degree of resolution and would seem justifiable if $T_2'$ were known to be smaller than $T_2$ (see

Section 2-2). Are we fooling ourselves by doing this, or does this kind of analysis really increase the spectral information content? There seems no answer to this question that will please everyone.

Much more sophisticated and very computer-intensive mathematical procedures have been devised for the analysis of FIDs in terms of the same variables that we have used in the analyses shown in Figures 5-21 and 5-22. One of these methods is known as the **maximum-entropy method**[6] and is a minimalistic approach in that it seeks to take into account the least number of variables which will satisfy some desired criterion of goodness of fit. The association with entropy comes from the connection between **entropy and information content**.

The concepts of the maximum-entropy procedure appear to have application in many fields of research, especially radio astronomy where it is used for image enhancement. The initial application of the maximum-entropy method to $^{13}C$ spectra,[6] implied that the Holy Grail of sensitivity improvement had indeed located and easily available to all; see Figure 5-23. However, inspection of the results suggests that removal of noise from the spectrum is likely to be the result of a **non-linear** filtering process. This can be deduced from the fact that the stronger a peak is in the original spectrum, the more prominent it is in the filtered spectrum. It should be clear that a non-linear filter which accentuates the stronger resonances relative to the weaker resonances (and noise) will appear to achieve greater sensitivity, provided that we define sensitivity in terms of gross noise reduction, as has been achieved in Figure 5-23. However, this is not the same as sensitivity defined as recovering very weak signals from spectral noise. Indeed, there is no hard evidence that the maximum-entropy method is the answer to this latter statement of the sensitivity problem.

The maximum-entropy procedure is discussed in more knowing detail by Freeman.[7] It has very substantial utility in clarifying some of the complex representations associated with two-dimensional NMR spectroscopy, some simple types of which will discussed later. Another, and quite different, method of estimating the parameters responsible for the complex decays exhibited by FIDs is called **linear prediction**[8] which can produce beautifully detailed and clear spectra

---

[6]S. Sibisi, J. Skilling, R.G. Brereton, E.D. Laue and J. Staunton, *Nature* **311**, 446-447 (1984); E.D. Laue, J. Skilling, J. Staunton, S. Sibisi, and R.G. Brereton, *J. Magn. Reson.* **62**, 437-452 (1985).
[7]R. Freeman, *A Handbook of Nuclear Magnetic Resonance*, Addison Wesley Longman, Essex, Second Edition. 1997, pp. 133-136.
[8]H. Barkhuisen, R. de Beer, W.M.M.J. Boveé and D. van Ormondt, *J. Magn. Reson.* **61**, 465-481 (1985).

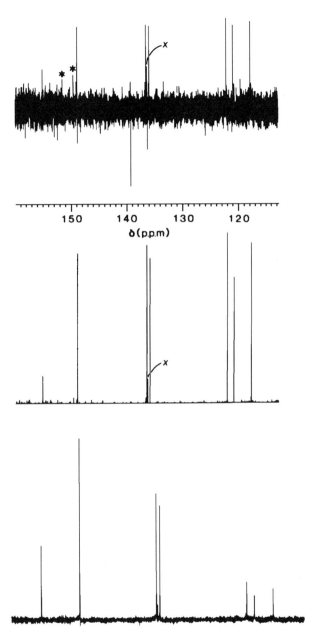

**Figure 5-23** Improvement in a FT-NMR $^{13}$C spectrum of 2-vinylpyridine by use of maximum-entropy signal processing.[6] For the upper spectrum, a short pulse of 1.6 μs gave a low signal-to-noise ratio. The lower spectrum (on a different horizontal scale) was obtained with a 13-μs pulse. The middle spectrum uses the spectral parameters obtained for the top spectrum by maximum-entropy analysis. Compare the ratios of the peak heights in the middle and lower spectra, because these are suggestive of a measure of non-linear filtering by the maximum-entropy procedure. The peaks marked with x are instrumental artifacts; those with asterisks are noise peaks. Reprinted by permission of the authors from *Nature* Vol. 311, p. 446, Copyright (c) 1984 Macmillan Magazines, Ltd.

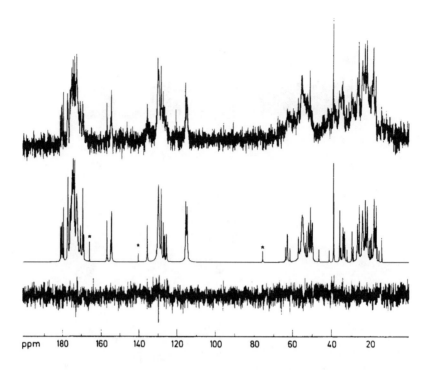

**Figure 5-24** The top spectrum is a conventional FT $^{13}$C spectrum of porcine insulin at 67.9 MHz, taken with time averaging over 24 h. The middle spectrum is the result of a linear-prediction analysis which required 90 h on a minicomputer. The lower curve is the imaginary component of a FT transform of the residuals from the difference between the experimental and calculated FIDs. Reproduced by permission of the *Journal of Magnetic Resonance* and the authors[9]

from a very complex mixture of signals and noise as shown by the example of the $^{13}$C spectrum of porcine insulin[9] in Figure 5-24. Like the maximum-entropy method, linear prediction is very computer intensive; thus, the calculation of 8192 data points of Figure 5-24 required some 90 hours on a minicomputer. Figure 5-24 has several peaks marked with * that have a degree of statistical uncertainty even though, because there is no noise, these peaks look quite real. This is the prime difficulty with all of the mathematical procedures that remove noise from spectra. Despite their substantial inconvenience, noisy spectra do provide standards for comparison. If a sharp peak is anywhere close in magnitude to the extremes of the noise, you will be justified in regarding it with some measure of suspicion, no matter what computer smoothing and computer analysis show.

---

[9]H. Gesmar and J.J. Led, *J. Magn. Reson.* **76**, 183-192 (1988).

# Exercises

**Exercise 5-1 a.** Show the sequence from left to right of the carbon resonance assignments you would make to account for the peaks in the proton-decoupled $^{13}C$ spectrum of monosodium glutamate, $Na^+ \ ^-O_2CCH_2CH_2CH(NH_3^+)CO_2^-$, shown in Figure 5-1.
**b.** Sketch the appearance you would expect for a **proton-coupled** $^{13}C$ spectrum of monosodium glutamate, taking account only the couplings expected for hydrogens directly attached to carbon.

**Exercise 5-2** The first commercial NMR spectrometers using massive electromagnets operated at 30 MHz for protons, moved rapidly to 40 MHz and then to 60 MHz which was the standard for several years. Although 100-MHz electromagnet proton spectrometers were in use for some time, the advent of superconducting magnets rapidly made them obsolete. Then, at the low-field end, commonly 60 and 90-MHz, permanent-magnet spectrometers came on line that are robust and inexpensive to run. Compare 60-MHz and 600-MHz pulse FT spectrometers; if one used a 10-µs 90° proton pulse how many complete waveform cycles would there be for each over the pulse period? What would be the range in frequencies and in ppm that each could cover if the condition were imposed that the highest and lowest frequencies would be no more than +30° or -30°, respectively, out-of-phase with the oscillator frequency at the end of the pulse when using quadrature detection? Review Section 2-5 and show your procedure.

**Exercise 5-3** What is the Nyquist frequency for a spectrometer with 16,384 data registers with an acquisition time of 5 s?

**Exercise 5-4** How many data registers would you need to do a **fast** Fourier transform covering a spectral range of 50,000 Hz with 2 Hz per point? How reliable would you expect a 10.0 Hz difference in peak maxima to be measured with these settings? Explain.

**Exercise 5-5** Determination of $^{13}C$-$^{13}C$ coupling constants is difficult at the natural-abundance level because the probability of a $^{13}C$-$^{13}C$ linkage in a two-carbon chain is about 0.02%. If we wanted to measure the $^{13}C$-$^{13}C$ coupling at the natural-abundance level in ethanol, we would use complete proton decoupling to get the strongest possible peaks corresponding the $^{13}C$-$^{13}C$ coupling. The chemical shift of the $CH_3$

carbon of ethanol is 16.4 ppm and that of the $CH_2OH$ carbon is 58.4 ppm relative to the carbon resonances of TMS.

**a.** Sketch the natural-abundance $^{13}C$ spectrum you would expect to see with proton decoupling and showing the positions of the resonances in Hz on the assumption of a 500-MHz (proton) spectrometer, as well as knowing that the $^{13}C$-$^{13}C$ coupling constant is expected to be about 40 Hz.

**b.** Consider the relative advantages of measuring this coupling by the FT procedure or by taking CW spectra over just the necessary sweep width, first at the $CH_3$ resonance and then as a check at the -$CH_2OH$ resonance. Time averaging would be necessary for both techniques.

**c.** Are there any differences in the data-register overflow problem for FT and CW spectra for determining this $^{13}C$-$^{13}C$ coupling? Explain.

**d.** Suppose you measured the $^{13}C$-$^{13}C$ coupling in ethanol and 2,2,3,3-tetramethylbutane at equimolar concentrations. How would the ease of detecting the signals corresponding to the various couplings compare for the two compounds? Explain.

**Exercise 5-6** A proton spectrum is taken with the pulse frequency 1000 Hz from the resonance frequency of tetramethylsilane (TMS) and, after transform (no quadrature detection), peaks appear on a 1000-Hz spectral width at 625, 593, 120 and 0 Hz (the last being the protons of the TMS in the sample). Now, suppose the pulse frequency is moved to 500 Hz from TMS and, after transform of the resulting FID, the frequency scale is set with the TMS peak again at zero. Sketch the proton spectrum that you would expect to see now for a 1000-Hz spectral width. Explain.

**Exercise 5-7** You will see from Fig. 5-10, that $A(\omega)$ coming out of quadrature detection is a mixture of $u$- and $v$-mode signals. Suppose one had a spectrum with a single resonance, how could you tell from $A(\omega)$, the composite FID, or its transform, whether the pulse frequency was to the + or the - side of the frequency of the resulting resonance signal? Suggestion: review Sections 4-3 and 4-6.

**Exercise 5-8** For monosodium glutamate (see Exercise 5-1 and Figure 5-1) explain how, if the $T_1$ relaxation times of the carboxyl carbon nuclei are longer than those of the $CH_2$ or CH carbon nuclei, the choice of parameters for signal averaging could either lead: to 1) equal intensities of the carboxyl carbon resonances of monosodium glutamate relative to the $CH_2$ and CH carbon resonances or 2) substantially smaller intensities for the carboxyl carbons.

# Exercises

**Exercise 5-9 a.** Calculate the Ernst angle, Equation 5-7, for time interval between pulses of 1 s and having nuclei with respective $T_1$ values of 0.75, 1.25 and 5 s.
**b.** Use Figure 5-13 to determine the pulse angle that gives the optimum value of $G$ for each value of $T_1$ and also your best estimate for a compromise over the spread of $T_1$, values, recognizing that it is desirable for the peaks to have comparable intensities and assuming $T_2$ is short enough to make the curves of Figure 5-13 reliable.

**Exercise 5-10** Consider again the $^{13}C$ spectrum of monosodium glutamate shown in Figure 5-1. Assume that this spectrum was taken with time averaging using a $t_p$ of 5 s and a 90° pulse angle. Assume further that the $T_1$ values of the $^{13}C$ in the $CH_2$ groups are about 3 s. Show how you can use the spectrum of Figure 5-1 and the curves of Figure 5-13 to guesstimate the $T_1$ values of the CH and carbonyl groups. What are likely to be large unknowns in the application of this procedure?

**Exercise 5-11** Referring to Figure 5-14, calculate the phase-angle change for $M_{xy}$ that would occur when a short 90° pulse is applied to a $M$ vector making an angle of 15° to the Z axis and having $M_{xy}$ in the rotating frame of reference making 30° clockwise angle from the +Y axis.

**Exercise 5-12** What values of $A$ and $B$ would you try using with Equation 5-8 and the spin system of Figures 5-17 to 5-20 to provide the best value of S/N without caring much about the resolution achievable? What would limit your choices in the real world of NMR spectra? Explain.

**Exercise 5-13** Exponential multiplication can widen or narrow lines, for example see Figure 5-17 and 5-18. Earlier, we said linewidths are determined by $T_2$. Explain what is going on here.

**Exercise 5-14** Estimate the S/N ratios of the upper and lower spectra of Figure 5-23 by using the procedure described in Exercise 3-6. Are the relative noise levels of these two spectra consistent with the pulse times stated to be used to obtain these spectra? Would you expect them to be? Explain.

**Exercise 5-15** The inverse-gated proton decoupled (no NOE, see Section 5-2-13) $^{13}C$ spectrum of polyisobutylene, $(CH_3)_3C[CH_2-C(CH_3)_2]_nCH_2-C(CH_3)=CH_2$, (along with

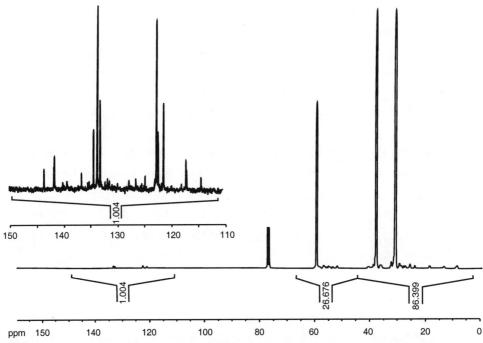

**Figure 5-25** A $^{13}$C NMR spectrum of polyisobutylene dissolved in CDCl$_3$. See Exercise 5-15

isomers having different positions of the double bond in the C4 unit that terminates the chain) is shown in Figure 5-25. The solvent CDCl$_3$ gives a closely spaced triplet resonance with each peak having the same intensity at about 78 ppm. Integrals are supplied below the baseline that encompass (reading from left to right) the weak resonances from about 145-110 ppm (1.004) and the stronger resonances from about 67-44 ppm (26.676), and 44 to about 3 ppm (86.399). The region from 145-110 ppm is also shown with a 2x expanded horizontal scale and a much expanded vertical scale.

**a.** Which functional groups of polyisobutylene would you assign to each of the principal $^{13}$C peaks shown in the spectrum of Figure 5-25? Show your reasoning.

**b.** Use the integrals to calculate the number-average molecular weight of the polymer. Show your procedure.

**c.** Why is the $^{13}$C resonance of the solvent comprised of three equally spaced lines with equal intensities?

**5-16** Consider the top FID of Figure 5-22. Its Fourier transform shows noise that was present in the original spectrum, but in the transform of the "Roberts" FID near the bottom of the Figure, there is no noise. Explain why these two spectra differ with respect to noise level and explain how the manner that the lower spectrum is obtained might be changed so that it would show noise in more the same way as the upper spectrum. What would the theoretical limit be for the "frequency" of noise in this spectrum?

**5-17** Calculate the least-squares fit and correlation coefficients for a 5-Hz sin wave sampled at 10 points over 1 s with a 4-Hz sin wave sampled over the same points: **a.** when the curves start off in phase and **b.** when they start out 45° out-of-phase with one another.

# 6
# Relaxation and the Nuclear Overhauser Effect

We have been dealing extensively with $T_1$ and $T_2$ relaxations in the last few chapters without concern for the mechanisms by which they occur. $T_2$ relaxation, at least when $T_2 \ll T_1$, generally has a very simple mechanism; namely, loss of phase coherence in the X, Y plane, however caused. When $T_2 \gg T_2'$, the loss of phase coherence, that causes the signal to disappear, is normally associated with magnetic-field inhomogeneities (Section 2-2). Our purpose in this chapter is to look into mechanisms of $T_2$ relaxation that are inherent to the sample rather than resulting from imperfections in the spectrometer. The $T_1$ relaxations involve transfer of energy, supplied from the $B_1$ field to the nuclei, back to the surroundings in the form of thermal motions and this is much more complex process than simple loss of phase coherence.

## 6-1 Some General Considerations of $T_2$ Relaxation

When the applied field is highly homogeneous and $T_2 \sim T_2'$, one possible situation is that there is no loss of phase coherence in the X, Y plane and that the decay of $M_y$ corresponds to the approach of $M_z$ to its equilibrium value as in Figure 2-13. In this circumstance, relaxation will be by a $T_1$ mechanism and $T_2$ will equal $T_1$. Quite a different situation is where $T_1 > T_2$ and the loss of phase coherence is a property of the sample. This is most often associated with viscous media.

We can see how viscosity can be important from a simple example. Suppose we have a molecule with two normally equivalent protons arranged so that, at any given instant, the locus within the medium is likely to be different for one H than the locus for the other H, but these loci can be interchanged by molecular motions (see Figure 6-1).

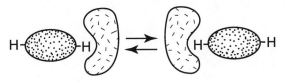

**Figure 6-1** Schematic representation of how respective local environments of otherwise equivalent hydrogens could lead to differences in precession frequencies in a viscous medium

## 6-2 General Considerations of $T_1$ Relaxation.

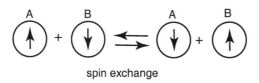

spin exchange

**Figure 6-2** Schematic representation of spin exchange, where nuclei A and B change their orientations simultaneously

Now, suppose the magnetic vectors of both nuclei are turned into the X, Y plane by a 90° pulse. There will be a difference in chemical shift for the H's in the different loci; but, if the lifetimes of the states are short, and the interchange rate is fast relative to $2\pi\Delta v$, where $\Delta v$ is the frequency difference (see the discussion in Section 1-2), then there will be no significant effect on $T_2$. However, if the interchange rate is slow, as would be expected for a highly viscous medium wherein the motion of the molecules relative to one another is slow, then the differences in precession frequencies that are characteristic of the differences in chemical shifts in the different loci will cause the two magnetic vectors to lose phase coherence and make the signal intensity diminish more rapidly than if there were no difference in chemical shift. Anything that causes the signal to diminish more rapidly shortens $T_2$. This loss of phase coherence is effectively due to magnetic-field inhomogeneities at the molecular level. However, in this situation, unlike what happens in the Carr-Purcell sequence (Section 2-3), the loss of coherence is **not** recoverable in the form of an echo by a 180° pulse. Why not? Because the inhomogeneities are internal to the sample and therefore will be subject to random fluctuations as the result of the molecular motions. The situation is rather like the one discussed earlier with respect to effects of diffusion on $T_2$ measurements (Section 2-3).

Nuclei can also lose phase coherence by **spin exchange** (Figure 6-2). Two nuclei in close proximity to one another can exchange spin orientations and this can cause phase changes, either as the result of the transition, or as the result of their chemical-shift differences. Obviously, the exchange process will be more favorable if the nuclei are held in close proximity for an extended period. Consequently, increasing the viscosity will help to facilitate such exchange. Spin exchange causes $T_2$, but not $T_1$ relaxation.

### 6-2 General Considerations of $T_1$ Relaxation.

Relaxation involving $T_1$ changes the energy content of a nuclear spin system by changing the degree of magnetization along the Z axis. It is not so difficult to visualize the process whereby the Z magnetization is turned by a 90° rf pulse into the X, Y plane.

Likewise, loss of phase coherence leading to $T_2$ relaxation is not a very complex idea. It is more difficult to see how the increase in energy associated with reducing the component of Z magnetization, is later dissipated to the surroundings (known to physicists as the **lattice**, hence **spin-lattice relaxation**). The nuclei are suspended in the electric field of the electrons and there is no mechanical way that their excess magnetic energy can be transferred to the lattice by collisions. What is needed is for the nuclei to act as **rf oscillators** and transfer their excess magnetic energy to suitable receivers in the lattice. Of course, for that to happen, there will have to be receivers with at least some degree of receptivity, or tuning, to the precession frequencies of the nuclei with excess energy.

What is often not appreciated is that when we use a receiver coil to detect a signal from a sample after a pulse, the coherently precessing nuclei can themselves induce a current in the receiver coil which then acts to turn the nuclear vectors back toward the +Z axis. This is called **radiation damping** and is sample, spectrometer and spectral parameter dependent. It can cause serious errors in relaxation-time measurements. Radiation damping is discussed in knowing detail by Freeman.[1]

The usual mechanisms of $T_1$ relaxation can generally be divided into two different groups, **intermolecular** and **intramolecular**. The **intermolecular** mechanisms usually involve motions of unpaired electrons in paramagnetic substances such as paramagnetic transition-metal ions in the neighborhood of the magnetic nucleus undergoing relaxation, or else, and less efficiently, corresponding motions of neighboring magnetic nuclei.

## 6-3 Intermolecular Mechanisms of $T_1$ Relaxation

Let us consider first some possible intermolecular mechanisms and how they might operate.

**1. Relaxation Induced by Unpaired Electrons.** We can transmit rf energy to a nuclear spin system by moving electrons through an oscillator coil. The same process can be achieved by the motion of substances with unpaired electrons, such as paramagnetic ions, in the neighborhood of a magnetic nucleus. Here, the energy can be transferred in either direction, until equilibrium is attained. You may wonder how a more or less random motion of a paramagnetic ion can generate enough of, or even the semblance of, a coherent rf field to cause a nucleus to flip between its magnetic states. A

---

[1] R. Freeman, *A Handbook of Nuclear Magnetic Resonance*, Addison Wesley Longman, Essex, Second Edition, 1997, pp. 203-206.

## 6-3 Intermolecular Mechanisms of $T_1$ Relaxation

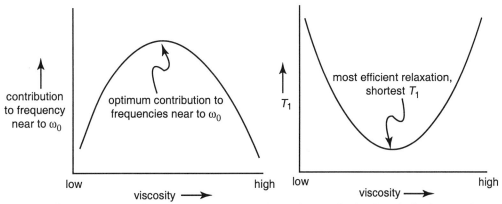

**Figure 6-3** On the left is a schematic representation of how the amplitude of particular frequencies near to $\omega_0$ of some particular kind of nucleus, produced by moving unpaired electrons or nuclear magnets would be expected to have a maximum value at some viscosity. On the right is shown the expected variation of $T_1$ for that nucleus with viscosity

major assist is provided by the fact that an unpaired electron has a large magnetic moment (~1000 times greater than that of a proton) and can usually approach on the order of molecular dimensions to an atom containing a nucleus to be relaxed. To provide the magnetic field is one thing, to provide the appropriate frequency is another matter altogether.

If we consider the motion of a paramagnetic ion relative to a particular nucleus at ordinary temperatures, we would expect it to have some sort of a distribution of velocities with low probability at zero, or very slow velocities, and also low probability at very high velocities. The ion will move about in a Brownian motion in various directions. Such motions of a magnet **in the time domain** can be analyzed by the Fourier method to determine their **frequency distribution in the frequency domain**. The corresponding plot of amplitude versus frequency is expected to have a maximum value corresponding to the maximum value of the motional spectrum (Figure 6-3). The dance of a paramagnetic ion around a nucleus undergoing relaxation is the equivalent of a very broad-band rf oscillator or receiver and there may, or may not be, a substantial amplitude in the relatively narrow range of frequencies that the nuclei can absorb efficiently to change their magnetic state. Obviously, if the amplitude maximum corresponds to the resonance frequency, the process will be as efficient as it can be with these ingredients. Thus, the efficiency of transfer of energy back and forth as the result of the motion of paramagnetic ions relative to the magnetic nuclei should be a function of viscosity. Consequently, we expect that there could be some value of the viscosity that would give a distribution of paramagnetic-ion velocities with the amplitude of its frequency maximum at the resonance frequency of the nuclei and thus cause relaxation, even if not with high efficiency.

If the paramagnetic ion can complex with an unshared electron pair in a molecule containing a nucleus undergoing relaxation, such complexation can be very effective in producing relaxation. It is especially important in observing $^{15}$N and $^{31}$P nuclei of atoms with unshared electron pairs that often complex strongly with paramagnetic ions. Oxygen also has enough paramagnetism to bring about substantial changes in $T_1$ and if you want to measure $T_1$ in a way that is intrinsic of the sample itself, oxygen should be rigorously excluded. Oxygen has a substantial affinity for aromatic compounds and it has been reported that the proton $T_1$ of benzene in dilute, oxygen-free, carbon disulfide solution is about 60 s. In the presence of oxygen, the relaxation time drops to about 3 s!

**2. Relaxation Induced by Extramolecular Magnetic Nuclei.** Clearly, if the motions of paramagnetic ions can induce relaxation, motions of the magnetic dipoles of atomic nuclei should do the same, but because of the difference in magnitude of electron and nuclear moments, the efficiency is expected to be far less. The existence of this kind of dipole-dipole interaction can be illustrated by the relaxation of protons in water. When H$_2$O is diluted with pure D$_2$O, with D having about 1/7 the magnetic moment of H, the relaxation rate of the protons in the mixture decreases as expected for less efficient relaxation by the dancing deuteron dipoles.

## 6-4 Intramolecular Mechanisms of $T_1$ Relaxation

There are several **intramolecular** $T_1$ relaxation mechanisms that have varying degrees of importance depending on the molecular structure, the magnetic field, and the medium in which the spectra are taken. These mechanisms include **dipole-dipole, chemical-shift anisotropy, spin-rotation, scalar-coupling** and **quadrupolar relaxation**. Remembering that the reciprocal of the relaxation time is the first-order rate constant for relaxation we can write:

$$k_1 = \frac{1}{T_1} = \frac{1}{T_{1\text{intermolecular}}} + \frac{1}{T_{1DD}} + \frac{1}{T_{1CSA}} + \frac{1}{T_{1SR}} + \frac{1}{T_{1SC}} + \frac{1}{T_{1Q}} \qquad \text{Eqn. 6-1}$$

where the last five terms on the right-hand side represent the several **intramolecular** contributions to the overall relaxation time. Equation 6-1 by no means universally applicable because not all of its terms are wholly independent of one another. However, it does provide a basis for discussion of relaxation mechanisms. Nonetheless, you should realize that relaxation is a very complex phenomena and much of what will be pre-

## 6-4 Intramolecular Mechanisms of $T_1$ Relaxation

sented in this Chapter is substantially oversimplified.

**1. Quadrupolar Relaxation.** Of all of the intramolecular $T_1$ relaxations, quadrupolar is probably simplest to grasp and illustrates some generally important principles. To begin with, quadrupolar relaxation requires that the nuclei involved have a spin $I > 1/2$. When a nucleus has its spin zero, as for $^{12}C$ and $^{16}O$, the nucleus has no magnetic properties and is important to NMR spectroscopy only in the way it modifies the chemical shifts or spin-spin couplings of magnetic nuclei to which it is attached, or is close enough to influence. A nucleus with $I > 0$ acts as though its nuclear charge is circulating and thereby generating a magnetic moment.[2] Nuclei with $I = 1/2$ behave as if their nuclear charge is distributed with **spherical symmetry**. The result is that no torque is exerted on such nuclei by motion of an asymmetric electric field in their neighborhood. Such a torque would act to reorient the nuclei and tend to change their magnetic quantum states.

When $I > 1/2$, the nuclei have magnetic moments and also act as though their nuclear charge is spinning in a non-spherically symmetric manner. Such nuclei can usefully be thought of as having charge distributions corresponding to either spinning prolate or oblate ellipsoids. When such a nucleus is surrounded by an asymmetric electric field and the molecule containing the nucleus turns over, then the **electric-field gradient** moves relative to the orientation of the nuclear spin axis. This creates a torque on the nucleus that will tend to reorient it and cause it to change its magnetic quantum number and relax. The parameters that determine the rate of this kind of relaxation include the $I$ value of the nucleus, $\tau_c$ (the **molecular correlation time** one definition of which is the average time for the molecule to rotate one radian as the result of molecular collisions), $eQ$ (**the magnitude of the quadrupole moment** of the nucleus), which is a measure of the degree of electric dissymmetry of the nucleus and, finally, the magnitude of the electric-field gradient that is acting on the nucleus. As with relaxation associated with intermolecular interactions arising from motions of nuclear moments, or of unshared electrons (such as of paramagnetic ions or free radicals), we expect that there will be optimal values of $\tau_c$, the correlation time, for producing relaxation, as shown in Figure 6-3.

Quadrupolar relaxation generally causes line broadening, if the NMR resonance

---

[2]Physicists **talk** about "spin" as though it were spin. But when asked to explain what they mean by "spin", they quickly point out that while the electron has "spin", its size is so small that the charge would have to be circulating faster than the speed of light to generate the rather large magnetic moment of the electron. So "spin" and nuclear magnetism turn out to be intrinsic nuclear properties, irrespective of how conceptually satisfying it is to think of them as arising from spinning nuclei.

being observed is also that of the quadrupolar nucleus. The line broadening is often small or negligible when the electron distribution around the nucleus is spherically symmetrical, as for $^{35}Cl^-$, $^{35}ClO_4^-$, $^{14}NH_4^+$ or $^{14}N(CH_3)_4^+$. The line broadening that arises from quadrupole relaxation can be a useful probe of complex formation when something like $^{35}Cl^-$ complexes rapidly and reversibly with a Lewis acid. During its brief period of complexation, if such an entity goes from a symmetrical to an unsymmetrical electrical environment, molecular tumbling will usually cause some relaxation, which will be evident as line broadening of the resonance of the uncomplexed $^{35}Cl^-$.[3]

**2. Dipole-Dipole Relaxation.** When observing $^{13}C$ or $^{15}N$, the principal mode of relaxation is usually by dipole-dipole relaxation involving directly attached or close-by protons. This is a very important intramolecular mechanism and it is most easily described in terms of the possible quantum states. Thus, it may be helpful for you to review the concepts involved in the quantum-mechanical description of nuclear magnetic energy states in Section 1-1 before charging ahead.

We will use for an example the two-spin system provided by methanoate (formate) anion labeled with $^{13}C$ in its carbon $^1H$-$^{13}CO_2^-$. Both $^1H$ and $^{13}C$ have spin $I = 1/2$ and each will have two magnetic states with the magnetic quantum numbers $+1/2$ and $-1/2$. The easiest way to deal with the energy changes possible for a two-spin system is to work with the total magnetic energy of the system, as in Figure 6-4. If we designate the proton as nucleus (1) and the carbon-13 as nucleus (2), we can devise a composite description of the system when both nuclei are in the $+1/2$ state (here taken, as before, as the more stable state with both nuclei having the same sign for their nuclear moments), thus $+1/2(1)$, $+1/2(2)$. If we let $\alpha$ represent $+1/2$ and $\beta$ represent $-1/2$, we might also write $\alpha(1)\beta(2)$. A further convenient abbreviation, particularly with many spins in the system, is to leave out (1) and (2) and simply write $\alpha\alpha$, $\alpha\beta$, $\beta\alpha$ and $\beta\beta$ for $\alpha(1)\alpha(2)$, $\alpha(1)\beta(2)$, $\beta(1)\alpha(2)$ and $\beta(1)\beta(2)$, respectively. The extension of writing $\alpha\beta\beta\alpha$ for $\alpha(1)\beta(2)\beta(3)\alpha(4)$ should be obvious. Because the resonance frequency of protons is about four times that of $^{13}C$ in the same magnetic field, we can construct an energy diagram for 600-MHz proton and 150-MHz $^{13}C$ frequencies that will have $\alpha\alpha$, the most stable state, 375 MHz below the zero of energy and $\beta\beta$, the least stable state, 375 MHz above zero. The states $\alpha\beta$ and $\beta\alpha$ will obviously have intermediate $\pm$ 225-MHz energies.

There will be four possible transitions in which only **one** nucleus changes its quantum number (**single-quantum transitions**). As shown in Figure 6-4, there are two

---

[3] For practical applications of this phenomenon to the study of proteins, see T.R. Stengle and J.D. Baldeschweiler, *J. Am. Chem. Soc.*, **89**, 3045-3050 (1967)

## 6-4 Intramolecular Mechanisms of $T_1$ Relaxation

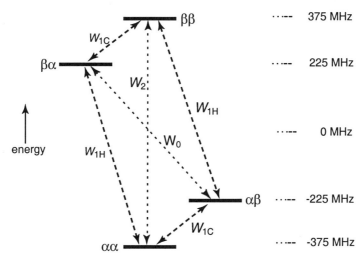

**Figure 6-4** Energy diagram for relaxation of the nuclear magnetic states of $^{13}C$-labeled methanoate anion H-$^{13}CO_2^-$. At a 600-MHz magnetic field for protons, the $^{13}C$ resonance will be about 150 MHz so that the separation between the $\alpha\alpha$ and $\beta\beta$ magnetic states will be a total of 750 MHz. The transitions labeled $W_{1H}$ and $W_{1C}$ are all single-quantum transitions, in that the total quantum number in the change is ±1, i.e. +1/2, +1/2 $\rightleftarrows$ +1/2, -1/2; +1/2, -1/2 $\rightleftarrows$ +1/2, +1/2; and so on. In contrast, $W_2$ is a double-quantum transition with the total magnetic quantum number changed by two, while $W_0$ is a zero-quantum transition

single-quantum proton transitions labeled $W_{1H}$, as well as two-single quantum $^{13}C$ transitions labeled $W_{1C}$. The transitions $\alpha\alpha \rightleftarrows \beta\beta$ labeled $W_2$, are **double-quantum transitions** and these are **forbidden** transitions, as far as the usual spectrometric equipment for NMR excitation is concerned. However, "forbidden" is a relative word when spectroscopy is involved and should not be taken as an absolute. The transitions $\alpha\beta \rightleftarrows \beta\alpha$ involve no change in total magnetic quantum number and are also "forbidden" transitions. Both $\alpha\alpha \rightleftarrows \beta\beta$ and $\alpha\beta \rightleftarrows \beta\alpha$ can also be classified as **combination transitions,** in which **two** nuclei change their quantum states **simultaneously**. These transitions are forbidden for the usual energy input from a rf oscillator because the total magnetic quantum number changes are +2 and 0, respectively. It turns out that combination transitions can be written for three or more nuclei which would have the usual allowed ±1 change in total magnetic quantum number. An example is $\alpha\beta\alpha \rightleftarrows \beta\alpha\beta$. However, such transitions are also forbidden.

Next to come is the relevance of all of this to relaxation. After excitation, establishment of $T_1$ equilibrium for H-$^{13}CO_2^-$ occurs by dipole-dipole relaxation, in the absence of significant relaxation by unpaired electrons or extramolecular magnetic nuclei. Dipole-dipole relaxation is connected with molecular motions in something like the same way as the relaxation that is associated with extramolecular motion of nuclear magnets (Section 6-3-1). What is different and surprising is that the multiple-quantum transitions that are "forbidden" and are marked in Figure 6-4 as $W_0$ and $W_2$ can be **very important**. For each relaxation path, there is the expected maximum relaxation rate that is a function of the molecular correlation time and the strength of the applied magnetic field.[4] For very rapid motions, the efficiency of relaxation is greatest for the largest frequency difference between the states (see Figure 6-4), so that $W_2$ is favored. For long correlation times (slow motions), $W_0$ will become more important than $W_2$. Only the **intramolecular** $W_2$ and $W_0$ dipole-dipole relaxations are sizable, because the efficiency of dipole-dipole relaxations depends on the reciprocal of the sixth power of the distance between the pairs of nuclei involved $r^{-6}$.

An important consideration is that, while relaxation by the $W_{1H}$ or $W_{1C}$ routes can be enormously influenced by paramagnetic ions and other relaxation modes, $W_0$ and $W_2$ are not similarly affected and are only associated with the intramolecular dipole-dipole mechanism. This has important consequences for the nuclear Overhauser effect as we will see shortly.

**3. Decoupling of Nuclear Spins and the Nuclear Overhauser Effect.** One of the most important techniques in using NMR spectroscopy for structural studies or reaction rates involves the simplification of spectra by removing spin-spin splittings by decoupling. In the simple explanation for how one kind of spin 1/2 nuclei splits another kind's resonance, we say that one or the other of the two possible magnetic states of the splitting nuclei either augments, or decreases, the magnetic field at the nuclei whose resonances are being split and so cause those nuclei to have one or the other of two resonance frequencies. If we call the increments to the field at the second group of nuclei $+J/2$ and $-J/2$, the observed splitting between the two resonances will be $J$. Such splittings are usually reported in Hz and, unlike the chemical shift, are independent of the magnetic field. The reason for the independence on field strength is simply because the increment to the field depends on the magnitude of the nuclear moment, not on the shieldings or populations associated with the α and β states. But surely, the populations make a difference in something. Indeed, they do, but are not expected to be observable in our usual simple spectra, except at very low temperatures. We know that the lower state (call it α) will be more populated than the β state, but as we have shown earlier (Section 1-2), the excess of the population of α over that of β is only 0.012% even for protons at room temperature in a 750-MHz spectrometer! What this means is that

---

[4]See H.O. Kalinowski, S. Berger and S. Braun, *Carbon-13 NMR Spectroscopy*, John Wiley & Sons, New York, 1996, Chapter 5 for discussion and references.

## 6-4 Intramolecular Mechanisms of $T_1$ Relaxation

one of the two lines in the doublet produced by the splitting will only be about 1.00012 times more intense than for the other. This is hardly observable at the present state of the art of NMR spectrometers. However, at 1° K, the ratio of the lines would be 1.037, which would be observable, if the lines were sharp enough (not likely).

In this simple picture of spin-spin splitting, we can consider the process of decoupling as being the selective irradiation of the first kind of nuclei (the splitting nucleus) so that transitions occur rapidly enough between the magnetic states of that nucleus so as to average $+J/2$ and $-J/2$ to zero. In the electromagnetic vector picture, we can envision decoupling as applying a continuous rf frequency at $\omega_1$, the frequency of the splitting nuclei, to produce a continuous rotation of the magnetization vector of the splitting nuclei around the $X$ axis, which if fast enough would average the magnetization vector of nucleus (1) to zero with respect to the nucleus (2).

The above simple picture suffices for many purposes, but does not account for the fact that irradiation of the proton of H-$^{13}CO_2^-$ not only causes the $^{13}C$ doublet ($J$ = 195 Hz) to collapse to a singlet, but gives a singlet with more than twice the intensity of the sum of the separate intensities of the doublet lines. This enhancement of the signal, we call the **Nuclear Overhauser Effect** or **NOE**. The influence of the NOE on the practice of $^{13}C$ NMR in organic compounds is discussed in Section 5-1 and the references cited there. To understand decoupling and the NOE, we need to modify our energy diagram of Figure 6-4 to reflect the energy changes that are involved in spin-spin splitting.

Now, let us consider that $\alpha\alpha$ represents two spin 1/2 nuclei with their moments pointing in the same direction, something like two bar magnets side by side that have their two north poles and their two south poles in juxtaposition.

We know that two bar magnets arranged in this way will repel each other and we might guess that the two-nucleus system arranged similarly will be less stable than if they were a long way from one another, in which circumstances their influence on one another would be negligibly small. Of course, $\beta\beta$ would be just like $\alpha\alpha$ in having repulsion between its magnetic dipoles. Contrariwise, we would expect that $\alpha\beta$ or $\beta\alpha$ would be more stable than two nuclei far apart, because now there would be an attractive force between the magnetic dipoles.

This picture involving magnetic dipoles is very simple and has the virtue that it is actually possible to use electromagnetic theory to calculate the repulsion or attraction between the dipoles at the distance they would be expected to have in molecules and thus make it possible to determine whether these repulsive or attractive energies correspond in any reasonable way to the measured $J$ values for the splittings. The short answer is that they do not. The calculated energies are $10^3$-$10^4$ times too large for two nuclei treated as simple bar magnets aligned as above. Here, we have treated the nuclei as having a completely fixed relationship to one another. But this is unrealistic when each nucleus can have only two magnetic quantum states relative to the field direction. Consider what happens when the two nuclei are separated by a distance $r$ in a diatomic molecule and that molecule is tumbling about the midpoint of $r$.

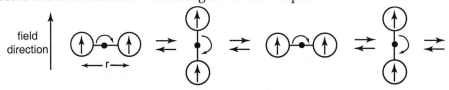

If tumbling occurs freely, it turns out that the dipole effect averages to zero. This is dismaying because it shows that our simple picture of coupling cannot be correct. But there is a residual splitting that, as we have said, is $10^3$-$10^4$ smaller than predicted. What is its origin? What is going on here? The answer is that, although the dipolar mechanism does not operate for freely tumbling molecules, what we observe is a splitting interaction that is transmitted from one magnetic nucleus to the other by the bonding electrons. The way in which this operates for a simple molecule like H-D (remember that $H_2$ will only show one resonance line, because the chemical shifts of the two protons are the same) is for each nuclei to contribute a small magnetic polarization to the bonding electrons. You should not expect it to be a wholly foregone conclusion that such an interaction would invariably cause αα and ββ to be destabilized, while αβ and βα are stabilized, but this is indeed the way that it turns out for the one-bond coupling of $H_2$. Interestingly for molecules where there is a two-bond geminal H-D coupling, as for $Cl_2C=CHD$, it is quite usual that the reverse is true, αβ and βα being destabilized while αα and ββ are stabilized!

Now assume that αα and ββ are destabilized and αβ and βα are stabilized for the $^{13}C$-methanoate anion. Our energy diagram (Figure 6-4) will require correction, αα and ββ each being raised by $J/4$, while αβ and βα are each decreased by $J/4$, where $J = 195$

## 6-4 Intramolecular Mechanisms of $T_1$ Relaxation

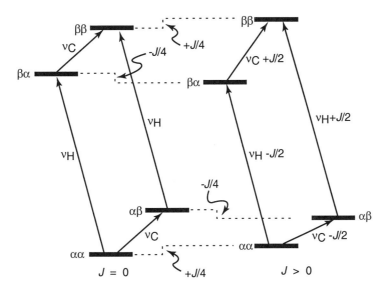

**Figure 6-5** Energy diagram for the observable transitions for the nuclear magnetic states of $^{13}$C-labeled methanoate anion, H-$^{13}$CO$_2^-$, for the cases where $J$ is zero and when $J$ is greater than zero. The states $\alpha\alpha$ and $\beta\beta$ are raised in energy by $J/4$, while the states $\alpha\beta$ and $\beta\alpha$ have their energies lowered by $J/4$. The transitions shown are all single-quantum transitions, where $\nu_H$ and $\nu_C$ represent the energy changes that correspond to the frequencies of the proton and $^{13}$C transitions, respectively. The energy diagram shows that, when $J$ is different from zero, a four-line spectrum will result with doublets separated by $J$ at the proton and the carbon frequencies, see Figure 6-6

Hz, as shown in Figure 6-5. Why $J/4$? Because it will be seen that, with $\alpha\alpha$ raised in energy by $J/4$ and $\beta\alpha$ lowered by $J/4$, the $\alpha\alpha \to \beta\alpha$ transition will be decreased by $J/2$ and the $\beta\alpha \to \beta\beta$ transition increased by $J/2$. Thus, we see that $^{13}$C resonance will be a doublet separated by the coupling constant, $J$. The two lines of the doublet will appear to be equally intense by the argument given earlier. The same procedures applied to $\alpha\alpha \to \beta\alpha$ and $\alpha\beta \to \beta\beta$ will lead to two proton resonances separated by $J$, see Figure 6-6. Note that the $J/4$ spacings in Figure 6-6 are wildly exaggerated with respect to the shift differences by a factor of ~ $10^7$, but no principles are sacrificed thereby.

If you go back through the analysis of Figure 6-5, you will find that it does not matter whether $J$ is positive or negative; either sign produces a four-line spectrum with the same line spacings. The only difference is that the position of the transitions $\alpha\alpha \rightleftarrows \alpha\beta$ and $\beta\alpha \rightleftarrows \beta\beta$ as well as $\alpha\alpha \rightleftarrows \beta\alpha$ and $\alpha\beta \rightleftarrows \beta\beta$ are interchanged. We could tell whether $J$ is positive or negative if we could measure the difference in the peak intensity between the doublets in each pair of transitions. However, as pointed out earlier in this Section, this is not a very practical proposition, because the predicted differences are quite small. Nonetheless, it has been possible to determine the absolute sign of $J$ for

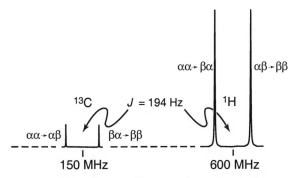

**Figure 6-6** Schematic representation of the transitions corresponding to the energy diagram of Figure 6-5. The $^1$H and $^{13}$C resonance intensities are not to scale. Because of the smaller magnetic moment of $^{13}$C, the intensity of the $^{13}$C resonances will be 1/60 of the $^1$H resonances for $^1$H-$^{13}$CO$_2^-$, see Table 1-1

several cases and subsequently the signs of many other $J$ values have been related to these absolute determinations.

We can understand how proton decoupling collapses the $^{13}$C splitting in the system of Figure 6-5 and 6-6, by causing sufficiently rapid transitions between the proton magnetic states, so that the relatively small coupling (194 Hz compared to the 600-MHz difference between the proton states) is averaged to zero. Understanding the line intensities resulting from the nuclear Overhauser effect (NOE) is less simple.

The intensity of a given NMR transition will depend on the populations of the magnetic states involved. If the lower-energy state has a higher population than normal relative to an upper state, we can expect that the absorption transition will have an enhanced probability. Conversely, if the upper state is more populated than the lower state, then we can expect an **emission signal** to be observed. This may not seem reasonable, but we have seen how a 180° pulse can give negative magnetization and, in the quantum-mechanical view, the 180° pulse inverts the normal population distribution. Another way to look at the population effect is in terms of laser action. Increases in population of one state relative to another from the normal can lead to **stimulated absorption** or **stimulated emission.**

The point of the above is that, if particular transitions are enhanced, the likely reasons are differences from the normal population distributions of particular magnetic states. Figure 6-7 shows that irradiation at the proton frequencies of the αα ⇌ βα and αβ ⇌ ββ will initially tend to increase the populations of states βα and ββ. What happens next will depend on how relaxation occurs. This is a complex situation but, if the $^1$H resonance lines are saturated, the $T_2$'s are long, the correlation times short, and equilibrium is established, it turns out that the various relaxation pathways have relative rates of $W_2:W_1:W_0 = 12:3:2$. That $W_2$ dominates is very important because $W_2$ provides a

## 6-4 Intramolecular Mechanisms of $T_1$ Relaxation

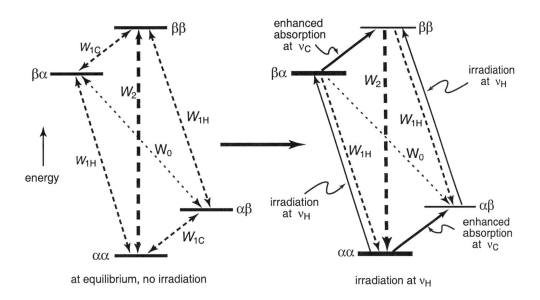

**Figure 6-7** Energy diagram for the nuclear magnetic states of $^{13}$C-labeled methanoate anion H-$^{13}$CO$_2^-$ with fast molecular motions. On the left, where the system is at equilibrium and the oscillators are off, relaxation occurs by the dipolar mechanisms; i.e., $W_0$ and $W_2$ along with the one-quantum mechanisms, $W_{1H}$ and $W_{1C}$. The transition arrows are drawn weighted to show that $W_2$ is most favored and $W_0$ is least favored. On irradiation at $\nu_H$, the populations of $\beta\beta$ and $\beta\alpha$ are expected at first to tend to increase, but then the predominant $W_2$ relaxation reduces the population of $\beta\beta$, thereby increasing the population of $\alpha\alpha$. When equilibrium is reached among all of the pathways in the presence of the radiation field, the transition at $\nu_C$, the $^{13}$C transitions, have enhanced probabilities, because of the increases in the populations of $\alpha\alpha$ and $\beta\alpha$. The heavy horizontal lines at $\alpha\alpha$ and $\beta\alpha$ reflect increases in the populations above normal, while the lighter lines are used for $\beta\beta$ and $\alpha\beta$ with subnormal populations. The enhanced absorptions are manifestations of the ($^{13}$C, $^1$H) nuclear Overhauser effect

pathway for preferentially depleting $\beta\beta$ and replenishing $\alpha\alpha$. The overall result is that $\beta\alpha$ becomes relatively overpopulated and $\beta\beta$ becomes depleted. The same is true for $\alpha\alpha$ and $\alpha\beta$. With this population distribution, the $\alpha\alpha \rightarrow \alpha\beta$ and $\beta\alpha \rightarrow \beta\beta$ transitions are enhanced and a NOE is observed. However, if the relaxation is not exclusively dipolar, then the relaxation via $W_2$ becomes relatively less efficient with respect to the $W_1$'s and the NOE diminishes. For $^{13}$C and protons, as in $^{13}$C-methanoate anion, the signal strength depends on the relative magnitudes of $T_{1\text{dipolar}}$ and $T_{1\text{other}}$ by:

$$\text{signal} = 1 + \text{NOE}(^{13}\text{C},^1\text{H}) = 1 + \frac{(\gamma_{1_H}/\gamma_{13_C})}{2} \cdot (1/T_{1DD})/(1/T_{1DD} + 1/T_{1\text{other}}) \qquad \text{Eqn. 6-2}$$

where $\gamma_{1_H}/\gamma_{13_C} \sim 4$. If $T_2$'s are short and the lines broad, computation of the NOE is more involved, but will generally be smaller. Be that as it may, the maximum signal

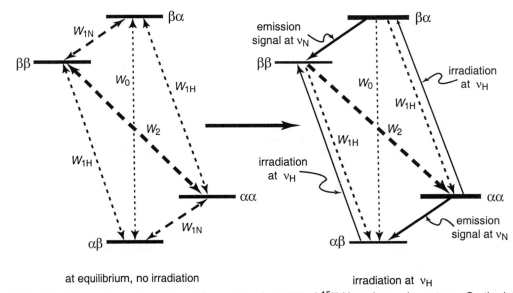

**Figure 6-8** Energy diagram for the nuclear magnetic states of $^{15}$N-H nuclear spin system. On the left, where the system is at equilibrium and the oscillators are off, the system undergoes relaxation through the dipolar mechanisms, $W_0$ and $W_2$ as well as the one-quantum mechanisms $W_{1H}$ and $W_{1N}$. Notice how the states shown in Figure 6-3 are now rearranged because of the negative magnetic moment of $^{15}$N. Also note that the relative positions of the $W_0$ and $W_2$ pathways for relaxation have been interchanged. The transition lines are again weighted in the figure to show that $W_2$ is most favored and $W_0$ is least favored. On irradiation at $v_H$ the populations of both states $\beta\beta$ and $\alpha\alpha$ are expected to increase, but the predominant $W_2$ relaxation reduces the population of $\beta\beta$, thereby increasing the population of $\alpha\alpha$. When equilibrium is reached in the presence of the radiation field, the transitions at $v_N$, the $^{15}$N transitions, have *emission* probabilities, because of the increases in the populations of $\alpha\alpha$ and $\beta\alpha$ The heavy lines at $\alpha\alpha$ and $\beta\alpha$ reflect these increases in the populations, while the lighter lines are used for $\beta\beta$ and $\alpha\beta$. The emission signals are the manifestations of the negative ($^{15}$N,$^1$H) nuclear Overhauser effect

strength will be 3 when $T_{1other} = 0$. It should be clear that, if one were to irradiate $^{13}$C and observe $^1$H, the NOE would be much smaller, because $\gamma_{13C}/\gamma_{1H}$ is about 0.25.

With $^{15}$N, the NOE takes an unusual turn in that **emission signals** are observed for the $^{15}$N resonances when $^1$H is irradiated and dipolar relaxation dominates. The underlying reason for this behavior is that the nuclear moment of $^{15}$N is negative, while that of $^{13}$C and most of the other nuclei of importance to organic chemistry are positive. The result of the difference in sign is to make $\alpha\beta$ the most stable state and $\beta\alpha$ the least stable state. Thus, Figure 6-8 shows the resulting arrangement of energies. Now, if we pump on the proton transitions by irradiating at the proton frequencies, the dominant $W_2$ relaxation will make the $\beta\alpha$ and $\alpha\alpha$ states abnormally populated, respectively, relative to $\alpha\beta$ and $\beta\beta$ and, consequently, emission signals will be observed for the $^{15}$N transitions. Because $\gamma_{1H}/\gamma_{15N} \sim -10$, the maximum NOE signal (when $1/T_{1other} = 0$) for $^{15}$N is given by Equation 6-3.

## 6-4 Intramolecular Mechanisms of $T_1$ Relaxation

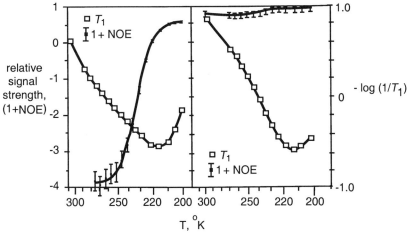

**Figure 6-9** On the left is shown how $T_1$ and the $^{15}$N-signal strength (as 1 + NOE) at 50 MHz changes for urea as a function of temperature in water-dimethyl sulfoxide solution where the principal effect of decreasing temperature is to increase the viscosity (and molecular-correlation times). A full NOE, giving about -4 for the signal strength, will only be observed when dipolar relaxation is the important mechanism at low viscosities. When the NOE is no longer effective, then the signal strength approaches the relative intensity of +1. On the right, is the same kind of plot for the $^{15}$N resonance of azabenzene (pyridine). Here dipolar relaxation is ineffective and as $T_1$ decreases with increasing viscosity only a small change takes place in the NOE[5]

$$\text{signal} = 1 + \text{NOE}(^{15}\text{N}, ^1\text{H}) = 1 + \frac{(\gamma_{^1\text{H}}/\gamma_{^{15}\text{N}})}{2} \cdot (1/T_{1DD})/(1/T_{1DD} + 1/T_{1\text{other}}) \sim -4 \qquad \text{Eqn. 6-3}$$

Clearly as $1/T_{1\text{other}}$ increases, the fraction of dipolar relaxation decreases, the $^{15}$N NOE will decrease and the strong emission signal expected for the maximum NOE will diminish in intensity, pass through zero and finally reach the positive value expected for no NOE. The left side of Figure 6-9 shows how this comes about for the $^{15}$N of the nitrogens of urea, $NH_2C(=O)NH_2$, dissolved in water-dimethyl sulfoxide. The viscosity of the solution is increased by lowering the temperature 100° from about 300° K to 200° K. You can see that the signal strength (measured as $1+\text{NOE}(^{15}\text{N},^1\text{H})$) begins to diminish where dipolar relaxation becomes less efficient relative to other modes of $T_1$ relaxation and $T_1$ is approaching its minimum value. A different situation is shown in the right-side of Figure 6-9 for azabenzene (pyridine) for which dipole-dipole relaxation is quite

azabenzene

---

[5] A. Wei, M.K. Raymond and J. D. Roberts, *J. Am. Chem. Soc.* **119**, 2915 (1997). A. Wei, Senior Thesis, California Institute of Technology, 1989; M.K. Raymond, Senior Thesis, Reed College, 1991.

inefficient, because there is no hydrogen directly attached to the nitrogen. Again, there is a $T_1$ minimum with increasing viscosity, but the signal strength curve shows that the NOE is small across the experimental range in temperature.

Unlike ¹³C, a little NOE for ¹⁵N can be much worse than no NOE at all. In using ¹³C for structural determinations, you are likely to be mostly interested in chemical-shift information, so you cheerfully take whatever NOE you get, along with proton decoupling. Differential NOEs for different ¹³C resonances will make quantitative comparisons of peak intensities virtually useless (see Section 5-2-**13**). However, all of the peaks will be there with, or without, NOE. With ¹⁵N, differential NOEs can be a disaster, because some peaks may be weakened, or even nulled, by the NOE, while others may be strong emission signals. Thus, attempts to take advantage of the ¹⁵N NOE for signal enhancement must take into account this possibility, especially when there are reasons to suspect that $T_{1\text{other}}$ may be comparable to $T_{1\text{dipolar}}$. This problem has special cogency for spectra run on water solutions, where there is the possibility of contamination by paramagnetic metal ions. Such ions, as we have mentioned before (Section 6-3-**1**), can be especially important in promoting $T_1$ relaxation of nitrogens carrying unshared electron pairs.

In using ¹⁵N spectroscopy for structural elucidation, it turns out that ¹⁵N $T_1$ values are often very long. This is a particular problem for small molecules, that have no hydrogens on nitrogen so that ¹H-¹⁵N dipolar relaxation is inefficient. Small molecules are particularly affected because their rates of tumbling are fast and, in the usual NMR solvents, the tumbling rates will be far from the $T_1$ minimum on the $T_1$ vs. correlation time curves (Figure 6-3). As one example, the ¹⁵N $T_1$ for 1-methyl-1-azacyclopentane (*N*-methylpyrrolidine) is on the order of 1000 s and, if we were to wait for 90% or

1-methyl-1-azacyclopentane

more of $T_1$ relaxation to occur in a pulse-FT procedure, the time between pulses would be about an hour. The obvious and simple way to alleviate this problem is to decrease $T_1$ by adding paramagnetic substances. A favorite is the chromium salt of 2,4-pentanedione (chromium *tris*-acetylacetonate). However, this procedure sacrifices any possible advantage to be gained from the NOE because it increases $T_{1\text{other}}$.

An alternative way to shorten $T_1$, without necessarily diminishing the NOE is to decrease the rate of tumbling and thus make the dipolar relaxation more effective, at least down to the point where other relaxation mechanisms become important. There

## 6-4 Intramolecular Mechanisms for $T_1$ Relaxation

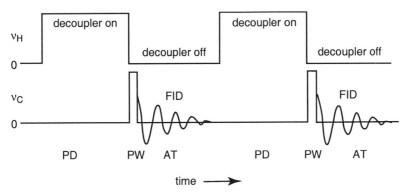

**Figure 6-10** Gated pulse sequence for producing a $^{13}$C-proton NOE and, at the same time, allowing the $^{13}$C-proton splittings to be observed. The principle here is that both the buildup of excess populations by proton decoupling during the pulse-delay period PD and relaxation are relatively slow processes, but the spin-spin splittings can be detected by a suitable short pulse PW and taking the FID during AT when the decoupler is off

are various ways of increasing the viscosity of the solutions to slow the tumbling rates, as by adding soluble high-polymeric materials. A sometimes convenient procedure (illustrated in Figure 6-9) is to decrease the temperature well below 0° C and this can be done, even with water solutions, if sufficient dimethyl sulfoxide is added to prevent freezing.[5]

There are a number of tricks by which the NOE or decoupling can be manipulated to obtain greater sensitivity, or line intensities that more accurately reflect relative nuclear concentrations. One such procedure is **gated decoupling** that allows a spectrum to show a NOE while still retaining coupling information. This may seem contrary to everything we've said about the NOE and decoupling, but remember that the NOE depends on differences of populations of states, while coupling depends on the states being mixed up by irradiation at the time the spectral information is being collected. Referring to the energy diagrams of Figures 6-5 and 6-7, we will show how we could have substantial excess population of particular states without sacrificing the coupling information. The important thing is to build up the abnormal populations **before** gathering the spectral information. This can be done for our $^{13}$C-methanoate example by pumping up the abnormal populations and then turning off the decoupler, immediately applying a $^{13}$C pulse (which could be 90°) and then observing the decay of the $^{13}$C FID with the decoupler off. The process can then be repeated, if spectral accumulation is necessary.

Diagrams, such as Figure 6-10 have been developed for this kind of sequence that illustrate in a simple way the temporal relationships involved. Here, the sequence is

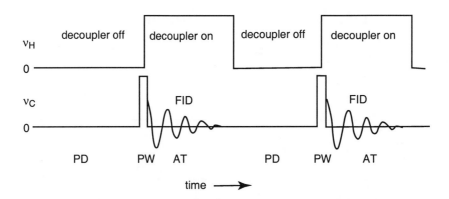

**Figure 6-11** Inverse gated pulse sequence for producing a decoupled $^{13}$C spectrum with no $^{13}$C-proton NOE. Here, in contrast to the sequence of Figure 6-10, the decoupler is left off during the pulse delay period PD to allow any excess populations produced by proton decoupling during the time period AT to return to normal. Then, following the short pulse PW, the spin-spin splittings are removed by proton decoupling while observing the FID. Before repeating the process, it is necessary to wait long enough during PD to allow any excess population growth during AT to decay away with the decoupler off

pulse delay PD, a $^{13}$C excitation pulse PW and acquisition time AT. Note that the period of $\nu_H$ irradiation is rather long; this is because it takes a significant time for the system to reach the equilibrium state represented by right-hand side of Figure 6-7.

The crux of this so-called **gated decoupling procedure** is that normally $T_1$ is substantially longer than the period over which the FID is observed, i.e. $T_1 > T_2$. Therefore, the abnormal population distribution can be expected to be maintained quite well during the FID, especially in the early part when the signal is strongest.

The opposite result can be obtained with **inverse gated decoupling**, whereby the decoupled spectrum is obtained without NOE. This sequence involves having the decoupler on only during the acquisition period; the build up of NOE is now slow compared to the time of taking the FID, Figure 6-11. Here, if repetitive scans are used, the pulse delay should be quite long compared to $T_1$ to ensure that any buildup of abnormal populations, during the irradiation period with $\nu_H$, is returned to near normal before the sequence is restarted.

The principal purpose of inverse-gated decoupling is to allow accurate measurements of peak intensities in the absence of NOE, see Section 5-2-13. Another use is to quantify the NOE from the ratio of two peak intensities; one intensity being that obtained with as much NOE as is possible with particular peaks and the other intensity being that obtained by inverse-gated decoupling.

Another useful technique for spectral analysis, particularly for $^{13}$C spectra is to use **off-resonance $^1$H decoupling**. The idea here is to use a low-enough power level to resolve the small secondary couplings involving protons and carbons separated by more than one bond and let the larger one-bond couplings remain, albeit with somewhat reduced magnitudes. In this connection, **off-resonance** means that the coupling frequency is not the same as the resonance frequencies of the protons that are being decoupled. (We have seen earlier in Section 2-5 how the degree of excitation depends on the rf power levels and the difference between the exciting frequency and the nuclei to be excited). Because the range of $^{13}C$ chemical shifts may be large, problems will be encountered when trying to use a single proton frequency to decouple all of the secondary couplings in a uniform way. Substantial improvement is possible with a noise-modulated proton frequency that can apply a substantial range of proton frequencies to the sample. Reasonably strong, but not complete, **noise decoupling** is very useful for structural analysis, because it will broaden the primary couplings in $^{13}C$ resonances of methine and methyl structural units and give sharp lines for the resonances of methylene and quaternary carbons.

It also may be very helpful to do **on-resonance single frequency decoupling** to identify particular proton resonances with particular carbons. However, this kind of information can be achieved much more generally by the so-called 2D method, which will be described later.

**4. Relaxation by Chemical-Shift Anisotropy.** We have seen how $T_1$ relaxation can be produced by a fluctuating magnetic field at a nucleus which has frequency components close to $\omega_0$. **Chemical-shift anisotropy** has the proper characteristics to produce this kind of $T_1$ relaxation. What is chemical-shift anisotropy and how does it work? Molecular anisotropy means that a property is not the same in all directions in the molecule. Normally, we think of the chemical shift of nuclei as being single-valued, but this is because we observe the shifts of tumbling molecules, whereby any directional differences in shifts within the molecule will be averaged to a single value. That there are directional differences should not come as a surprise. Consider benzene; we know, or should expect, that the magnetic susceptibility will be different perpendicular to, or along the plane of, the benzene molecule. This means that if we have a way to orient the molecule edgewise, and then broadside, to the field direction, we will expect both the carbon shifts and the proton shifts to be different for the two orientations. In many cases, the differences in shift corresponding to different orientations of molecules in the magnetic field, which are usually quantified by **chemical-shift tensors,** may be more than 100 ppm. This fact can make qualitative theorizing about comparative chemical shifts rather hazardous, because the various tensors are known to change drastically with changes in molecular structure and yet the changes often occur in ways that may make only small changes in the measured **average shift** values.

How chemical-shift anisotropy enters as a factor in $T_1$ relaxation should be clear provided we recognize that, as an anisotropic molecule tumbles, the magnetic field at the nucleus will change because of the directional differences in chemical shift. Early on, it was believed that chemical-shift anisotropy would probably not be very important, except at very high magnetic fields, because the effect is expected to depend on the square of the applied magnetic field. However, there are many important contrary examples. Consider the relaxation of $^{15}$N in azabenzene (pyridine). The electrical dissymmetry of this molecule suggests a substantial chemical-shift anisotropy. There is no proton directly attached to nitrogen and so, by token of the $r^{-6}$ effect, dipolar relaxation should not be very efficient. We can easily tell what fraction of the $T_1$ relaxation time can be attributed to dipolar relaxation from the $^{15}$N-$^1$H NOE. With the $^{15}$N of azabenzene coming into resonance at 18 MHz, $T_1$ at room temperature is about 90 s, but even so there is not a full - 4 NOE. As shown in Figure 6-9, when the spectra are taken at 50 MHz, $T_1$ drops to 16 s and the NOE becomes almost negligible. The change in $T_1$ and NOE is the result of more efficient relaxation by $T_{1\text{other}}$ and, because this is associated with a change in magnetic field, chemical-shift anisotropy is indicated.

It is interesting to examine the behavior of the $^{15}$N signals of protonated azabenzene (in the form of the salt of azabenzene with trifluoroacetic acid) dissolved in water-dimethyl sulfoxide with changes in temperature as shown in Figure 6-12. With this

protonated azabenzene

substance, dipolar relaxation should be more efficient than with azabenzene itself, because there is now a hydrogen on nitrogen, but there is still the electrical dissymmetry that makes relaxation by $T_{1\text{CSA}}$ an important contributor to $T_1$. The consequence is that, at near room temperature (about 300° K) and a 50-MHz $^{15}$N frequency, there is some NOE, as can be seen in the left side of Figure 6-12. You will also see that $^{15}$N dipolar relaxation becomes more efficient as one lowers the temperature, because the NOE increases significantly. The reason is that the dependence of $T_{1\text{CSA}}$ on the molecular correlation time is different from that of $T_{1\text{DD}}$. The NOE now increases until the tumbling rate becomes slow, $T_1$ approaches its minimum, and dipolar relaxation starts to become less important, with a concomitant decrease in the NOE. When the same experiment is carried out at a 30-MHz $^{15}$N frequency (right side of Figure 6-12), $T_{1\text{CSA}}$ is

# Exercises

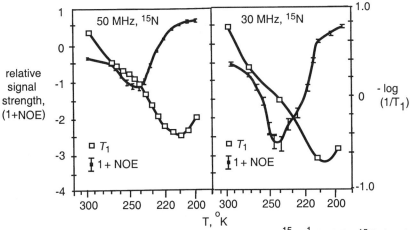

**Figure 6-12** Variations of the $T_1$ and NOE, measured as 1 + NOE($^{15}$N, $^1$H) of the $^{15}$N signals of protonated azabenzene (in the form of the salt of azabenzene with trifluoroacetic acid) dissolved in water-dimethyl sulfoxide with changes in temperature at 50 MHz (left) and 30 MHz (right) for the $^{15}$N resonance frequency[4]

less important, dipolar relaxation becomes more efficient and the NOE becomes more negative with decreasing temperature, until the correlation time becomes long.

**5. Spin-Rotation Relaxation.** If a methyl group twirls around the bond connecting it to the rest of a molecule, the nuclei and bonding electrons circulate around the bond axis and thereby generate a local magnetic field. The field will fluctuate as the result of collisions and thereby produce frequency components that can induce relaxation. This relaxation mechanism is called **spin rotation** and is most important for small molecules. Because fast motions are required to produce significant effects, $T_{1SR}$ normally decreases with temperature, which is opposite in behavior to the other forms of relaxation we are considering. Spin rotation is usually important only for small molecules in nonviscous solvents.

**6. Scalar-Coupling Relaxation.** If a nucleus is coupled to a quadrupolar nucleus and that coupling is modulated by the rate of relaxation of the quadrupolar nucleus then there is a magnetic-field fluctuation that leads to variety of what is called **scalar coupling relaxation**. Another example would be modulation of a coupling, such as might be expected for a compound with a $^{15}$N-$^1$H situation, when undergoing chemical exchange.

## Exercises

**Exercise 6-1** Consider the system shown in Figure 6-1. Will an increase in the magnetic field used in the spectrometer be expected to increase or decrease $T_2$, if all other things remain the same? Give your reasoning.

**Exercise 6-2**  Explain how paramagnetic ions can assist $T_2$ relaxation, particularly if the nucleus undergoing relaxation is in an atom that has an unshared electron pair.

**Exercise 6-3**  The proton NMR spectrum of an acidic solution of methylammonium chloride in water shows in succession, starting from TMS at zero: a clean 1:3:3:1 quartet, then a large water peak and, finally, three broad featureless peaks separated from one another by about 80 Hz. Sketch out the spectrum and explain why it appears the way it does.

**Exercise 6-4**  Consider observing the complex $^{13}C$ satellites on time-averaged proton spectra of benzene in oxygen-free carbon disulfide solution, where the proton $T_1$ relaxation time is about 60 s (see Section 6-3, **2**) and use Figure 5-13 to estimate the optimum interval between pulses for both 20° and 90° pulse angles. Suppose to decrease $T_1$, we exposed the sample air so that $T_1$ dropped to 3 s. Estimate what the optimum interval between pulses for both 20° and 90° pulse angles should be now. What possible NMR disadvantage might there be to allowing the sample to be exposed to air?

**Exercise 6-5**  The role of paramagnetic substances in inducing $T_1$ relaxation in water solution can hardly be overestimated. Consider a solution of calcium methanoate (formate) in ordinary deuterium oxide ($D_2O$) and note that natural-abundance calcium nuclei are comprised of more than 99% of two isotopes, $^{40}Ca$ and $^{44}Ca$, both having spin $I = 0$. A commercial sample of calcium methanoate in $D_2O$ showed a $T_1$ of 1.30 s for the methanoate proton. Calcium methanoate prepared from ultrapure materials with the solvent run through ion-exchange resins raised $T_1$ to about 14.5 s and removal of oxygen raised it further to about 30 s. Suppose now the solvent were changed from $D_2O$ to water ($H_2O$) free of paramagnetic ions and oxygen. What would you expect to happen to $T_1$? Explain.

**Exercise 6-6**  Explain whether you think that doubling the magnetic field would, or would not, change the viscosity required to give the minimum value of $T_1$ shown in the right side of Figure 6-3.

**Exercise 6-7**  It is generally found that when you take similar organic substances that differ primarily only in molecular weight that $T_1$ usually becomes smaller as you increase the molecular weight. However, at some point, if you continue to increase the molecular weight, $T_1$ starts to increase. Explain.

**Exercise 6-8** If you make $^{13}$C $T_1$ measurements of successive carbons along the chain of something like a ten-carbon straight-chain alcohol, there are as much as five- to ten-fold differences. If you had a five-carbon alcohol which carbons would you expect to have the largest and which the smallest $^{13}$C $T_1$ values? Give your reasoning.

**Exercise 6-9** It has been reported that, when aminoethanoic acid (glycine, isoelectric point, pH 6.4) was dissolved in commercial deuterium oxide (almost always contains traces of copper ions), the $^{13}$C $T_1$ of the carboxyl group was found to be 16 s at pD 3.7 s at pD 7.3 and 14 s at pD 10.0. When the glycine was sublimed and the deuterium oxide was distilled, then extracted with a powerful copper-complexing solution, the $T_1$ observed was 75 s at pD 3.1, 86 s at pD 5.8 and 83 s at pD 9.6. Then, when the deuterium oxide was replaced by highly purified water as solvent, the $T_1$ dropped to 44 s. The $^{13}$C-proton NOEs ranged from 2.2 to 2.0 for the purified materials in deuterium oxide and 2.7 in water. Before purification, the $^{15}$N-proton NOE was -4.87 in acidic solution and about -2.1 at pH 6. The NOE values are probably accurate to ± 15% and the $T_1$ values to ± 10%. Explain what went on in these experiments, touching on the pD and solvent dependences of $T_1$, as well as the NOE values.

**Exercise 6-10** Suppose the different intensities of the resonances (as judged by the peak heights) in the $^{13}$C proton-decoupled spectrum of monosodium glutamate (Figure 5-1) were determined only by the efficiency of the $^{13}$C-proton NOE and there is a full $^{13}$C NOE for the $CH_2$ carbons. Estimate the NOE for the carboxyl carbons. Give your reasons and any assumptions. Which terms in Equation 6-1 (Section 6-4) might contribute more, or less, to $T_1$ relaxation for the carboxyl carbons than to the $CH_2$ carbons? Explain your choices.

**Exercise 6-11** Calculate what you would expect the NOE to be in the proton spectrum of $^{13}CH_3OH$ if we were to irradiate the $^{13}$C. Show your method.

**Exercise 6-12** Consider an ordinary (no proton decoupling) $^{15}$N NMR signal from a substituted amide R-(CO)-$^{15}$NH-R' group (in a solvent where the hydrogens are not rapidly exchanging).

a. How much greater would the resulting $^{15}$N resonance be than the absolute intensity of one line of the doublet signal, if proton decoupling produced a full $^{15}$N-proton NOE? What would be the saving of time, when time averaging, to attain the same S/N ratio for the decoupled spectrum compared to the ordinary spectrum?

b. What value of the $^{15}$N-proton NOE would give a null signal? If the $T_1$ for the null signal for dipole-dipole induced $^{15}$N relaxation is 4 s, what is the value of $T_{1\text{other}}$?

c. What would you expect for the appearance and line intensities of the amide $^{15}$N resonances with gated proton decoupling and with inverse-gated proton decoupling? Give your reasoning.

**Exercise 6-13** Suppose that you were taking the proton-decoupled $^{13}$C NMR of benzene in a very slightly viscous solution (conditions of the left side of the left-hand diagram in Figure 6-3). Of the various contributions to $T_1$ shown in Equation 6-1 (Section 6-4), explain qualitatively how you would expect each to change in direction on increasing the magnet field of the spectrometer. One of these contributions to $T_1$ should be especially strongly dependent on the magnetic-field strength. Which one is that and why?

**Exercise 6-14** Why does partial proton noise decoupling tend to give broader $^{13}$C resonances from methine and methyl carbons than from methylene or quaternary carbons, the latter two of which actually tend to sharpen up on partial proton decoupling?

**Exercise 6-15** What would be the maximum value of a **proton-proton** NOE (irradiate one set of protons, observe another set)? Show your method.

**Exercise 6-16** When one takes the $^{15}$N spectrum of lysozyme, a mucolytic enzyme of MW about 15,000 made up of many different amino acids connected together by peptide linkages, some of the $^{15}$N resonance peaks are weakly upward from the base line, while others are strongly downward. What general kinds of nitrogen would you expect to give upward peaks and which of the natural amino acids (consult an organic textbook for the list) would you expect to have one or more nitrogens that would give strong downward peaks. How might you expect the spectrum to change if it were taken at a higher temperature (but not so high as to cause the enzyme to be denatured)?

# 7
# Pulse Experiments in One-Dimensional NMR Spectra

The use of pulse excitation followed by free-induction decay has wreaked a revolution in NMR spectroscopy by allowing the possibility of manipulating the decaying FID magnetization in a variety of ways not easily possible with the classic CW spectra. Some of the procedures derive from ones originally developed for CW spectroscopy, such as decoupling to simplify spectra, or to produce nuclear Overhauser enhancement (Chapter 6). Others are quite different and can increase sensitivity as well as simplify greatly the analysis of complex spectra.

There are serious difficulties in trying to make these procedures even qualitatively intelligible. The problem is that the vector diagrams we used in Chapter 2 to show how spin echoes can be utilized in the measurement of $T_2$ and how multiple pulses are employed for measurement of $T_1$ by inversion recovery, are not helpful in many situations, such as when pulses are applied at the resonance frequencies of one nucleus of a spin-coupled pair. The same situations also cause difficult problems for applying the Bloch equations in the simple way that we used them in Chapter 4. Nonetheless, much help in qualitative understanding can come from consideration of energy-level diagrams and changes in energy-level populations as was done in the discussion of nuclear Overhauser effect in Chapter 6. However, there will be aspects of advanced NMR pulse techniques that essentially defy qualitative explanation and are dealt with quantitatively, by those expert in such matters, using such exotica as "product operators", density matrices and the like. Let us start with a specific example that illustrates something of what we can and cannot do.

## 7-1 Selective Magnetization Transfer

We have discussed the elements of the nuclear Overhauser effect in Chapter 6. It is a rather complex phenomenon intimately related to energy-level populations and mechanisms of spin-lattice relaxation. It is also of great value in increasing the intensities of $^{13}C$ and $^{15}N$ resonances. Now, we consider a much simpler technique that is usually called **selective magnetization transfer**. However, it is often known also as **selective polarization transfer, selective population transfer** or **coherence transfer**.

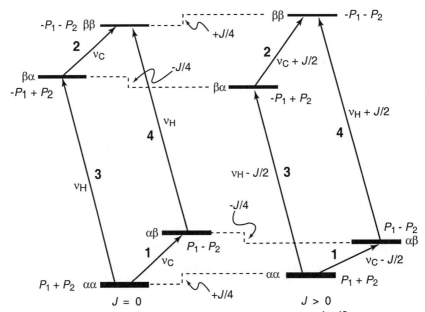

**Figure 7-1** Energy diagram and equilibrium populations for a two-nucleus $^1$H-$^{13}$CR$_3$ system. The rationale for the energy levels is given in Section 6-4-3. The procedure for deriving the population differences ($P_1 + P_2$, $P_1 - P_2$ and so on) is described in the text below

Consider a two-nucleus system such as afforded by $^1$H-$^{13}$CR$_3$ where the R's are to be taken to be some variety of appropriate nonmagnetic groups whose effect we can ignore for the purposes of our discussion. Following the lead of Section 6-3, we construct an energy-level diagram with upward transitions as shown on the left side of Figure 7-1. At equilibrium, the $\alpha\alpha$ and $\alpha\beta$ states will each have slightly more than 50% of the proton population in the favored $\alpha$ orientation with the field, while $\beta\alpha$ and $\beta\beta$ will have slightly less. Let us denote the excess population of $\alpha(1)$ as $P_1$. The same approach for the $^{13}$C nuclei would give $P_2$ as the excess population with the $\alpha$ orientation of the $^{13}$C nuclei in $\alpha\alpha$ and $\beta\alpha$. In Figure 7-1, the population excess, or deficit, is written beside each energy level in terms of $P_1$ and $P_2$. If we subtract the population of $\alpha\beta$ from $\alpha\alpha$, we get $2P_2$ for what we might call the normal population difference for the ordinary $^{13}$C transition, labeled **1**. Almost exactly the same population difference $2P_2$ is expected for transition **2** and therefore we expect that transitions **1** and **2** should have almost exactly the same intensity.

There is usually a spin-spin coupling of perhaps 125-200 Hz between the proton and the directly attached $^{13}$C. Assuming the coupling, which we designate as $^1J_{\text{H}^{13}\text{C}}$ (the superscript denoting a **one-bond** coupling and the subscripts denoting a coupling between a **proton and $^{13}$C**) is positive, the energy-level diagram will change, as was

## 7-1 Selective Magnetization Transfer    145

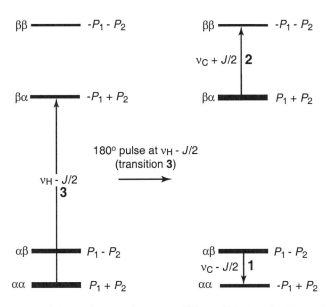

**Figure 7-2** Changes in populations of magnetic states of Figure 7-1 attendant to applying a 180° pulse at the frequency of proton transition **3** with the purpose of interchanging the populations of αα and βα

worked out for Figure 6-5, and is repeated on the right of Figure 7-1, except that now the population information is included for each of the various levels. Perturbations of the populations by changes in the energy levels as the result of the spin-spin couplings are not shown in Figure 7-1. They are expected to be negligible because the couplings are so small compared to the chemical-shift differences between the +1/2 and -1/2 states. When account is taken of the spin-spin coupling, we expect two $^{13}$C resonances separated by $^1J_{H^{13}C}$ and two proton resonances similarly separated.

The key step in selective magnetization transfer is to apply a **selective** 180° pulse to one resonance of the coupled system and here we use the proton frequency of transition 3 (see Figure 7-2). This 180° pulse will **interchange the populations of βα and αα**, so that immediately after the pulse, βα will have the population $P_1 + P_2$ and αα will have the population $-P_1 + P_2$. Now, we can apply a 90° pulse to the $^{13}$C nuclei, collect and transform the FID in the usual way. If we compute the population differences between the levels as before, transition **2** will correspond to $2P_1 + 2P_2$, while transition **1** will correspond to $-2P_1 + 2P_2$. Thus, we would expect the $^{13}$C transition **2** to have more intensity than the normal, and transition **1** to be less intense than normal. How large will the population differences be? We know that the excess population in the lower magnetic energy state depends on γ, the temperature and the strength of the magnetic field (see Section 1-2). For a given temperature and magnetic field, the

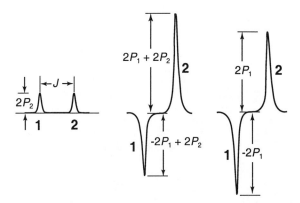

**Figure 7-3** On the left, representation of the normal $^{13}$C doublet expected for the AX system of Figure 7-1 with $J_{H-13C} \neq 0$. In the center are the resonances and intensities expected if the system is energized with a 180° pulse at the proton frequency corresponding to transition 3 to give the population differences shown in Figure 7-2. The far-right group of peaks is the result of point-by-point subtraction of the normal doublet on the left from the center group of resonances and represents the actual enhancement, both positive and negative, of the $^{13}$C resonances by the magnetization transfer from $\alpha\alpha$ to $\beta\alpha$

proportional gain in intensity over a normal $^{13}$C spectral line will be $P_1/P_2 = \gamma_{1_H}/\gamma_{13_C} \sim 4$ for transition **2**. The diminution in the population of $\alpha\alpha$ by the 180° pulse means that transition **1** will be favored in the **downward** direction and we will observe an emission line with a relative intensity of $-2P_1 + 2P_2$.

Figure 7-3 represents the spectrum that would result if everything worked ideally. Left is the normal $^{13}$C spectrum with the resonance split by $^1J_{H13C}$ and, in the middle, we see the result of taking a $^{13}$C spectrum immediately after a 180° pulse applied at the frequency of the proton transition **3**. On the right is the result of a point-by-point subtraction of the left-hand doublet from the middle spectrum and thus represents the ideal positive and negative enhancements of the normal $^{13}$C doublet.

The sequence will be seen to allow for the possibility of substantially increasing the intensity of $^{13}$C signals and, as a one-pulse experiment, the sequence is more efficient than the nuclear Overhauser effect (NOE). With $^{13}$C, the NOE only adds to the signal strength by a factor of two (see Section 6-4-3). Why then use the NOE at all? There are a number of reasons. Clearly, if you want to use this sequence to transfer the $\alpha\alpha$ magnetization cleanly to $\beta\alpha$ via transition **3**, and also do time averaging, you will need to wait at least four times the longer of the $T_1$ values of the proton or the carbon before starting a new sequence. Furthermore, with $^{13}$C at its natural-abundance level, finding the frequency of the proper **proton** transition to apply the 180° pulse may not be easy and also the power level, as well as the pulse width, must be adjusted so that transition **3** is where energy absorption occurs, with essentially no excitation of transition **4** (see Section 2-5 and you could look ahead to Section 7-9). Added to all of these strictures, you will not often be interested in selectively enhancing just one $^{13}$C resonance in a molecule with a multiplicity of carbons.

So why bring the selective magnetization transfer experiment up at all? There are some important points that it illustrates. Magnetization transfer, in one form or the other, is a feature of many advanced NMR techniques. Also, techniques for enhancing weak resonances, by any means, can be extremely important in practical spectroscopy and, finally, there is the question of understanding what is going on in such experiments. The explanation of magnetization transfer may seem straightforward when it is considered as a simple interchange of the populations of the $\alpha\alpha$ and $\beta\alpha$ states as the result of a 180° pulse. What is harder to explain is the overall result on the basis of the magnetic-vector model we have used for other pulse techniques. Why is this a problem? What is difficult to understand is the role of the $J$ coupling in the magnetization transfer. If $J$ is zero, transitions **3** and **4** coincide and selective pulses are not possible. Also, if the proton and $^{13}$C are in different molecules, one can construct an energy diagram similar to the one on the left side of Figure 7-1, and yet a pulse at the frequency of transition **3** will not enhance the intensity of transition **2**. Our simple vector models work well where the nuclei are independent of one another, but not when the $J$ coupling is significant. We need to understand why $J$ is important and that is not easy. For now, we will just say that it is important and it has to be different from zero.

It should be clear that there is little advantage to be gained by enhancing the **proton signals** through selective magnetization transfer from $\alpha\alpha$ to $\alpha\beta$ by a 180° pulse applied to the system of Figure 7-1 at the frequency of the $^{13}$C transition **1**. If this is done, transitions **3** and **4** will have relative intensities of $2P_1 + 2P_2$ and $2P_1 - 2P_2$. Now, if we make a point-by-point subtraction of the normal proton doublet, the resultant will be an absorption line with intensity $2P_2$ and an emission line of intensity $-2P_2$. Each of these lines would have only $\pm (\gamma_{13C}/\gamma_H) \sim \pm 0.25$ times the intensity of the lines of the original proton doublet.

Let us move now to more useful and (of course) more complex applications of these ideas to the enhancement of signals of nuclei now with smaller $\gamma$ values than protons, but again necessarily where $J$ couplings are significant.

## 7-2 An Inept Preparation for INEPT

In virtually every field of science, there is unending growth of acronyms to save space for journals and authors from writers' cramp or CTS (carpal tunnel syndrome).

NMR is no exception and, besides NMR itself, we have been using FT, CW, FID, S/N, FFT and NOE rather shamelessly because these are quite common terms of wide applicability. Now, as we begin to venture into the more arcane techniques of NMR, we find their practitioners using acronyms that are contrived to be easy to pronounce, often whimsical (so as to be easy to remember), but not so often easy to translate to their basic descriptions. We will begin to deal now with INEPT, but there remains for later DEPT, COSY, NOESY, INADEQUATE and so on to contend with. You and I may not like these acronyms, but it will be hard for you to wreak a revolution in nomenclature unless you become far more of a mover and shaker in the field than I have ever been. Fortunately, many NMR acronyms have not received wide acceptance and particularly, one that was wonderfully contrived to form a word not often used in drawing-room conversation (but made newsworthy, if not more respectable, by the Bobbitt family).

INEPT is an acronym for **Insensitive Nuclei Enhanced by Polarization Transfer** and may sound related to the selective magnetization transfer described in the previous section. Such a surmise would be correct, the difference is removal of need for selectivity of the pulses used. The requirement for the $J$ coupling still remains critical.

Let us prepare for INEPT by considering some ways we can manipulate our system $H^{13}CR_3$ (where the R's are non-magnetic groups) with special emphasis on the effects of coupling and decoupling. Assuming relaxation of the magnetic nuclei is relatively slow (on the order of seconds), we can expect that there will be four, rather long-lived, magnetic varieties of the $H^{13}CR_3$ molecules in almost equal concentrations, each corresponding to one of the states of Figure 7-1; namely $\alpha\alpha$, $\beta\alpha$, $\alpha\beta$ and $\beta\beta$. These represent what we call **symmetric** states (to be discussed later, but if you are impatient, read Section 9-2). These symmetric states interact with pulses at either the proton or $^{13}C$ frequencies. However, only the net magnetizations arising from the tiny excesses of nuclei in the lower magnetic quantum states are observable. At room temperature, the positive $Y$ magnetization produced by a 90° pulse on the nuclei in the +1/2 state is almost, but not quite, canceled by the negative $Y$ magnetization resulting from the 90° pulse on the nuclei in the -1/2 state.

Taking a very simple view, we can regard those excess $^{13}C$ nuclei in the +1/2 state as being either on a molecule with a +1/2 proton $\alpha$ (1) or a molecule with a -1/2 proton $\beta$ (1). From Figure 7-1, we can see that the $H^{13}C$ coupling (assumed to be positive) will increase the difference in energy between the $\beta\alpha$ and $\beta\beta$ states and thus increase the $^{13}C$ resonance frequency of transition **2** associated with the proton state $\beta$ (1). The coupling will operate in the opposite direction to decrease the frequency of transition **1** associated with the proton state $\alpha$ (1). The respective rates of rotation of the

## 7-2 An Inept Preparation for INEPT

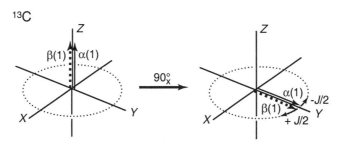

**Figure 7-4** A 90° pulse applied to a system H$^{13}$CR$_3$ ( the R's are nonmagnetic groups) and we identify the $^{13}$C nuclei that are associated with H$^{13}$CR$_3$ molecules having the proton in the +1/2 state α(1) or in the -1/2 state β(1). The curved arrows represent the direction of movement of these vectors in the rotating reference frame at the rates of ± J/2

$^{13}$C vectors in a rotating reference frame, with the frame rotation rate equal to the average $^{13}$C resonance frequency, will be $\pm (^1J_{H^{13}C})/2$. Application of a short $90°_x$ pulse at the $^{13}$C frequency will move the vectors corresponding to each of these species to the X, Y plane (see Figure 7-4). If we now turn on the receiver, collect and transform the FID, we will get two $^1J_{H^{13}C}$ signals of equal intensity separated by J, which will be understood here as $^1J_{H^{13}C}$.

In the next part of our discussion, we assume both the proton and $^{13}$C $T_1$ values are large compared to what else happens, so we will not be concerned with the rate of change of the magnetization along the Z axis. We start with the nuclei in a rotating reference frame having its frequency ω equal to the average $^{13}$C frequency just after the 90° pulse of Figure 7-4. As we have said, with J > 0, the nuclear magnetic vector corresponding to the molecules with the -1/2 proton will rotate around the Z axis at the rate J/2, while the other nuclear vector will go in the opposite direction at -J/2 (look again at the right side of Figure 7-1). These vectors will be 180° out-of-phase with each other along the X axis in 1/2 J$^{-1}$ s, and in-phase again after J$^{-1}$ s, along the -Y axis. You should be sure that you understand this before proceeding further.

Now, if we apply a 180° $^{13}$C pulse after τ s (Figure 7-5), there will be a negative echo after another τ s and, if data acquisition is started then, the FID will consist of two negative signals with frequencies of $\omega_{13C} \pm J/2$. The overall sequence here is just the same as the Carr-Purcell sequence described earlier in Section 2-3.

A variation on this theme is to carry on broad-band proton decoupling after the 90°-τ-180°-τ sequence; that is, **beginning at the start of acquisition of the negative echo**. The timing diagram, analogous to those shown in Figures 6-10 and 6-11, is shown in Figure 7-6 for 90°-τ-180°-τ. Decoupling eliminates the J coupling and hence removes the frequency difference between the rotating vectors. For this reason, the transform

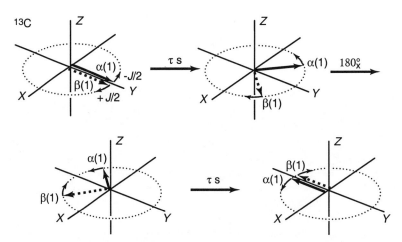

**Figure 7-5** Result of a 90°-τ-180°-τ $^{13}C$ pulse sequence applied to a $H^{13}CR_3$ (where the R's are nonmagnetic groups). At this point, data acquisition and transform will yield two negative signals separated by $J$ Hz. The sequence here is exactly that of the Carr-Purcell sequence discussed earlier

of the FID taken starting at the echo point and decoupled will produce a single negative resonance signal.

It is also useful to consider what happens when we start the broad-band proton decoupling **immediately following the 180° pulse** (Figure 7-7). Don't forget that decoupling averages $J$ to zero and, as soon as decoupling is started, the two magnetic vectors will have **no frequency difference relative to one another** and, if the $^{13}C$ resonance frequency is equal to ω, the rotating-frame frequency, then until something else is done to them, these vectors will remain stationary in the rotating frame, except for $T_1$ and $T_2$ relaxation (see Figure 7-8). As a result, any phase difference between the vectors that was allowed to build up in the first τ waiting period remains constant.

Clearly, if the phase difference remains constant and if $T_2^* \sim T_2$ (a highly homogenous field), there will be no echo at the end of the second τ period, even though

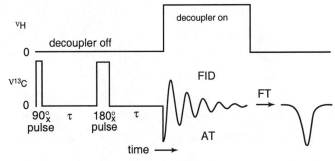

**Figure 7-6** Timing diagram for proton decoupling in a Carr-Purcell sequence, where decoupling is on only during the FID which is taken starting at the echo point

## 7-2 An Inept Preparation for INEPT

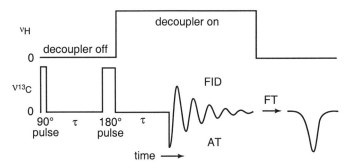

**Figure 7-7** Timing diagram for proton decoupling in a Carr-Purcell sequence, where the decoupler is turned on immediately after the 180° pulse

the magnetic vectors will be flipped by the 180° pulse. Here, the second $\tau$ period might seem superfluous and, as a consequence, data acquisition could be started immediately after the 180° pulse, provided that $T_2^* \sim T_2$. However, there are two important reasons for inclusion of the 180° pulse. The first is the obvious one and that is, if the field is inhomogenous, the $^{13}$C $M_+$ and $M_-$ vectors will each start to fan out during the first $\tau$ period, because each is made up of a spread of frequencies $\pm J/2 \pm \Delta$ where $\Delta$ is a measure of the field inhomogeneities. Therefore, the 180° pulse will be useful to refocus the individual vectors and eliminate the spread in frequencies (see Section 2-3). Use of 180° refocusing pulses, for the purpose of correcting for field inhomogeneities is a common feature of many pulse sequences.

The other important reason for having the 180° pulse is to take care of phase errors arising from chemical shifts or, in the particular case we are discussing, the possibility that the rotating-frame frequency is not the same as the proton-decoupled $^{13}$C

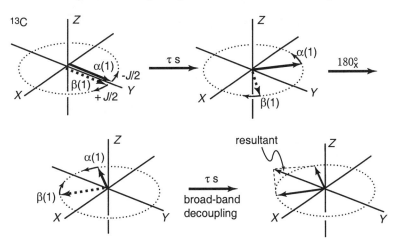

**Figure 7-8** Vector diagram corresponding to the pulse-timing diagram of Figure 7-7

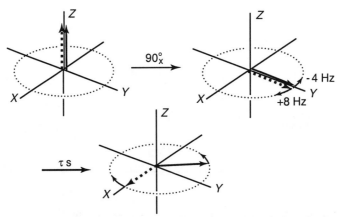

**Figure 7-9** Diagrams showing the effect after a τ period of 1/32 s of a chemical-shift difference on two uncoupled nuclei differing in frequency by 12 Hz, one moving 8 Hz faster than the rotating frame and the other 4 Hz slower

frequency. What these possibilities do is to introduce phase problems that can be corrected by the 180° pulse. The rationale is not difficult. Consider first the simple case of two uncoupled nuclei ($J = 0$), one with a frequency 4 Hz less than rotating frame and the other with a frequency 8 Hz greater that the rotating frame. Suppose now, we take τ to be 1/32 s. In the first τ period following the initial 90° pulse, the faster vector will have progressed 90° and will lie along the +X axis, while the other vector will have lagged behind by 45° in the rotating frame, see Figure 7-9.

If we now begin data acquisition, the faster vector would by itself produce a FID that would transform to a dispersion signal, because, at the start of data acquisition, it would be 90° out-of-phase. The slower vector by itself would give a FID that would transform to a phase-distorted absorption signal. Transform of the combined FID

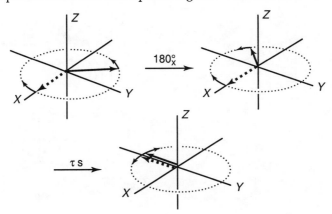

**Figure 7-10** Diagrams showing the effect of a refocusing 180° pulse on the vectors produced by the sequence of Figure 7-9

## 7-2 An Inept Preparation for INEPT

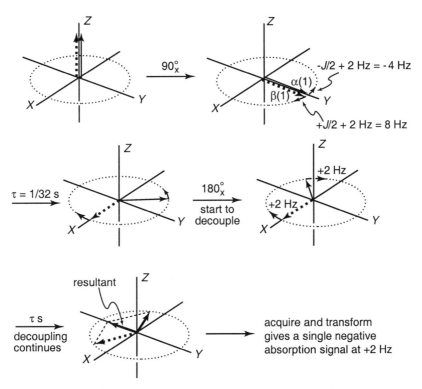

**Figure 7-11** How the 180° pulse refocuses the magnetic vectors corresponding to a doublet with a 12 Hz coupling constant, where the average frequency of the doublet is 2 Hz greater than of the rotating-frame and τ is 1/32 s

would clearly give a complex signal. However, that can be avoided by use of the 180° pulse. If you look at Figure 7-10, you will see that the 180° pulse will have no effect on the faster vector, because it lies along the X axis, but will bring the slower vector around so it is in the X, Y plane, 45° away from the -Y axis. Clearly, after another τ period, the vectors will be back in phase along the -Y axis and this then is an appropriate time to begin data acquisition. The resulting FID, taken with quadrature detection and suitably transformed, will afford two negative absorption signals at -4 and +8 Hz, respectively. So, from this, you can see how the 180° pulse is important to obtaining proper phase relationships at the start of the acquisition period.

The same kind of argument can be applied to a heteronuclear AX case with a coupling constant of 12 Hz, where the rotating frame frequency is at a 2-Hz different frequency than the decoupled doublet frequency (see Figure 7-11). Again, we will use τ = 1/32 s. Now, one of the doublet lines will have a frequency 4 Hz less than that of the rotating frame while the other will have a 8 Hz greater frequency. If the decoupler is

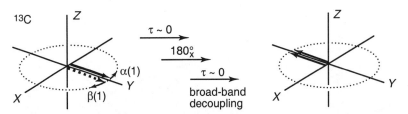

**Figure 7-12** The result of the sequence of Figure 7-11 when $\tau$ is very short and the effect of the $J$ coupling is removed by decoupling after the 180° pulse. The result here is a single negative resonance

turned on immediately after the 180° pulse; then, at the end of a first 1/32 s $\tau$ period, the vectors will remain at a fixed angle with respect to one another but will differ on the average from the rotating-frame frequency by $(8 - 4)/2 = +2$ Hz. In the second $\tau$ period, these vectors will move in the rotating frame to become symmetrically disposed with respect to the $Y$ axis. Their resultant vector sum will be just 180° out-of-phase with the frame frequency when data acquisition begins. Transform of the resulting FID will give a negative absorption peak at +2 Hz with respect to the reference frequency. Of course, this peak will have a diminished absolute intensity, because the vector sum at the start of acquisition is for an angle between the vectors greater than zero.

Applying the lessons of Figure 7-11 to our example of $H^{13}CR_3$, we will see that, if decoupling begins just after the 180° pulse, the intensity of the transformed signal will depend on $\tau$. If $\tau$ is very short, so that no phase difference builds up before the 180° pulse, then transform of the resulting FID, will give a single negative signal of maximum intensity, see Figure 7-12.

On the other hand, if $\tau$ is such as to have the phase angle between the vectors become 180°, at the time of the $180°_x$ pulse and the start of decoupling, the $M_+$ and $M_-$ vectors will produce **no net signal** in a receiver coil oriented to detect magnetization along the $Y$ axis. The idea is that two equal-magnitude vectors, 180° out-of-phase with one another, will cancel each other and thus give no FID, see Figure 7-13.

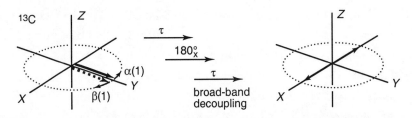

**Figure 7-13** The result of the sequence of Figure 7-12 when $\tau$ is such as to have the vectors come to be 180° out-of-phase with one another and then the effect of $J$ coupling is removed by decoupling after a 180° pulse. The result here is no resonance signal at all, the two vectors canceling each other

## 7-3 INEPT

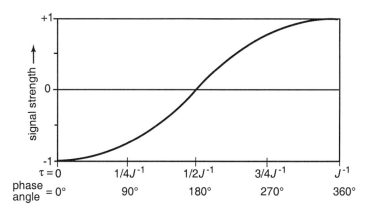

**Figure 7-14** Signal strength of the transformed FIDs obtained as a function of τ (or the phase angle) for H-$^{13}$CR$_3$ using a 90°-τ-180°-τ pulse sequence with appropriate decoupling at the start of the second τ period and continuing through acquisition of the FID. Any possible buildup of NOE as the result of decoupling is ignored

We can see from these examples that, when one uses the sequence of Figure 7-6, the signal intensity is a function of τ, or the phase angle, and follows the curve of Figure 7-14. Here, we ignore any contribution to the signal intensity by buildup of NOE because of proton decoupling. Further, it should be clear that analogous behavior would be observed, if you were to carry out the pulses at the proton frequency and decouple the $^{13}$C nuclei following the 180° pulse. The changes in these spectra, as τ is changed, are clearly a function of τ and it is common to designate such spectra as **J-modulated**.

### 7-3 INEPT

As we mentioned earlier, the basis for INEPT is magnetization transfer to enhance the signal of the less-sensitive nucleus with the aid of **non-selective pulses**, that is pulses that do not have to be applied at particular frequencies. The sequence is not simple because it requires pulses to be applied to both the $^1$H and $^{13}$C channels. We start with the nuclei at equilibrium and first give a nonselective $90°_x$ pulse to the protons (see Figure 7-15). Now we wait for $\tau = 1/4J^{-1}$ s, this period allows the proton vectors to get just 90° out-of-phase. Obviously to do this properly we need to know the value of $J_{H^{13}C}$, a problem we will address later. At this point, $180°_x$ pulses are applied to the $^1$H channel and then to the $^{13}$C channel. You may remember from the selective magnetization transfer (Section 7-1) that 180° pulses will interchange the populations of the states involved, so the $^{13}$C pulse causes the proton vectors in the X, Y plane to

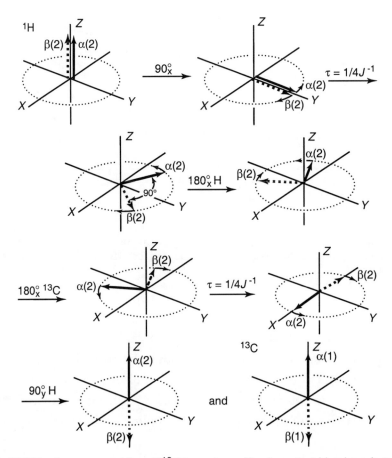

**Figure 7-15** INEPT pulse sequence for a $H^{13}CR_3$ system. The important idea here is to achieve the same pattern of magnetization that results from the selective magnetization-transfer sequence (see Section 7-1) by using non-selective pulses. This is largely brought about with pulses at the proton frequency, with a key 180° $^{13}C$ pulse that changes the direction of rotation of the proton magnetization vectors. This step is followed by a $\tau$ waiting period and a 90° H pulse along the $Y$ axis that leads to the desired $^{13}C$ magnetic state shown at the bottom right of the figure. In this example, we do not assume that the $^1H$ frequency is the same as that of the rotating frame

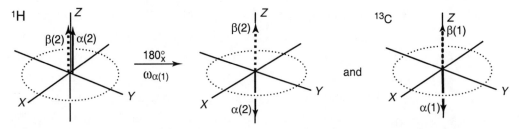

**Figure 7-16** The result of a proton 180° pulse applied at the frequency of transition **3** in Figure 7-1. Note that the orientation is opposite to the last configuration of Figure 7-15, which corresponds to excitation of transition **4**

## 7-3 INEPT

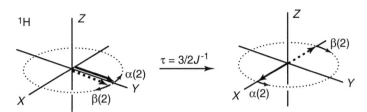

**Figure 7-17** Alternative mode of achieving the next-to-last configuration of Figure 7-15 without using a 180° pulse but one that leads to other problems

interchange their positions. Now, after another τ period, the vectors will be 180° out-of-phase. At this point, a $90°_y$ proton pulse (applied along Y axis) turns these vectors so that now the one labeled β(2) is in the -Z direction while the one labeled α(2) is back where it started at the beginning of the sequence.

So where does all this get us? To answer that question look at what happened in the selective magnetization-transfer experiment. There, we applied a selective 180° pulse at the frequency of one resonance of the proton doublet of $H^{13}CR_3$. Translated into a vector diagram, and choosing the α(2) component (transition 3 of Figure 7-1), we have the result shown in Figure 7-16. Clearly, the more elaborate sequence of Figure 7-15 has achieved the same kind of transformation with nonselective pulses. Therefore, the observed $^{13}C$ spectrum, after we apply a $90°_x$ $^{13}C$ pulse, is the same as in Figure 7-3, except that now it corresponds to excitation of transition 4 of Figure 7-1. But, there are caveats. In one case, we had to apply a selective frequency to either the +J/2 or the -J/2 component of the doublet and, in the other, we had to know the value of $J_{H^{13}C}$ in order to adjust τ to produce the desired, reasonably precise, 90° phase difference. Despite this, the great advantage of INEPT is that it can be used to enhance simultaneously the resonances of $^{13}C$ carbons with different chemical shifts.

You might ask why not achieve the same result, after the initial 90° pulse, by simply adjusting τ to $3/2J^{-1}$, so that the proton vectors come to the next-to-last configuration of Figure 7-15 and then apply the $90°_y$ proton pulse, see Figure 7-17.

Again, the answer is, on the one hand, to take advantage of the capability of the $180°_x$ proton pulse to reduce line broadening by refocusing any spread in frequencies arising from field inhomogeneities and, on the other hand, to avoid phase problems that would come with different chemical shifts, because, for τ = $3/2J^{-1}$, not all of the pairs of vectors corresponding to different chemical shifts would come to lie at opposite ends of the X axis after the τ waiting period (see discussion in Section 7-2).

Adjustment of τ to accord with expected values of $J_{H^{13}C}$ can be troublesome if a molecule has several quite different kinds of C-H bonds. The range of the one-bond

158        7 Pulse Experiments in One-Dimensional NMR Spectra

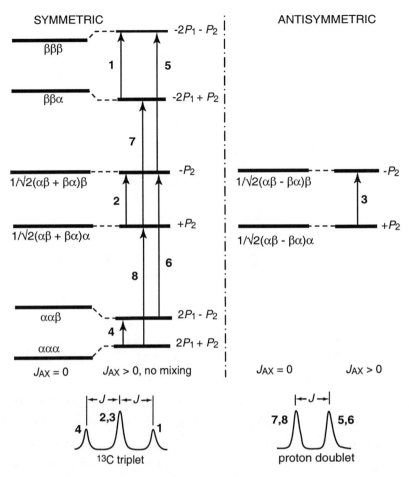

**Figure 7-18** Energy diagram for a system of three nuclei ($A_2X$) with two different chemical shifts, i.e. $\nu_A$ and $\nu_X$. On the left are the symmetric states with $J_{AX} = 0$. On the right side are the antisymmetric states. The expected doublet and triplet peaks in the normal spectrum are shown below

$J_{H^{13}C}$ couplings is a factor of two from ethane (125 Hz) to ethyne (249 Hz). Obviously, INEPT with a molecule such as propyne would have to be carried out with $\tau$ adjusted for two very different one-bond proton-carbon $J$ values. However, molecules, such as steroids or carbohydrates, with many saturated carbons having about the same $^1J_{H^{13}C}$ values are expected to give good results with a single reasonably well-selected $\tau$.

Some aspects of the application of INEPT to carbons carrying different numbers of hydrogens can be worked out according to the same principles that we used for our H-CR$_3$ system in Figure 7-15.

Suppose first that there is no hydrogen on the carbon. Clearly nothing will happen during the proton pulses, but the $180°_x$ $^{13}C$ pulse of Figure 7-15, followed

## 7-3 INEPT

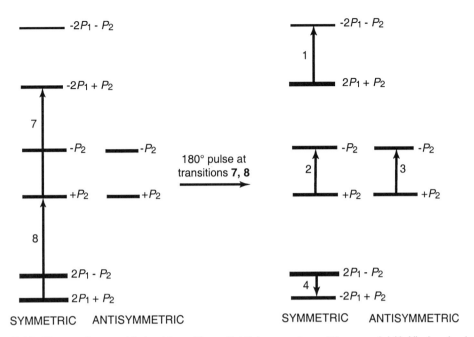

**Figure 7-19** Energy diagram (derived from Figure 7-18) for a system of three nuclei ($A_2X$) showing how the populations are perturbed by a 180° pulse applied to the A transitions 7 and 8 that have the same frequency. The X transitions that result from a 90° pulse to the X nucleus and acquisition are shown in Figure 7-20

subsequently by a $90°_x$ pulse, will give a single negative resonance with no enhancement. With a $H_2{}^{13}CR_2$ arrangement, having **equivalent** protons (a so-called $A_2X$ case), there will be four essentially equally probable proton states. The nature of these will be discussed in detail in Chapter 9 (see for example Section 9-2). For now, we will designate these as **symmetric** corresponding to the functions; $\alpha\alpha$, $1/\sqrt{2}(\alpha\beta + \beta\alpha)$ and $\beta\beta$. The fourth designated as **antisymmetric** corresponds to $1/\sqrt{2}(\alpha\beta - \beta\alpha)$. Of these, only the symmetric states will be involved with the proton pulses, the antisymmetric protons are nonmagnetic and this means that one-fourth of the total $^{13}C$ will act just like a quaternary carbon, as described above, and thus give an unenhanced negative signal. There will be molecules with protons corresponding to each of the symmetric states; $\alpha\alpha\alpha$, $\alpha\alpha\beta$, $1/\sqrt{2}(\alpha\beta + \beta\alpha)\alpha$, $1/\sqrt{2}(\alpha\beta + \beta\alpha)\beta$, $\beta\beta\alpha$ and $\beta\beta\beta$, where nucleus (3) is $^{13}C$. If INEPT works the same way as selective magnetization transfer for $H^{13}CR_3$, then we can see what the result would be by simply applying a 180° pulse to one line of the $H_2{}^{13}CR_2$ proton doublet. We can now construct Figure 7-18 taking account of $^1J_{H^{13}C}$.

One half of the proton doublet shown in Figure 7-18 is a composite of the equal transitions **5** and **6**, the other half of transitions **7** and **8**. Suppose we apply a 180° pulse

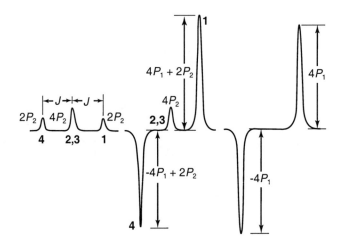

**Figure 7-20** On the left, representation of the normal $^{13}$C triplet expected for H$_2$$^{13}$CR$_2$ (an A$_2$X system, as in Figure 7-18) with $J_{H13C} \neq 0$. In the center are the resonances and intensities expected if the system is energized with a 180° pulse at the proton frequency corresponding transitions **7** and **8** of Figure 7-19. The far-right group of peaks is the result of point-by-point subtraction of the normal triplet on the left from the center group of resonances and represents the net enhancement, both positive and negative, of the $^{13}$C resonances by the magnetization transfer shown in Figure 7-19

to transitions **7** and **8**. The result is that we exchange the populations of ααα and of $1/\sqrt{2}(\alpha\beta + \beta\alpha)\alpha$ as well as of $1/\sqrt{2}(\alpha\beta + \beta\alpha)\alpha$ and ββα. Because transitions **7** and **8** are linked, this is equivalent to exchanging the populations of ααα and ββα (see Figure 7-20). As a result, the $^{13}$C spectrum will show transitions **2** and **3** as a composite unenhanced absorption, **4** will show enhanced emission, and **1** will show enhanced absorption (see Figure 7-20).

Rationalizing the course of INEPT for a methyl group as of H$_3$$^{13}$CR is complicated because the symmetric and antisymmetric proton energy states of H$_3$$^{13}$CR are more involved than those of H$_2$$^{13}$CR$_2$. When these are known, extension of Figures 7-18 and 7-19 to H$_3$$^{13}$CR offers no great difficulties. However, rather than work through this here, I only show the expected results for H$_3$$^{13}$CR in Figure 7-21 (see Exercise 7-9).

## 7-4 Phase Cycling

The right-hand sides of Figures 7-3, 7-20 and 7-21 for the INEPT spectra of H$^{13}$CR$_3$, H$_2$$^{13}$CR$_2$ and H$^{13}$CR$_3$, include point-by-point subtraction of the normal spectrum from the "raw" INEPT spectrum to produce plots of the actual INEPT enhancements (positive or negative). It is possible to do this subtraction of the digitized spectra as described, but an alternative process is to use **phase cycling** to remove the contribu-

## 7-4 Phase Cycling

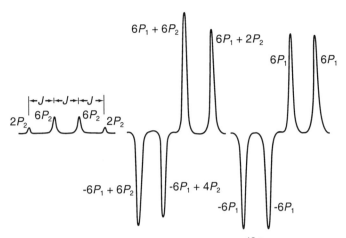

**Figure 7-21** On the left, schematic representation of the normal $^{13}$C quartet expected for H$_3$$^{13}$CR (an A$_3$X system) with $J_H{}^{13}{}_C \neq 0$. In the center are the resonances and intensities expected if the system is energized with a 180° pulse at one of the frequencies of the proton doublet. The far-right group of peaks is the result of point-by-point subtraction of the normal quartet on the left from the center group of resonances and represents the net enhancement, both positive and negative, of the $^{13}$C resonances by the magnetization transfer

tions of the "natural" resonances. Phase cycling can be applied when taking a time-averaged INEPT spectrum by changing the phases of both the $90°_y$ H pulse in the last step of the sequence of Figure 7-15 and of the receiver in the analyzing step ($90°_x$ $^{13}$C pulse-acquire) that follows after the sequence of Figure 7-15.

In thinking about what phase cycling does, it is convenient to divide a transition, for example, transition **1** of Figure 7-2, into two parts. The first part corresponds to the emission signal resulting from the final configuration of Figure 7-15 with a relative intensity of $-2P_1$ and the second part corresponds to the normal "natural" upward transition with intensity $+2P_2$. At the end of the sequence of Figure 7-15, the natural magnetization, as the result of the $180°_x$ $^{13}$C pulse, will lie along the -Z axis and the 90° $^{13}$C analyzing pulse will direct it along the -Y axis, as in Figure 7-22, so that it will diminish the vector $\beta(1)$ and augment the vector $\alpha(1)$. If there is no phase cycling, with,

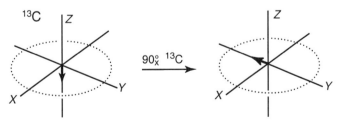

**Figure 7-22** The "natural" $^{13}$C magnetization after the $180°_x$ $^{13}$C pulse of Figure 7-15 and then after the $90°_x$ $^{13}$C analyzing pulse

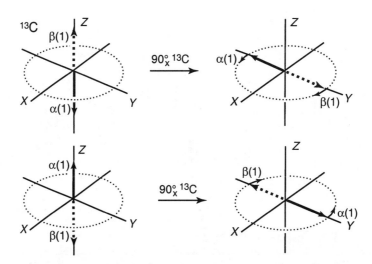

**Figure 7-23** Diagrams showing the two different $^{13}C$ magnetizations identified with the α(1) and β(1) proton states formed by 90° $^1H$ pulses along the ±Y axis, and how these can give the same signal after the 90° $^{13}C$ pulses if the receiver phase is switched by 180°. The procedure is called **phase cycling**

or without, time averaging, this will lead to a spectrum similar to that shown in the center of Figure 7-3. However, if in the second time-averaging cycle, we substitute for the $90°_{+y}$ $^1H$ pulse, a $90°_{-y}$ $^1H$ pulse, then the natural magnetization will augment vector β(1) and diminish vector α(1). Continuation of the sequence of alternating between $90°_{+y}$ and $90°_{-y}$ $^1H$ pulses will average the effects of the natural magnetization to zero.

But we will need to do more and that will be to alternate the receiver phase. This is important for the INEPT, because the sequence of alternating $90°_{\pm y}$ $^1H$ pulses will give alternating orientations of the magnetization shown at the bottom of Figure 7-15. The key to the cancellation of the natural magnetization, while retaining the INEPT magnetization is the alternation of receiver phase. You will see from Figure 7-23, that the different vector orientations produced by the $90°_{\pm y}$ $^1H$ pulses give different orientations after the $90°_x$ $^{13}C$ pulse, but if we switch the receiver phase 180°, then the receiver will be looking at those orientations from two different directions, down the +Y and up the -Y axis toward the origin. As far as the alternating INEPT magnetizations are concerned those views will be identical. This phase-cycling procedure and its variations have many applications in complex pulse sequences. As one very simple example, phase cycling will cancel out any stray constant voltages introduced from the electronics into the receiver channel. We will assume in our further discussions that phase cycling will be used to simplify INEPT spectra by removing the contributions to the resonances arising from the natural $^{13}C$ magnetizations.

# 7-5 Refocused INEPT

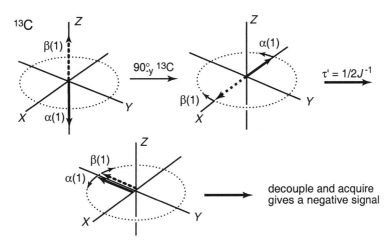

**Figure 7-24** Result of a 90° $^{13}$C pulse along the Y axis and a τ' waiting period starting with the last configuration of Figure 7-15 and using H$^{13}$CR$_3$

## 7-5 Refocused INEPT

INEPT spectra offer substantial enhancement, but with large molecules having many carbons and hydrogens, there may be a confusing melange of upright and downward peaks to sort out. If the aim is simply sensitivity enhancement, there are advantages to being able to use proton decoupling to remove the many C-H splittings. This is not feasible with INEPT in the form we have described it, because decoupling during acquisition of the FID would simply average the positive and negative enhancements and leave behind a normal decoupled $^{13}$C spectrum. It is possible to achieve the desired result by further manipulation of the pulse sequence with retention of the enhancement with the help of a technique called "**refocusing**." Because INEPT is not usually the method of choice for taking $^{13}$C spectra, unless substantial enhancement is needed (and there are other methods of achieving that), there may not seem much point in discussing the finer details of refocusing; which, as we will see, does not work equally well with CH, CH$_2$ and CH$_3$ groupings. Still there is intellectual satisfaction in looking at what has to be added to the pulse sequence to solve this kind of problem and the results have application to other pulse sequences. Let us start again with our simple H$^{13}$CR$_3$ case.

The last step of INEPT (see Figure 7-15) before a $90°_x$ $^{13}$C pulse and taking the FID was a $90°_y$ $^1$H pulse that completed the exchange of the populations of β(2) and α(2). Starting from here (see Figure 7-24), we now apply a $90°_{-y}$ $^{13}$C pulse instead of the $90°_x$

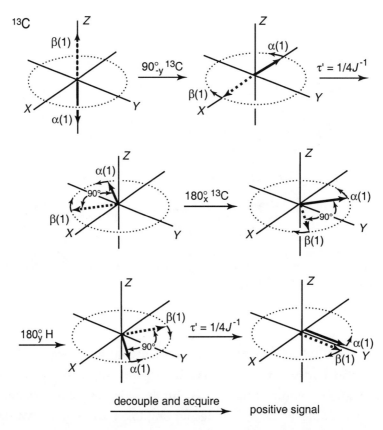

**Figure 7-25** Pulse sequence for INEPT to allow $^{13}C$ signals to be refocused and decoupled to give single upright enhanced resonance peaks from $H^{13}CR_3$ and the inner lines of $H_3^{13}CR$ type compounds. The formation of the starting magnetic configuration in the upper left is shown in Figure 7-15

$^{13}C$ pulse and consider the fate of the $^{13}C$ magnetism. Following the $90°_{-y}$ $^{13}C$ pulse, the $^{13}C$ vectors will have different frequencies relative to the frame of $\pm J/2$, which means that these vectors will each take $2J^{-1}$ s to go 360° in the rotating frame. At the outset, the vectors will move so as to come into phase, where they will be augmenting each other, rather than being directed in opposite directions. We could just let them come into phase by waiting for $\tau' = 1/2J^{-1}$, turn the decoupler on and take the FID. As we have discussed before (Section 7-2), this loses the advantage of a 180° refocusing and has problems when there are different $^{13}C$ chemical shifts. The better way is to wait for $\tau' = 1/4J^{-1}$, which will bring the vectors to a phase angle of 90°, and then do a $180°_x$ $^{13}C$ pulse (see Figure 7-25). The $180°_x$ $^{13}C$ pulse will refocus, but along the X axis and then be out-of-phase with the receiver. The way to avoid this is to immediately follow the $180°_x$ $^{13}C$ pulse with a $180°_y$ H pulse. What this does, without causing a population exchange

## 7-5 Refocused INEPT

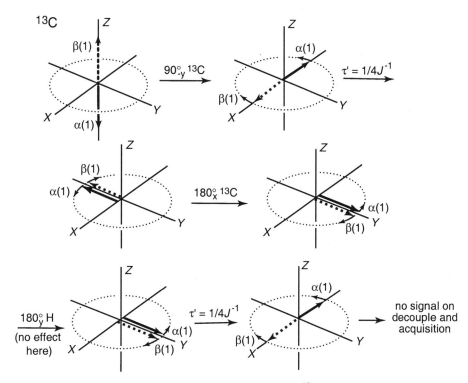

**Figure 7-26** Application of the pulse sequence for INEPT to a $H_2{}^{13}CR_2$ type compound with the use of $\tau'$ = $1/4J^{-1}$. Formation of the starting magnetic configuration in the upper left is shown in Figure 7-15. With $\tau'$ = $1/4J^{-1}$, it will be seen that there will be no resonance signal on decouple and acquire for $H_2{}^{13}CR_2$

(there is no change in magnetization along the Z axis) is to interchange α(1) and β(1) so that the vectors change direction and refocus along the Y axis after the second time τ'. Now, transform of the FID will give two upright enhanced signals or, if the decoupler is turned on, a single enhanced upright decoupled peak. Clever, these INEPT spectroscopists!

The sequence that allows refocused INEPT to give a single decoupled $^{13}C$ resonance for H-$^{13}CR_3$ leads to disaster with $H_2{}^{13}CR_2$ because the signal disappears. Let us see why, and we will start again at the end point of Figure 7-15, where selective magnetization transfer was achieved by the INEPT sequence of non-selective pulses (see Figure 7-26). The key issue is what happens in the τ' waiting period when τ' = $1/4J^{-1}$, as was used with $H^{13}CR_3$. Here, in the $^{13}C$-frequency domain, things are not as they are in the proton domain. What is different? Remember that the proton resonances of $H^{13}CR_3$, $H_2{}^{13}CR_2$ and $H_3{}^{13}CR$ are all split into a doublet by the $^{13}C$. Assuming the J values are about the same (a reasonable assumption), the rates of rotation of the proton

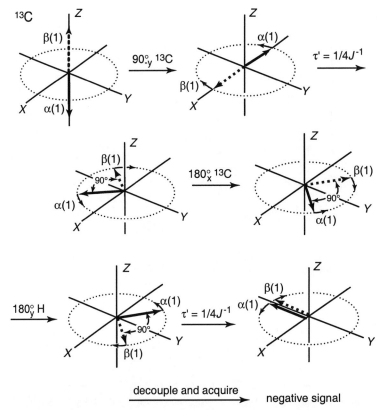

**Figure 7-27** Application of the refocused INEPT pulse sequence to the outer lines of $H_3{}^{13}CR$ with $\tau' = 1/4J^{-1}$. The rates of rotation of the vectors in the rotating frame are $\pm 3J/2$. Formation of the starting magnetic configuration in the upper left is shown in Figure 7-15

vectors in the rotating frame will all be $\pm J/2$, for $^{13}CH$, $^{13}CH_2$ or $^{13}CH_3$ during the $\tau$ periods of Figure 7-15. This will not be true in the $^{13}C$-frequency domain. The $^{13}C$ vectors of $H_2{}^{13}CR_2$ will rotate at $\pm J$ and not at all, while those of $H_3{}^{13}CR$, rotate at $\pm 3J/2$ and $\pm J/2$. So we can see in Figure 7-26 that, when the sequence with $\tau' = 1/4J^{-1}$ s is used at the end of the INEPT series of pulses to $H_2{}^{13}CR_2$, then the start of the decouple and acquisition period will find the $\alpha(1)$ and $\beta(1)$ $^{13}C$ vectors 180° out-of-phase (**antiphase**, some say) along the X axis. As stated before, the non-rotating vector, corresponding to the center line of the $^{13}C$ doublet of $H_2{}^{13}CR_2$, is not enhanced by INEPT, so for practical purposes, all of the $^{13}C$ resonances of $H_2{}^{13}CR_2$ should disappear (assuming phase cycling, see Section 7-3).

Depending on your point of view when you do such an experiment, this can be good or bad. If you compare a conventional proton-noise decoupled $^{13}C$ spectrum with a refocused and decoupled INEPT spectrum, you can distinguish the resonance of $^{13}CH$

## 7-5 Refocused INEPT

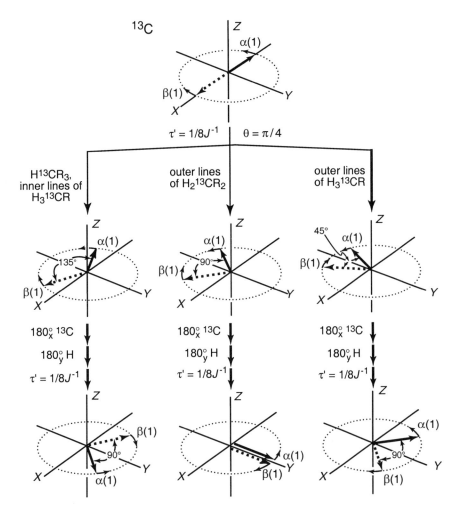

**Figure 7-28** The expected $^{13}C$ signals from refocused INEPT with $\tau' = 1/8J^{-1}$ ($\theta = \pi/4$). Formation of the starting magnetic configuration at the top is shown in Figure 7-27. Only the results of the $\tau'$ evolution periods are shown. The changes in magnetization resulting from the key refocusing steps involving 180° $^{13}C$ and H pulses are not shown, but can be worked out as in Figure 7-25 and 7-26. If there is a 90° final phase angle, the signal will have 0.707 of the intensity of a 0° final phase angle

groups from those of $^{13}CH_2$ groups, because the $^{13}CH_2$ signals are **edited out** by the INEPT sequence. Obviously, if your intention was only to enhance all of the $^{13}C$ peaks, you will not like the result. **Spectral editing** by techniques of this type can be an important part of the analysis of complex NMR spectra.

What about $H_3{}^{13}CR$ in the same experiment? Clearly the two inner lines of the expected $^{13}C$ quartet will respond just like the $^{13}C$ doublet of the outer lines of $H_2{}^{13}CR_2$ and give an upright enhanced signal. The rates of rotation of the vectors corresponding

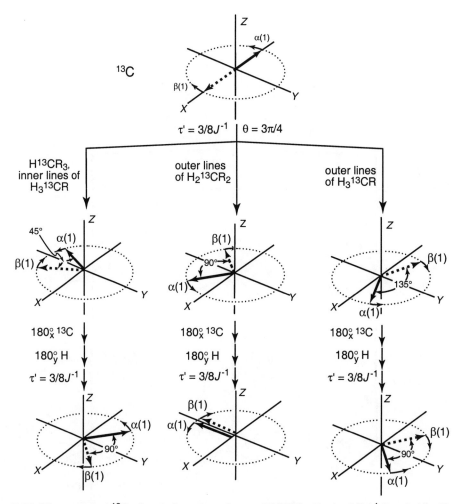

**Figure 7-29** The expected $^{13}C$ signals from the refocused INEPT with $\tau' = 3/8J^{-1}$ ($\theta = 3\pi/4$). Formation of the starting magnetic configuration at the top is shown in Figure 7-27. Only the results of the $\tau'$ evolution periods are shown. The changes in magnetization resulting from the key refocusing steps involving 180° $^{13}C$ and H pulses are not shown, but can be worked out as in Figures 7-25 and 7-26

to outer lines of the quartet are three times faster than of the inner lines so that when the inner-line vectors make a 90° phase angle, the outer-line vectors each make a phase change of 135° that puts them at a 90° angle. The sequence as shown in Figure 7-27 gives a negative signal for the inner lines.

You will remember that INEPT for $H_3^{13}CR$ was shown in Figure 7-21 to produce two negative and two positive $^{13}C$ resonances of equal absolute intensity, spaced $J$ Hz apart. Figure 7-25 and 7-27 show that the resonances arising from proton-decoupled refocused $H_3^{13}CR$ INEPT with $\tau' = 1/4J^{-1}$ will respectively be positive from the inner lines and negative from the outer lines, thereby averaging to zero; so here, then the $^{13}CH_3$ peaks will, like the $^{13}CH_2$ peaks, be edited out of the spectrum.

Useful control of spectral editing can be achieved by changing the $\tau'$ waiting period. It is common to define a quantity $\Delta$ that is equal to $2\tau'$ and corresponds to an angle $\theta = J\pi\Delta$ radians. The interesting values of $\theta$ are $\pi/4$, $\pi/2$ and $3\pi/4$. The value $\pi/2$ corresponds to $\tau' = 1/4J^{-1}$ that we have already covered, while $\pi/4$ corresponds to $\tau' = 1/8J^{-1}$ and $3\pi/4$ to $\tau' = 3/8J^{-1}$. Let us see what we can expect from $\pi/4$ and $3\pi/4$ for each of the kinds of the $^{13}CH_n$ systems. To make the diagrams somewhat more concise (Figures 7-28 and 7-29), we will start the sequences following the $90°_y$ $^{13}C$ pulse, and show only the effect of the first and last $\tau'$ period. You should be able to work out the magnetization changes in the intermediate steps. Try it!

It should be obvious that the value of $\tau'$ (or of $\theta$) has a very profound effect on the strength of the signals observed in **decoupled refocused** INEPT spectra. The editing possibilities associated with this can be very important in analyzing complex $^{13}C$ spectra. Enhanced peaks will not be observed for quaternary carbons, so the resonances of these carbons will be automatically edited out. All of the other $^{13}C$ resonances will be enhanced, to at least a substantial degree, with $\tau' = 1/8J^{-1}$ ($\theta = \pi/4$) (see Figure 7-28). The $^{13}C$ resonances of $CH_2$ and $CH_3$ groups are edited out when $\tau' = 1/4J^{-1}$ ($\theta = \pi/2$). The $^{13}C$ resonances of $CH_2$ will be distinguished from $CH_3$ or $CH$ when $\tau' = 3/8J^{-1}$ ($\theta = 3\pi/4$), because the resonances corresponding to $CH_2$ groupings will be enhanced, but will be negative going (see Figure 7-29). The changes of signal intensity as a function of $\theta$ for each kind of grouping is shown in Figure 7-30.

## 7-6 An Inept INEPT Summary

To summarize, INEPT can produce sizable enhancements of the resonances of less-sensitive nuclei, provided that there is a significant coupling to a more-sensitive nucleus. The resonances of those nuclei that have small, or negligible, couplings to the more-sensitive nuclei will only show unenhanced resonances, or none at all, if there is phase cycling or the normal spectrum is subtracted from the INEPT spectrum. Typically, an INEPT spectrum of a $^{13}C$-H group will give one strong $^{13}C$ absorption and one strong emission signal separated by $J_{H^{13}C}$. The $^{13}C$ resonances of a $^{13}CH_2$ group will also give two INEPT enhanced signals. One will be absorption, the other emission and the separation will be $^2J_{H^{13}C}$. With a $^{13}C$ methyl group, there will be two enhanced absorption and two enhanced emission signals with spacings of $J_{H^{13}C}$.

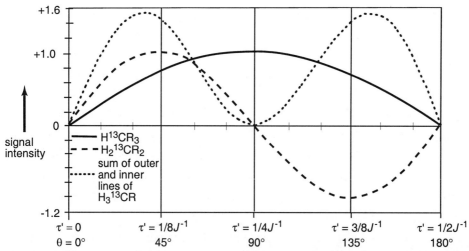

**Figure 7-30** Signal strengths of the transformed FIDs for the different groups of resonances as a function of τ' for $^{13}CH_nR$ using the refocused INEPT sequence with appropriate **decoupling** at the start of the acquisition of the FID. The vertical lines represent the values of τ' (or θ) that are usually selected in separate experiments to distinguish between CH, CH$_2$ and CH$_3$ groupings for INEPT or DEPT sequences

The "raw" intensities of the resonances for $^{13}CH_nR_{4-n}$ groupings generated by the INEPT procedure include contributions from the "natural" $^{13}C$ resonances (see Figures 7-3, 7-20 and 7-21). These contributions can be removed by digital subtraction of the normal spectrum from the INEPT spectrum, or by phase cycling prior to the data-collection process (Section 7-3). INEPT resonances corresponding to particular $^{13}C$ groups can be refocused and proton decoupled to give upright unsplit signals of the less-sensitive nuclei, but as can be seen in Figure 7-30, selection of the proper τ' ($\theta = 2\pi J\tau'$ rad) period is very important in determining the maximum upright signal strength for CH, CH$_2$ and CH$_3$ carbons. Figure 7-30 indicates that a good compromise value to give upright signals for all three groupings would have θ be a bit less than 45° degrees.

The degree of INEPT enhancement for the less-sensitive nucleus will depend on the ratio of the magnetogyric ratios of the two nuclei involved in the coupling $\gamma_1/\gamma_2$. For $^{15}N$, this will be the striking factor of 9.9 that represents, in the ideal case, a saving of a factor of 100 in observation time to produce the same S/N ratio.

## 7-7 The Role of J. Multiple-Quantum Coherences

A question we posed earlier was the importance of the spin-spin coupling in obtaining selective magnetization transfer (see Section 7-1). One can just accept this for a fact, or have concern about its origin. If you are willing to do the former, you might as well skip over to the next section. That could well be good advice, even if you would

## 7-7 The Role of J. Multiple-Quantum Coherences

like to understand, because what you will get in the following does not have much meat on it. "Where's the beef?" is a proper question.

In our example of selective magnetization transfer (see Figures 7-2, and 7-3) for a $^{13}$C-H system, we interchanged the populations of αα and βα by a 180° pulse applied at the frequency of the proton transition **3**. This would seem an unexceptional process if it were not for the fact that it results in enhancement of the intensity of transition **2** and an emission signal corresponding to transition **1**, both in the $^{13}$C-resonance regime. Clearly to have this happen, αα and βα must represent some sort of composite nuclear magnetic state that is not as simple as a system having a proton as α(1) on one molecule with no $^{13}$C, and a $^{13}$C as α(2) on a different molecule with no proton. In such circumstances (the left-side of Figure 7-1), the magnetic states would be independent of one another, with no spin-spin coupling and no positive or negative INEPT enhancements.

So far in our explanations of NMR phenomena we have relied on either energy-level diagrams or on simple magnetic vector models. While the energy-level diagrams have substantial utility in dealing with the ordinary NMR transitions with single-quantum transitions characterized by $\Delta m = \pm 1$, as we have used them, they take no account of phase relationships between the precessing magnetic moments, which we have seen in this chapter to be of critical importance when $J$ couplings are significant. On the other hand, the vector diagrams do take account of phase, but are deficient when we need to consider different quantum states. From this, perhaps you can understand the possibility of existence of NMR phenomena that can not be explained by either of the ways that we have used so far. The phenomena we have been dealing with almost exclusively up to now are the **single-quantum coherences**, the phenomena that we can not deal with so easily involve the **multiple-quantum coherences** of multiplicity 0, 2, 3, 4, .. We can think of the multiple quantum coherences as magnetic states populated by magnetic moments that we normally assign to two different energy levels that would be connected by $\Delta m = \pm 0, \pm 2, \pm 3, \pm 4$, and so on. Nuclear moments, when in these states, possess phase relationships of such character as to not produce detectable $X$, $Y$ magnetization and, what is more remarkable, not be subject to changing those phase relationships as the result of spin-spin interactions. The multiple-quantum states are transitory because, over time, they are subject to $T_1$ and $T_2$ relaxation. We rationalized a similar, appearing, but basically different situation in Section 7-3 (see Section 9-2 for more detail) for the two-nucleus antisymmetric state. How do they differ? Freeman[1] suggests that, in the multiple quantum state, the nuclear magnetic vectors can be

---

[1] R. Freeman, *Concepts in Magnetic Resonance*, **10**, 63-84 (1998)

thought of as antiparallel, but because of their special mode of preparation, they are locked into this kind of state by not being subject to changes in phase caused by spin-spin interactions of the kind illustrated by Figure 7-5.

A general basis for understanding the nature of multiple quantum coherences is to some degree supplied by consideration of the elements of the square symmetric **density matrix**, Equation 7-1, that represents a coupled two-nucleus system like the one represented by Figure 7-1.

$$\begin{vmatrix} \rho_{11} & \rho_{12} & \rho_{13} & \rho_{14} \\ \rho_{12} & \rho_{22} & \rho_{23} & \rho_{24} \\ \rho_{13} & \rho_{23} & \rho_{33} & \rho_{34} \\ \rho_{14} & \rho_{24} & \rho_{34} & \rho_{44} \end{vmatrix}$$

Eqn. 7-1

The elements of this matrix can be understood to reflect the average densities of populations of various possible magnetic states of the system. At equilibrium, with no applied $B_1$ field, all of the elements are zero, except for $\rho_{11}$, $\rho_{22}$, $\rho_{33}$, and $\rho_{44}$. These can be taken to correspond respectively to the states $\alpha\alpha$, $\alpha\beta$, $\beta\alpha$ and $\beta\beta$ of Figure 7-1 with respective populations: $P_1 + P_2$, $P_1 - P_2$, $-P_1 + P_2$ and $-P_1 - P_2$.

When we apply a pulse to the transition marked **3** in Figure 7-1, we perturb the density matrix and a significant population density will appear in $\rho_{13}$. This population density corresponds, on the one hand, to a state with single-quantum phase coherence of nuclear magnetic victors processing around the Z axis and, on the other, to a change in the population of the $\alpha\alpha$ and $\beta\alpha$ quantum states of Figure 7-1. It is this murky combination of vector and quantum concepts that is hard to describe other than mathematically. If nothing further is done to the system after the excitation of transition **3**, relaxation will cause the density matrix to return to its equilibrium condition. From this we see that, besides shifts and couplings, the pulse power, pulse duration and relaxation will be important to the evolution of the density matrix with time.

Each of the terms, $\rho_{12}$, $\rho_{13}$, $\beta_{24}$, and $\beta_{34}$ corresponds to excitation of a single-quantum transition and generation of a single-quantum coherence of detectable magnetization until that decays by $T_2$ relaxation. However, the density matrix will not reach the equilibrium condition until $T_1$ relaxation is complete.

The difficult-to-understand quantum coherences correspond to the density-matrix elements $\rho_{23}$ and $\rho_{14}$. Figure 7-1 shows that $\rho_{23}$ is associated with what we have thought of as forbidden transitions, $\alpha\beta \rightleftarrows \beta\alpha$ which have zero change in total quantum

## 7-7 The Role of J. Multiple-Quantum Coherences

number and thus are called zero-order coherences. On the other hand, $\rho_{14}$ is associated with "forbidden" transitions $\alpha\alpha \rightleftarrows \beta\beta$ and are designated as second-order coherences. Like the first-order coherences, these coherences are unstable and decay by relaxation.

During the lifetime of the coherences, they cannot themselves be observed but can be converted to detectable single-quantum coherences by a 90° pulse. What is harder to understand are the differences in sensitivity to magnetic field inhomogeneities and field gradients of the coherences as a function of their order. The differential sensitivity of the coherences in the this regard provides the most vivid evidence for their existence and is discussed in more detail in Section 8-3-**8**.

Quantum-mechanical procedures using the density-matrix formalism are available that allow one to calculate what happens in experiments like selective magnetization transfer and INEPT, as well as in other multipulse experiments, but they are abstruse and not easily susceptible to qualitative interpretation. Simpler procedures based on the "**product-operator**" formalism can be used to follow the effects of the different pulses and waiting periods in multipulse sequences. However, this approach does not lend itself to simple interpretation by our vector models either. Significantly, the product operators include terms that are of the form $I_1 \cdot I_2$ that represent for our case, interaction between the magnetizations of the $^{13}C$ and $^1H$ in $^{13}CH_n$ groupings.

Weird and wonderful things can be achieved by knowing how to manipulate multiple-quantum coherences. If you look at Figure 7-1, you will see that the maximum multiplicity possible for a two-nucleus system will correspond to the double-quantum transitions $\alpha\alpha \rightleftarrows \beta\beta$ while the minimum will correspond to the zero-quantum transitions $\alpha\beta \rightleftarrows \beta\alpha$. Proceeding in a similar way, Figure 7-19 will show that maximum multiplicity for a three-nucleus system corresponds to the three-quantum transitions $\alpha\alpha\alpha \rightleftarrows \beta\beta\beta$, and so on. By properly generating and manipulating the desired multiple-quantum coherences for given two, three, four or other nucleus groupings, for example, $^{13}CH$, $^{13}CH_2$ or $^{13}CH_3$, it is possible to manipulate the coherences to cause the particular resonances to essentially completely, as well as selectively be edited out! The procedures involve using what are called **multiple-quantum filters**.

As we proceed further, there will be occasions where we will be forced to hold up placards carrying the message "No user serviceable parts inside", which will mean that expertise in density-matrix or product-operator formalisms will be necessary to determine what happens in the particular sequence of pulses or evolution periods.[2] So,

---

[2] See R. Freeman, *Spin Choreography. Basic Steps in High-Resolution NMR*, Oxford University Press, New York, 1997, for a detailed and useful account of product-operator formalism, as well as of multiple-quantum coherences and wonderful clear descriptions of complex NMR experiments.

"Where's the beef?" Frustratingly evanescent, until you can turn it into something your spectrometer's rf receiver can detect.

## 7-8 DEPT and Its Variations

If you look back at Figures 7-2, 7-20 and 7-21 you will be reminded that INEPT produces coupling patterns for $^{13}$C resonances with intensity ratios of -1, +1 for CH; -1, 0, +1 for CH$_2$; and -1, -1, +1, +1 for CH$_3$ groups, respectively. Refocusing, without decoupling, using $\tau' \approx 1/8J^{-1}$, can invert the negative peaks, but in the absence of phase errors or large deviations in the $J_{H^{13}C}$ values, refocusing preserves the relative intensities of the multiplets. It is fair to say that the INEPT intensities are distorted relative to the 1:1, 1:2:1 and 1:3:3:1 found for the intensity patterns of ordinary spin-spin coupling of the $^{13}$C resonances of CH, CH$_2$ and CH$_3$ groups.

A solution to the distortion problem is supplied by DEPT (**Distortionless Enhanced Polarization Transfer**). This sequence operates in a rather similar way to INEPT, but gives normal ratios of coupling intensities, if not decoupled, in the acquisition part of the sequence. However, both coupled and decoupled peaks may have diminished total intensity, have zero intensity, or be negative going, in the same kind of manner as shown in Figure 7-30. The key parameter in this aspect of DEPT is the angle $\theta$ in INEPT, but applied in a very different way.

The INEPT refocusing sequence in the currently fashionable pulse-delay shorthand notation follows:

$^1$H   $(\pi/2)_x$   $1/4J^{-1}$   $\pi_x$   $1/4J^{-1}$   $(\pi/2)_{\pm y}$   $\theta$   $\pi_x$   decouple

$^{13}$C                     $\pi_x$            $(\pi/2)_x$        $\pi_x$   acquire ($\pm y$)

The early steps are shown in Figure 7-15 and it may be well for you to satisfy yourself how the notation corresponds to the diagrams we have used up to here. The $\pm y$ subscript on the $\pi/2$ H pulse denotes phase cycling (see Section 7-3) and is paired with the $\pm y$ notation appended to the acquire part of the sequence.

The DEPT sequence is:

$^1$H   $(\pi/2)_x$   $1/2J^{-1}$   $\pi_x$   $1/2J^{-1}$   $\theta_{\pm y}$   $1/2J^{-1}$   decouple

$^{13}$C                  $(\pi/2)_x$            $\pi_x$             acquire ($\pm y$)

The difference between INEPT and DEPT for $^{13}$CH$_n$ groups begins with the early evolution period of $1/4J^{-1}$ being replaced by $1/2J^{-1}$ and then with replacement of the $180°_x$ $^{13}$C with a $90°_x$ $^{13}$C pulse. This is then followed by the $\theta_{\pm y}$ H pulse, where $\theta = \pi \Delta J$ as before, but now, instead of corresponding to an evolution time for refocusing, it corresponds to the $^1$H pulse angle. The DEPT sequence depends on multiple-quantum coherences to produce its magic and no one seems to have succeeded so far in rationalizing this in terms of simple vector diagrams.

DEPT is very useful for differentiating CH, CH$_2$ and CH$_3$ groups in $^{13}$C spectra. The procedure involves changes in $\theta$ and, in this, the intensities follow the intensity plots of Figure 7-30. The customary choices of $\theta$ are 45°, 90° and 135° with results that are basically the same as for INEPT. Difficulties in selecting the $1/2J^{-1}$ interval because of variations of $J$ with chemical structure are common to both INEPT and DEPT.

There are a number of variations on DEPT, as well as other pulse sequences, that can be used to assist in one-dimensional analysis of spectra. However, it is not likely that further excursions that we might make in this area will improve our overall understanding and, after a brief excursion to look at "soft" pulses, we will turn to two-dimensional NMR.

## 7-9 Soft Pulses and Their Uses

Although we have not explicitly defined **soft pulses** we have considered several of the basic concepts involved in some detail previously. The purpose of soft pulses is to provide selective excitation of single transitions or groups of transitions with similar frequencies. This is in contrast to hard pulses which excite transitions of nuclei with a wide range of precession frequencies. In most of our discussions so far, the emphasis has been on what is required for non-selective excitations, although earlier in Section 7-2, we pointed out that selective pulses are needed for direct selective magnetization transfer.

An important criteria for setting up discriminatory soft pulses is the degree to which they can selectively activate one resonance relatively near to another resonance, here the pulse shape is important. In our earlier calculations (Section 2-5), we used a square-wave pulse with a rapid rise and rapid cutoff and we indicated that it made little difference what shape was given to the pulse, provided that it was short and powerful (Section 5-2-**3**).

Now, suppose we wish to produce a 180° pulse at the resonance frequency of one nucleus with little or no change in the magnetization of a second nucleus 100 Hz away. Numerical-integration calculations of the type we have discussed before show that when the frequencies to be discriminated are relatively close together and the pulse period short, then a 180° pulse at the frequency of one nucleus causes a change in

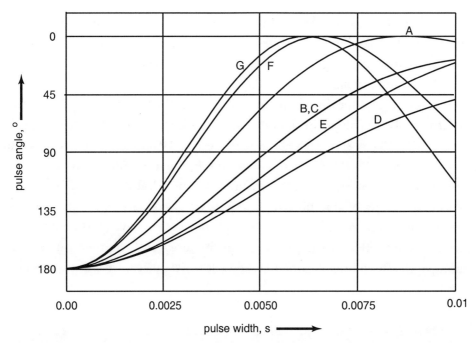

**Figure 7-31** Calculated pulse angles for a system of two nuclei, one of which has a frequency the same as the oscillator and the other 100 Hz away. The pulse time and amplitude are set up so that, for each different soft pulse, the "on-resonance" nuclei undergo an exact 180° pulse. The curves represent how the "off-resonance" nuclei respond as a function of total pulse time in s. Thus, Curve A (from a simple square-wave pulse) shows that when the pulse time is very short, both the "on-resonance" and the "off-resonance" nuclei will undergo a 180° change and the excitation is non selective. However, as the pulse time becomes longer, the excitation becomes more selective and tops out at 100% selectivity for the given 100-Hz frequency difference. The other curves represent different ways of applying the 180° pulse to the nuclei. B and C represent linear changes in pulse power, with B increasing from zero and C decreasing to zero. D represents an increasing Gaussian curve change in power. E has linearly decreasing power to the middle of the pulse and a linear increase thereafter. F is the same as E, except that it is inverted, with the maximum power in the middle of the pulse period. G involves two square wave pulses with one in the first quarter of the pulse period and the second in the last quarter

magnetization of the second nucleus that will be near to 180° if the pulse period is short as shown on the left side of Figure 7-31. As the pulse period is made longer, then the magnetization vector of the second nucleus gets out-of-phase with the pulse frequency and is turned upward toward its original value along the Z axis. How soon and how completely it regains its original value depends on the pulse shape. If the objective is to use a relatively short pulse time then one can use either an "on-off-on" or a "declining-ascending" pulse sequence, see Figure 7-31. These sequences are efficient in returning the second nucleus vector back to its original value, but require more precise timing than some of the other possibilities shown in Figure 7-31. In any case, you can see that the shape of the pulse is quite important with regard to excitation of "off-resonance"

## 7-8 Soft Pulses and Their Uses

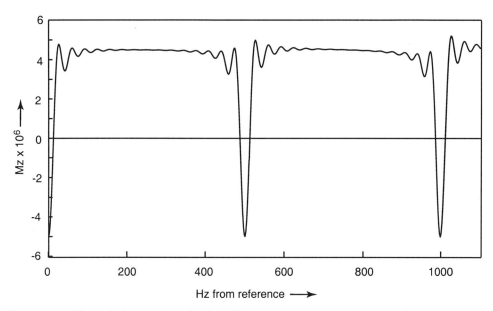

**Figure 7-32** The calculated effect of a DANTE sequence with an 0.001 s spacing between each of 32 pulses on nuclei having resonance frequencies ranging from 0 to 1100 Hz from the reference. The integrated intensity of the sequence produces a 180° pulse at the reference frequency. The effects were calculated assuming $M_{z0} = 5.0 \times 10^{-6}$ and used numerical integration of the Bloch equations

nuclei when the pulse period is longer than that normally used to produce a 90° pulse. In many cases, it is desirable to use a pulse that is shaped to a Gaussian curve.

One alternative to soft pulses for selective activation of nuclei are DANTE pulses (**Delays Alternated with Nutations for Tailored Excitations**).[3] These are short hard pulses with specified spacings between the pulses. The idea is to choose the proper spacing to activate nuclei with particular resonance frequencies. The procedure is the reverse of the swing analogy used at the end of Section 2-5. Now, we activate one swing, of several with different frequencies, by timing of our pushes to the frequency of that swing. The other swings are pushed one way or the other during the course of the activation and are not themselves activated. An example of the DANTE technique is shown in Figure 7-32, where the effect of 32 short pulses spaced 0.001 s apart is calculated for resonance frequencies ranging from 0 to 1100 Hz. The total intensity of the 32 pulses is taken to provide a 180° pulse for those nuclei whose precession frequency is exactly on resonance and it will be seen that the sequence also produces 180° pulses for nuclei whose precession frequencies are 500 and 1000 Hz off resonance.

---

[3]See R. Freeman, *A Handbook of Nuclear Magnetic Resonance*, Addison Wesley Longman, Essex 1997. Second Edition, p. 242 for the connection between the DANTE sequence and Dante's Purgatory.

However, the pulse sequence will not activate those nuclei with precession frequencies of 100-400 and 600-900 Hz.

**Exercises**

**Exercise 7-1** Show how you could modify the timing diagram of Figure 7-6 to give an upright signal as the result of AT, FT.

**Exercise 7-2** Show with vector diagrams how the pulse sequence of Figure 7-7 with the $\tau$ value of Figure 7-8 corrects for inhomogeneities in the magnetic field.

**Exercise 7-3** Consider substances representing the respective systems $^{13}CR_4$, $H^{13}CR_3$, $H_2{}^{13}CR_2$ and $H_3{}^{13}CR$, with non-magnetic R groups; and, for simplicity, assume that the $^{13}C$ resonance frequency of each equals the rotating-frame frequency. For each substance, make a plot of the signal intensities that you would expect for its $^{13}C$ resonance as a function of $\tau$ following a 90° $^{13}C$ pulse, a $\tau$ waiting period and then acquisition with decoupling only during the acquisition period (this amounts to the sequence of Figure 7-6 without the 180° pulse and the second $\tau$ period). Cover the range of $\tau$ from zero to $1/J$ and neglect any possible NOE resulting from the decoupling. Remember that the proton-decoupled $^{13}C$ resonances of $CH_2$ and $CH_3$ groups are composite resonances.

**Exercise 7-4** Investigate the evolution of the vectors of Figure 7-13 during an acquisition time AT of 1 s when the average frequency of the $^{13}C$ nuclei is 2 Hz different from the rotating frame and determine whether or not the signal is nulled, when the rotating-frame frequency is different from the average 13C frequency.

**Exercise 7-5** Show how phase cycling can cancel out a constant rf leakage voltage into the receiver coil.

**Exercise 7-6** Show how proton decoupling during the acquisition period of INEPT applied to the $^{13}C$ spectrum of $HCCl_3$ would remove the usual positive and negative enhancements of the $^{13}C$ signal characteristic of the INEPT sequence without refocusing.

**Exercise 7-7** Work out the changes in vector diagrams corresponding to each of the intermediate steps in Figures 7-28 and 7-29.

# Exercises

**Exercise 7-8** Show how you can derive the curves of Figure 7-30 by finding the points for 0°, 45°, 90°, 135° and 150°, with the procedures of Figures 7-25, 26, 27, 28 and 29.

**Exercise 7-9** Sketch the proton-decoupled and the various DEPT 75-MHz $^{13}$C spectra of 2,2-dimethyl-3-pentanol on the expectation that the shifts will be C1, 26 ppm; C2, 33 ppm; C3, 72 ppm; C4, 39 ppm and C5, 10 ppm; all with respect to tetramethylsilane.

**Exercise 7-10** In this exercise you are to work out why the intensities of the right-hand resonance lines in Figure 7-21 are -6P1, -6P1, +6P1 and +6 P1 on the basis of analogy with Figures 7-18 and 7-19, along with the following information. You will have to take into account that the protons of a $^1H_3{}^{13}CR$ system have four symmetric states with total magnetic quantum numbers of +3/2, +1/2, -1/2 and -3/2. There is also a pair of degenerate (equal-energy) antisymmetric states at +1/2 and another at -1/2. The energy-level setup is shown in Figure 7-33 with the allowed $^{13}$C transitions.
**a.** Use Figure 7-33 to construct the expected pattern of resonance lines for a $^1H_3{}^{13}CR$ system similar to the lower part of Figure 7-18 and the expected relative populations of the various magnetic states at equilibrium.
**b.** When a 180° pulse is applied to one line of the proton doublet, assume the populations of the states involved will be inverted from top to bottom. Use this information to derive the intensity pattern of Figure 7-21 on the basis of phase cycling as described in Section 7-4.

**Exercise 7-11** What problems would you foresee if you were to apply INEPT to a long-range $^2J_{H13C}$ coupling, for example in $R_2CH$-$^{13}CR_3$ (where the R's are nonmagnetic groups) and the coupling is 2 Hz compared to 100-200 Hz for H-$^{13}CR_3$? Explain.

**Exercise 7-12** Suppose that you had to excite two resonances 250 Hz apart, but not one in the middle between them. What sequence of DANTE pulses could you use for this purpose? Explain.

**Exercise 7-13** Sketch an energy diagram for a three-nucleus, spin 1/2 system with different chemical shifts and determine the number of zero-, one-, two- and three-quantum transitions.

**Exercise 7-14** The ordinary INEPT spectrum shown in Figure 7-34 is of a compound $C_8H_{10}$. Write a structure that fits with the spectrum and give your reasoning.

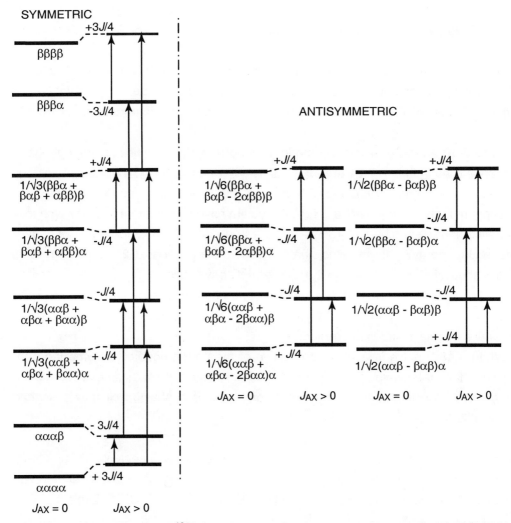

**Figure 7-33** Energy levels for a R$^{13}$CH$_3$ system, where R is a nonmagnetic group. See Exercise 7-10

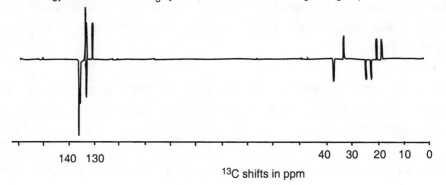

**Figure 7-34** INEPT spectrum for Exercise 7-14

Exercises                                                                                                          181

**Exercise 7-15** Sketch a $v$-mode FID or the FT of the $v$-mode FID for the final result of the last step in the given pulse sequences. You need not show the intermediate steps. But do indicate the scale of the time or of the frequency axes, as appropriate. Let $v_0$ be the rotating-frame frequency. For proton-$^{13}$C systems, $v_1$ will be the proton frequency relative to the proton $v_0$, and $v_2$ will be the $^{13}$C frequency relative to $v_0$ in the $^{13}$C rotating frame. The $\tau$ values are waiting periods where indicated, and AT is the acquisition time of the Acquire period. Assume that $T_1$ is long compared to $T_2$ throughout.

a. One nucleus, $v = 5$ Hz, $T_2 = 1$s, AT = 1 s; sequence = $90°_x$ - Acquire.
b. Same as **a.**, except that $T_2 = 0.25$ s; sequence = $90°_x$ - Acquire.
c. Same as **a.**; sequence = $90°_x$ - Acquire - FT. (Review Section 5-2-**12**)
d. Same as **b.**; sequence = $90°_x$ - Acquire - FT.
e. Same as **b.**; sequence = $90°_y$ - Acquire - FT.
f. Same as **b.**; sequence = $90°_y$ - $90°_x$ - $180°_y$ - Acquire - FT.
g. Two protons, $v_1 = -1$ Hz, $v_2 = 2$ Hz, $T_2 = 0.25$ s for both nuclei, AT = 1 s; sequence = $90°_x$ - Acquire - FT.
h. Same as **g.**, except that $\tau = 0.25$ s; sequence = $90°_x$ - $\tau$ - Acquire - FT.
i. Two nuclei, $v_1(H) = 2$ Hz, $v_2(^{13}C) = 0$ Hz, $T_2 = 0.25$ s for both $^1$H and $^{13}$C, $J^1_{H^{13}C} = 0$, $\tau = 0.25$ s; sequence = $90°_x$ (nonselective $^{13}$C) - $\tau$ (decouple $^1$H during $\tau$) - Acquire ($^{13}$C, decouple $^1$H) - FT.
j. Same as **i.**; except that $v_2(^{13}C) = +1$ Hz, $J_{1_H13_C} = 3$ Hz; sequence = $90°_x$ (nonselective $^{13}$C) - $\tau$ (decouple $^1$H during $\tau$) - Acquire ($^{13}$C, decouple $^1$H) - FT.
k. Same as **j.**; sequence = $90°_x$ (nonselective $^{13}$C) - $\tau$ - $180°_x$ - $\tau$ (decouple $^1$H during $\tau$) - Acquire (decouple $^1$H) - FT.
l. (CH$_3$)$_2$CH-CH$_2$OH, $v_2$ (C$_1$) = 7000 Hz, $v_2$ (C$_2$) = 3200 Hz, $v_2$ (C$_3$) = 2000 Hz, $^1J_{13CH} \sim 125$ Hz; let all $T_2$ values = 0.25 s, AT =1 s; sequence = $90°_x$ - Acquire ($^{13}$C) - FT.
m. One nucleus, $v = 0$ Hz, $T_2 = 0.25$ s, AT = 1 s, $\tau = 0.25$ s; sequence = $45°_x$ - $\tau$ - $45°_x$ - Acquire - FT.
n. Same as **m.**, except that $v = 2$ Hz; sequence = $45°_x$ - $\tau$ - $45°_x$ - Acquire - FT.

**Exercise 7-16** Devise a pulse sequence for $^{13}$CHCl$_3$ that will produce an anti-phase configuration of $\alpha$ and $\beta$ $^{13}$C moments without using proton pulses.

# 8
# NMR Spectroscopy in Two Dimensions

The modern art of two-dimensional (actually multidimensional) NMR is extraordinarily complex and continuously evolving. Our mission here is not to provide a handbook of the plethora of available pulse sequences that generate 2D spectra or their uses, but rather to provide an introduction to the principles on which 2D NMR is based and furnish a few examples of the kinds of information it can give you as well as supplying a basis for understanding the way that 2D NMR works. Excellent books are available that expand on the details of how 2D spectra are obtained and interpreted.

## 8-1  What is a 2D Spectrum?

Before answering that question, let us be clear as to what a 1D spectrum is. Ordinary NMR spectra are hardly one dimensional. We use two dimensions when we plot intensity against frequency. When we speak of 2D NMR we mean NMR with **two frequency dimensions**. At first, this may seem odd, but if you add, in one way or the other, a third dimension of intensity to two frequency dimensions you can see that there are all kinds of interesting information could be available in such plots. One example is COSY ("Correlated Spectroscopy"). COSY can provide almost at a glance, which nuclei are coupled to which, in a complex spectrum. Before proceeding further to show the advantages of COSY, we should say a word or two about graphical representations of 2D NMR spectra. The usual choices are of a simple topographic map showing the range of intensities with isointensity contour lines, or of a three-dimensional projection, preferably made with a hidden-line algorithm, that will show the terrain of the plot more clearly, especially with respect to the relative heights of the peaks, see Figure 8-1. Ordinary topographic maps in most atlases try to make relative heights clearer by using color-coded fill between the contour lines. Obviously, it takes many more points and much more programming skill to generate plots like those in Figure 8-1, than to plot a conventional 1D spectrum. If a conventional spectrum is carried on with 1024 spectral intensities, a corresponding 1024 x 1024 2D plot will have $(1,024)^2 = 1,048,576$ points and, if 16-bit precision is employed, then about 2 Mbytes of memory or disk space is needed. Greater precision and/or more points require more computer power.

## 8-1 What is a 2D Spectrum?

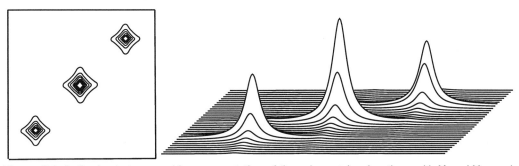

**Figure 8-1** Left, simple topographic representation of three Lorentzian functions with X and Y axes in frequency units. Right, three-dimensional representation of the same functions

Returning to COSY, look at Figure 8-2 where we have plotted an ethanol proton spectrum along both the $X$ and $Y$ axes of a contour plot. This plot has two in-plane frequency axes and the signal intensity is the out-of-plane dimension, with isointensity points connected to give contour lines. For COSY, the pulse sequence is designed so to give non-zero intensities at points that are along the diagonal where the chemical shifts intersect, and also at off-diagonal frequency intersection points, each of which represents a significant spin-spin splitting interaction. Figure 8-2 shows both of these features of COSY for ethanol molecules having slowly exchanging OH protons and methyl group protons, each coupled to the methylene group protons, but not coupled to each other. It should be clear that multiple-quantum coherences are likely to be

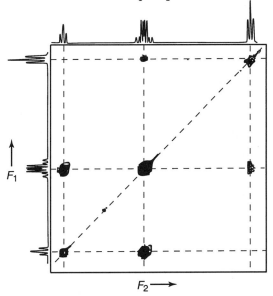

**Figure 8-2** Proton 2D COSY plot for slowly exchanging ethanol (spin-spin splitting observed between OH and $CH_2$ protons)

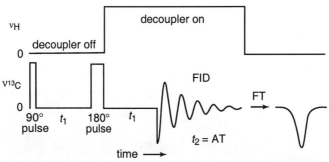

**Figure 8-3** Timing diagram for $^{13}C$ pulses for J-modulated spectra. The sequence here is the same as for Figure 7-7, except that the variable time is shown as $t_1$ and $t_2$ as the acquisition time AT

necessary to generate the off-diagonal peaks and the description of how the pulse sequence operates to give plots like Figure 8-2 will clearly be no less complicated than the descriptions already given of INEPT or DEPT. However, more straightforward aspects of 2D NMR exist to which we can give attention to first.

At the outset, let us consider how we can get a second frequency dimension spliced on to a conventional 1D spectrum. The seeds for doing this were planted in Chapter 7 where we looked into the time evolution of the magnetization of AX systems, with various waiting periods between pulses, using $^{13}CH_nR_{4-n}$ systems as examples. Whenever a quantity varies with time, it is a candidate for Fourier transform into a frequency domain by the procedures discussed in Chapter 3. This then is the way a second frequency dimension is obtained.

## 8-2 *J*-Modulated 2D Spectra

To illustrate how 2D spectra can be generated, we will choose a very simple example that both has some practical application and also is easy to understand, because it does not involve multiple-quantum effects. To do this and, at the same time, be consistent with much of the literature on the subject, we will need to change our notation slightly. Unfortunately, the change can make oral communication complex, because we will be dealing with $t_1$ and $T_1$, as well as $t_2$ and $T_2$. All I can say is that it was not my idea. Perhaps the experts who devised the notation only communicate with each other in writing. In any case, by convention, $t_1$ is a variable time that will lead us by transform to the second frequency dimension, quantified as $F_1$. The time that corresponds to the usual acquisition time of a FID is taken as $t_2$ and $F_2$ is then the corresponding frequency (see Figure 8-2).

Consider the *J*-modulated 1D spectra of the $H^{13}CR_3$ system that we discussed in

## 8-2  J-Modulated 2D Spectra

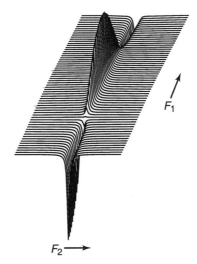

**Figure 8-4** Stack plot of signal strength of the transformed FIDs obtained as a function of $t_1$ for the $^{13}C$ resonances of H$^{13}$CR$_3$ using the 90°- $t_1$ -180°- $t_1$ sequence of Figure 8-3 with appropriate decoupling at the start of the second $t_1$ period and continuing through the acquisition of the FID

Section 7-2. Suppose that we take a series of $^{13}$C spectra using the timing diagram of Figure 8-3 and vary $t_1$. If we stack the plotted spectra with increasing $t_1$ to the rear, the result will be as in Figure 8-4, provided that $T_1$ and $T_2$ are both long compared to $t_1$. If we follow the center line of the stack up over the maximum, we expect the cross section to be as in Figure 8-5, which you will see is no more than Figure 7-14 extended to $2J^{-1}$s.

If we were to do the transform on just a few cycles of Figure 8-4, we would have a truncation problem, but that could be relieved by apodization and zero filling (see Chapter 5, particularly Sections 5-2-4 and 5-2-6)  However, the curve in Figure 8-4 is sinusoidal and could easily be extended to include many more cycles as shown in Figure 8-6. In this extension, $T_2$ causes the signal to decay and each member of the set of

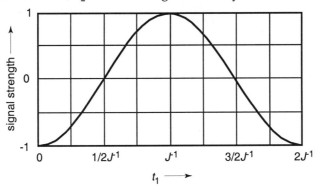

**Figure 8-5** Signal strength as a function of $t_1$ for H$^{13}$CR$_3$ along the midline of the stacked spectra of Figure 8-4

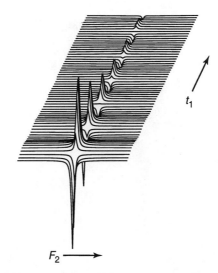

**Figure 8-6** Stacked plot of the same system as of Figure 8-4, except extended over several cycles with account taken of $T_2$ relaxation

decays in the $t_1$ direction will resemble a FID and can be transformed to the frequency domain. The decay that occurs along the center line has a negative first point and so an inverted absorption signal with a maximum at $J/2$ Hz is expected. Assuming that each spectrum in the stack plot of Figure 8-6 is made up of $n$ points, say 1024, and the stack also is 1024 deep, we would need to run 1024 transforms of 1024 points each on the cross sections parallel to the center section shown in Figure 8-6. The result will be a true 2D NMR spectrum with two-frequency dimensions (Figure 8-7). One dimension shows

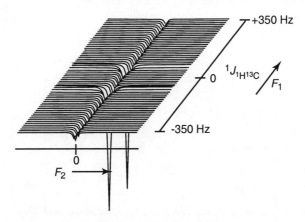

**Figure 8-7** A 2D *J*-modulated spectrum of an AX system that results from a Fourier transform of the data represented by Figure 8-6 along the $t_1$ direction. The two peaks are at ± $J/2$ Hz represent the AX spin-spin coupling. The peaks are negative because the center-line peak, when $t_1$ = zero, is negative in the stacked FT's of Figure 8-6. An upright plot of the data is shown in Figure 8-8

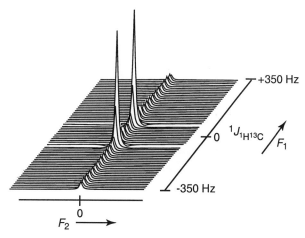

**Figure 8-8** Upright plot of the data represented by Figure 8-7

chemical shift, and the other dimension provides spin-spin coupling information. One can readily see from this example that 2D spectra are quite computer intensive and require resources that were not easily available in the early days of NMR.

To be sure, the negative peaks in Figure 8-7 do not look very attractive, but the plot can easily be inverted by a computer program, as in Figure 8-8, or by using a 180° phase shift in the receiver so that it is in effect looking down the $Y$ axis from the negative direction (see Sections 7-4 and 8-3-1). Another way to generate the plot is to make use of the magnitude spectrum (Section 5-2-11), where the intensity is $(x^2 + y^2)^{1/2}$. As noted before, for 1D spectra, this procedure reduces phase errors, but also broadens the bases of the peaks.

Consider now what will happen with a $H_2{}^{13}CR_2$ group run by the same pulse sequence. Let us look first at the case where $t_1 = (J/2)^{-1}$ that gives a null signal for $H^{13}CR_3$ (see Figure 7-24). After the initial 90° $^{13}C$ pulse, there will be three vectors we will have to consider. One corresponds to the center line of the $^{13}C$ $A_2X$ triplet and this will give a negative signal invariant with $t_1$ (except for $T_2$ decay). The two outside-line vectors will each precess in the rotating frame at $\pm J$ Hz and, if you look at Figure 7-30 you will see that when $t_1 = (J/2)^{-1}$, the outside-line vectors will lie along the $-Y$ axis. Then the 180° pulse, followed by another delay of $t_1 = (J/2)^{-1}$ will bring the vectors back along the $-Y$ axis, data acquisition and transform will lead to a negative signal. Again, if you look back at Figure 7-25, it should be clear that, for $t_1 = (J/4)^{-1}$, the outer $^{13}C$ triplet vectors will result in a null signal. Putting this information together with a $T_2$ decay, we will have $F_2$ signal intensity *vs.* $t_1$ diagram that will look like Figure 8-6. Apodization of this set of curves and Fourier transform into the second dimension gives three peaks,

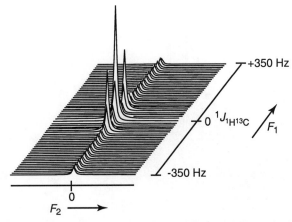

**Figure 8-9** An upright 2D J-modulated spectrum of an $A_2X$ system that results from a Fourier transform of the data like that of Figure 8-6 along the $t_1$ direction

one at 0, 0 and the others at 0, ± J, Figure 8-9. The 0, 0 peak arises from the center-line magnetic vector of the triplet while the 0, ± J peaks arises from the outer lines.

If you work through the same procedure for $H_3{}^{13}CR$, you will find peaks in the 2D spectrum at 0, ± J/2 and 0, ± 3J/2 that correspond the inner and outer pairs of the $^{13}C$ quartet ($A_3X$) with the peaks at 0, ± 3J/2 being one-third as intense as the others, as shown in Figure 8-10.

This may seem like a very cumbersome way indeed to get spin-spin coupling information. Why would anyone bother to make a 2D plot when the whole story would appear to be available by simply looking at the 1D spectrum, and picking out resonances of particular multiplicities and intensities? That procedure will indeed work, but if there are many carbons in the molecule having somewhat comparable chemical shifts

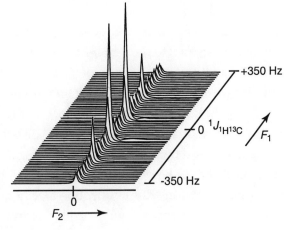

**Figure 8-10** An upright 2D J-modulated spectrum of an $A_3X$ system

and different numbers of attached hydrogens. The advantage of the 2D approach is that much more information can be shown without overlapping than is possible in a 1D plot. Thus, if the COSY method of Figure 8-2 were applied to the proton spectrum of a complex substance, there would be much more space available for showing the various couplings than along a one-dimensional frequency coordinate.

## 8-3 Some Techniques and Considerations for Multidimensional Spectra

I hope you are satisfied that the goal of generating multidimensional spectra is to gather information about the magnetization of the spin system as a function of time, in addition to that used for the simple FID, and then carry out a Fourier transform to convert the data that changes with this second function of time into a function of frequency. Most multidimensional spectra differ from the example given above in that the nuclei are massaged with a variety of pulses before the actual acquisition of the data to be transformed. Many different techniques are used and let us look at a few of the most basic, as well as some of the complications encountered. These individual techniques are widely used as building blocks in different forms of 2-D spectra.

**1. Phase Cycling.** We have encountered phase cycling earlier in Section 7-4 in connection with 1D spectra where it was used to cancel out the natural $^{13}C$ resonances when using the INEPT pulse sequence. The taking of multidimensional spectra, especially of large molecules, such as proteins, that can usually be studied only at low concentrations and so require very extensive time averaging, is a most severe test of the stability and sensitivity of NMR instrumentation. Often multidimensional spectra are plagued with "ghost" and "phantom" peaks that result from a variety of causes, one being the result of either differences in sensitivity of the two channels used for quadrature detection (see Section 5-2-9) or differences in their baseline output. Phase cycling can alleviate these problems through use of successive steps of 90° phase changes in the direction of application of the $B_1$ field in the $X, Y$ plane and appropriate related changes in the receiver phases along with switching between the two quadrature-detection channels.

Another use of phase cycling is in the simplest form of COSY in which the pulse sequence is $90°_x - t_1 - 90°_x - t_2$. Suppose we were to carry out this sequence on a group of nuclei with the same chemical shift, but one that does not correspond to the frequency of the rotating frame. For this case, we will take into account both $T_1$ and $T_2$ relaxations, a matter that we glossed over in Chapter 7. The vector diagrams of Figure 8-11 show a possible evolution of the magnetization in the time period $t_1$, where $T_2$ has diminished

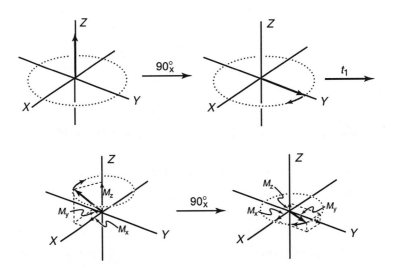

**Figure 8-11** The result of the $90°_x$ - $t_1$ - $90°_x$ part of the COSY pulse sequence. The point here is to show how as the result of $T_1$ relaxation and the second $90°_x$ pulse there will be a component of Z magnetization

the magnetization in the X, Y plane and $T_1$ has caused some growth of magnetization along the Z axis. The second $90°_x$ pulse results with an undesirable Z magnetization. Phase cycling in successive repeats of the COSY sequence reduces or eliminates the Z magnetization component, as described in Section 7-4. In many pulse sequences, the oscillator and receiver phases to be used in phase cycling will be listed in a table for a cycle of successive acquisitions when time averaging.

**2. Cross Relaxation.** In reading or hearing about 2D NMR (and 1D NMR as well), you may be confronted with allusions to **cross relaxation**. Fear not, because this topic has been covered in Section 6-4 with respect to the nuclear Overhauser effect (NOE). Figure 8-12 shows once again what is involved, ordinary relaxation in the two-spin H-$^{13}$C system occurs by the two $W_{1H}$ and the two $W_{1C}$ pathways, while the cross relaxation pathways are $W_2$ and $W_0$. These cross-relaxation routes are only implemented by way of dipolar interactions and, as we have discussed earlier, the NOE that is associated with them diminishes and finally disappears when other relaxation mechanisms begin to dominate. In the "extreme-narrowing mode" where $T_2$ is very long (short molecular correlation times), the fraction of cross relaxation of the total is given by Equation 8-1.

$$\frac{W_2 - W_0}{2W_{1C} + W_2 + W_0} \qquad \text{Eqn. 8-1}$$

## 8-3 Some Techniques and Considerations for Multidimensional Spectra

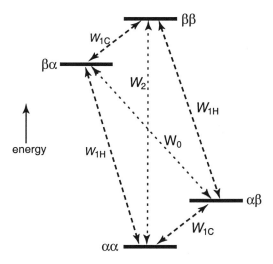

**Figure 8-12** Relaxation pathways for a two-nucleus system, illustrated here with a $^1H^{13}CR_3$ system, where the R's are nonmagnetic groups. The cross relaxation pathways are $W_0$ and $W_2$. Many of the important features of this graph were discussed in Section 6-4

When the relaxation is totally dipolar, protons are decoupled and $^{13}C$ is observed, then $W_2 : W_{1C} : W_0 = 12 : 3 : 2$.

If dipolar relaxation remains dominant, but the viscosity increases greatly, then a striking change takes place, because $W_0$ becomes the principal relaxation mechanism and this makes the NOE become negative.

**3. Spin Locking.** A number of multidimensional NMR experiments employ spin locking in different ways and for different purposes. The basic ideas behind spin locking are very similar to those used in the Carr-Purcell-Meiboom-Gill pulse sequence (Section 2-3). The sequence $90°_x - \tau - 180°_y - \tau$ is shown in Figure 8-13 and you should remember that this sequence provides a way to determine the $T_2$ of a sample with all positive echo peaks and minimal influences of diffusion. The important message for our purpose here is in the way that, with short $\tau$ times, the magnetization is held essentially along the Y axis independently of chemical-shift differences. To be sure, Figure 8-13 was worked out on the basis of $T_1 \gg T_2$ and we neglected relaxation along the Z axis. So, let us reverse field and neglect dephasing in the X, Y plane, but take into account $T_1$ relaxation. With these conditions, that is $T_1 \sim T_2$, relaxation in the rotating frame will occur after the $90°_x$ pulse, as shown in Figure 8-14. Now, a $180°_y$ pulse will turn the + Z magnetization into - Z magnetization and, after another $\tau$ period, there will be a sort of a "spin echo" when $T_1$ relaxation causes the magnetization to refocus along the Y axis. Because $T_1 \sim T_2$, we know that $T_2$ causes the X, Y magnetization to diminish, so that

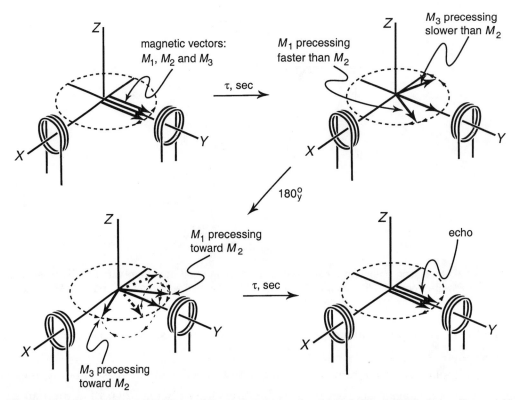

**Figure 8-13** The Carr-Purcell-Meiboom-Gill pulse sequence for determination of $T_2$ that was discussed in Section 2-3. It is illustrated here for three magnetic vectors $M_1$, $M_2$ and $M_3$ after a $90°_x$ pulse. The purpose of revisiting this sequence is because of its relation to spin locking

the echo-like effect will not be large, but you can see that there is some parallelism between dephasing and rephasing with a $180°_y$ pulse spins in the $X, Y$ plane and $T_1$ relaxation causing $-Z$ magnetization to relax upward to meet the $Y, Z$ plane.

The important point is that very short $\tau$ periods, combined with $180°_y$ pulses, will essentially lock the magnetization along the $Y$ axis. And what a series of short pulses can do is also possible with a strong continuous pulse - this is what we call a **spin-locking** pulse. The spin-lock sequence is the simple $90°_x$ - $180°_y$ sequence. We will see later how spin locking is used in some 2D NMR spectra.

One characteristic of the spin-locking condition that deserves comment is that when the nuclei are lined up along the $Y$ axis in the rotating frame, they undergo a $T_1$ type of relaxation along the $Y$ axis. Of course, this is really a $T_2$ relaxation in the laboratory frame of reference. What is noteworthy is that this relaxation occurs under the influence of the much smaller $B_1$ spin-locking field, than $T_1$ relaxation in the strong $B_0$ field along the $Z$ axis. That does not make much difference for most non-viscous liquids,

## 8-3 Some Techniques and Considerations for Multidimensional Spectra

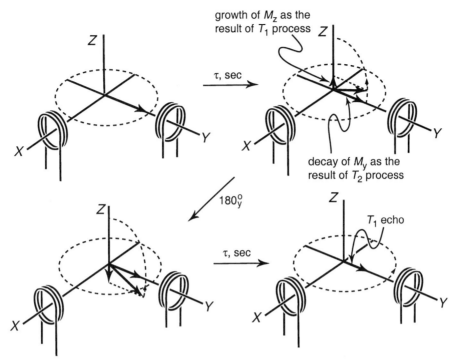

**Figure 8-14** Representation of the behavior of the $Z$ magnetization resulting from $T_1$ relaxation when spin locked. The $Z$ magnetization grows as the result of $T_1$ relaxation over some period $\tau$ and then is reversed by the $180°_y$ pulse. After that, the $Z$ magnetization "refocuses" along the $Y$ axis somewhat in analogy with refocusing in the Carr-Purcell sequence

but is important for situations where molecular motions are slow. We can see how this should be so if we return for a look at Sections 6-3 and 6-4 where we discussed qualitatively the dependence of relaxation on the magnetic field strength. Because of possible differences between the actual $T_2$ of the sample and spin-lock relaxation rate, the spin-lock relaxation rate is reported as $T_\rho$ or as $T_{1\rho}$.

**4. The Hartmann-Hahn Condition and Cross Polarization.** In quite a number of NMR experiments, there is a need to transfer energy between different sets of magnetic nuclei with quite **different precession** frequencies. We have seen how this can be done as the result of $J$-coupling in selective magnetization transfer in Section 7-2. An alternative procedure is by cross polarization using the magic of the **Hartmann-Hahn condition**. Transfer of energy from one set of spins to another is greatly facilitated if the nuclei have the same resonance frequency. Clearly, for transfer of energy from $^1H$ to $^{13}C$ nuclei there is an obstacle to overcome because their resonance frequencies are different by a factor of four. We can contrive a "thought experiment" to get the frequencies to be the same, whereby we place $^{13}C$ nuclei in a high magnetic field and the $^1H$ nuclei

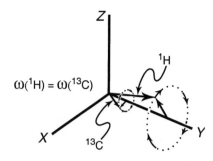

**Figure 8-15** A depiction of the Hartmann-Hahn condition for a proton-$^{13}$C system, where both sets of nuclei are precessing around the Y axis in their respective rotating frames and the pulse power on each is such to make $\gamma_{13C}B_1 = \gamma_{1H}B_1'$. The relative lengths of the proton and $^{13}$C vectors reflects the difference in the nuclear magnetic moments

in one-fourth as strong a field so that $\gamma_{13C}B_0 = \gamma_{1H}B_0'$. There are obvious difficulties with having two magnetic fields differing by a factor of four at the protons and the $^{13}$C nuclei in the **same molecule**, but these difficulties are just what Hartmann and Hahn overcame.

The first step in the Hartmann-Hahn process is to bring each set of nuclei to X, Y plane by a $90°_x$ pulse. Each set of nuclei, within its own rotating frame, can be spin-locked to the Y axis by either a series of $180°_y$ pulses or equivalent continuous pulses at the appropriate frequencies. The sleight-of-hand begins when the $B_1$ power at **each frequency** is adjusted, so that along the Y axis, we achieve the Hartmann-Hahn condition, which is $\gamma_{13C}B_1 = \gamma_{1H}B_1'$. When this is true, then both sets of nuclei are in effect, within their respective rotating frames, precessing with the **same** frequency about the Y axis, thereby providing an opportunity for exchange of energy. And it works!

A clearer idea about why it works can be seen by imagining a laboratory frame of reference taken along the Y axis. In that frame, the $^1$H and $^{13}$C spin-locked vectors precess around the Y axis, acting respectively to one another like the $B_1'$ rotating vectors in the ordinary rotating frame (see Figure 8-15). If the Hartmann-Hahn condition is met accurately, the transfer of energy from one set to the other is facilitated primarily by dipolar couplings in viscous media and solids, and by scalar J couplings in ordinary liquids.

If we were to use the Hartmann-Hahn condition to facilitate energy transfer between sets of nuclei in our "thought experiment", we would expect the equilibrium to be such that both sets of magnetic nuclei would have the same spin temperature. Things should be different in the real world when we start a Hartmann-Hahn experiment using protons and $^{13}$C, because we will expect that there is more energy stored in

### 8-3 Some Techniques and Considerations for Multidimensional Spectra

the protons than in the $^{13}$C nuclei. For one thing, unless the sample is enriched with $^{13}$C, there will likely be far more protons in the sample than $^{13}$C nuclei. The second reason is that the protons have a larger magnetic moment by a factor of about four than $^{13}$C. Consequently, using the Hartmann-Hahn condition with these nuclei in our sets, the original 90° pulse will, in effect, produce a set of nuclei with a relatively large magnetization aligned along the small $B_1'$ field. A **large** magnetization in a **low** field is possible only at **low spin temperatures** (see Section 1-2), so in effect, the spin-locked protons are very "cold". The spin-locked $^{13}$C nuclei will also be cold, but not as cold as the protons because their magnetization is smaller. As a result, the favored transfer of magnetization will be from protons to carbon to make the carbon spins colder. So, if we then turn off the $^{13}$C spin lock field, the $^{13}$C nuclei will give an FID at their frequency with a substantially enhanced signal (the maximum enhancement is $\gamma_{1H}/\gamma_{13C} \sim 4$). Because the cold proton reservoir is large, if time averaging is used and the $^{13}$C nuclei allowed to come to thermal equilibrium again, the $^{13}$C enhancement can, in principle, be repeated until the proton reservoir reaches the same spin temperature as the $^{13}$C nuclei. Then, to continue, the proton reservoir has to be recooled.

The Hartmann-Hahn condition can also be used for homonuclear systems. In such systems, a single $B_1$ field is all that is required and, because the γ values are the same, there will not be a significant temperature differences. In the absence of dipolar couplings, magnetization will be transferred between the sets of nuclei at rates in accord with the scalar $J$ couplings, modulated by the chemical-shift differences, because the latter represent deviations from the Hartmann-Hahn condition. One should remember that the spin-locking field along the Y axis is quite small compared to $B_0$. What this means is that the effective chemical-shift differences along Y are small, while the coupling constant remains unchanged. The result is that mixing of states such as αβ and βα (see Sections 7-3 and 9-2) becomes much easier than in the $B_0$ field and such mixing greatly facilitates magnetization transfer.

**5. Magnetic-Field Gradients.** In virtually everything we have discussed up to now, the emphasis has been on the deleterious effects of magnetic field inhomogeneities and how these can be canceled out by 180° refocusing pulses. The single exception has been with the possible use of magnetic-field gradients to measure diffusion constants (see Section 2-3). In recent times, magnetic-field gradients applied as pulses of varying length have become an almost indispensable part of multidimensional NMR spectroscopy and we will explore here some of the uses for which such gradients excel.

One form of magnetic gradient that has been used for a long time is called **homospoil**. It uses a simple coil that, when energized by passing a current through, generates an inhomogeneous local magnetic field that plays havoc with the homogeneity of the $B_o$ field, thus the name, homospoil. The purpose of homospoil is to cause rapid decay of magnetization in the $X, Y$ plane (transverse magnetization). A circumstance where this is quite desirable is in determining $T_1$ by the inversion-recovery procedure (see Section 2-4). You may remember that this procedure involves a 180° pulse, a recovery period and then a 90° analyzing pulse. If the 180° pulse is not really 180°, then some transverse magnetization will result and this would make the residual Z magnetization different from what would be expected, especially in the neighborhood of the time in which the Z component is near zero (for review, see Figure 2-26). Homospoil allows such transverse magnetization to be dephased easily and quickly.

Homospoil is hardly a precision procedure and not very comparable to today's precision magnet-field gradients, that have the capability of being applied for varying times, including as pulses. The technology is not trivial, because pulsed gradients can cause eddy currents in the magnet coils supplying the $B_0$ field. The need for producing pulsed gradients in magnetic resonance imaging (MRI) instrumentation spurred the development of shielded magnetic-field gradient pulses to reduce eddy currents and what was learned was transferred to spectroscopy procedures.

A precision-gradient pulse, along the Z axis, causes the nuclei across the sample to have continuously different frequencies with Z and, of course, become dephased during the pulse period - the extent of dephasing depending on the steepness of the gradient and the length of the pulse. As dephasing progresses, it causes the signal to decay and then disappear in just the manner we discussed for $T_2$ signal decay in Section 2-3. What is different now is that applying the **reverse** precision-gradient pulse, for the same time period as the original pulse, will bring the spins back into phase as though nothing had happened, except for the usual $T_1$ and $T_2$ processes inherent to the sample. It may seem strange to carry out such a refocusing sequence, but as we shall see later it has substantial power in certain 2D applications.

It should be recognized that refocusing with gradient pulses will be subject to the same limitation as with 180° pulses, in that diffusion can take nuclei from volume elements with a particular phase angle into another with a different phase angle. For Z-gradient pulses, diffusion will be important only when it occurs in the vertical direction.

Another use of gradient pulses is to improve resolution when the homogeneity of the magnet is less than perfect, but at the expense of overall signal strength. The key to doing this involves using a precision-gradient pulse to dephase the nuclei along the Z axis. What you do is to follow a 90° pulse with a positive gradient above some particular horizontal layer of the sample and a negative gradient below it. With these gradients, the spins in the upper and lower levels dephase, while the spins in the particular layer between the gradients will be unchanged in the rotating frame and, thus, will give a FID as usual, but from a magnetically homogeneous region. Of course, the observed resonances will be from a relatively small volume of the sample, which will cut down the signal strength.

**6. WALTZ Decoupling.** In many experiments, it is desirable to remove the splittings of some particular nucleus for an observed nucleus. The most common example is observing $^{13}$C and decoupling protons, some of the advantages of which were discussed in Section 5-1. Earlier, I did not make clear just how this decoupling, that is, **broad-band decoupling**, is carried out. The important thing is that all of the relevant protons be decoupled, but as the $B_0$ fields increase with improvements in magnet technology, the range of frequencies to be covered in decoupling goes up linearly with the field. Thus, 10-ppm proton shift at a 17.6 T (750-MHz) magnetic field is 7500 Hz compared to 600 Hz at the field of the 60-MHz magnets used in the "good old days". This range is rather small, compared to $^{13}$C with a range of some 37,500 Hz at 750 MHz (200 ppm) or $^{15}$N with a range of 67,500 Hz (900 ppm) at 17.6 T.

Single-frequency decoupling is useful, but far from efficient when broad-band decoupling is desired. The early alternative was to use "noise decoupling". What this involved was modulation of a single-frequency decoupling input by random noise so that, in effect, side bands were produced with a wide range of frequencies. Such procedures worked well at 60 MHz with a $^{13}$C observe frequency of about 15 MHz. Nevertheless, problems were sometimes encountered through the required power levels, particularly in water solutions, where a level of microwave dielectric heating could be involved. In one personal experience, the spectrum of an extraordinarily difficultly prepared sample of an enzyme, nitrogenase, complexed with $^{13}$C-labeled ethyne was taken with a debatably needed provision for proton decoupling. The result was conversion of the enzyme by thermal denaturation to the consistency of hard-boiled egg.

Better decoupling procedures became necessary as fields increased and these evolved in response to the need. It was true then, and is still true now, that the limits of human ingenuity and the flexibility of NMR capabilities have apparently not yet been reached. Consider, rather than decoupling with random noise, a sequence applied to the protons when one applies a repeated sequence such as $90^\circ_x - 180^\circ_y - 90^\circ_x$ (see Figure 8-16). In the $^{13}$C-$^{1}$H case, the aim is to efficiently cycle the protons through a rapid and continuous sequence of 180° inversions so as to cancel out their overall coupling effect. If there is a range of $^{1}$H shifts, these will be refocused by the $180^\circ_y$ pulse (compare the Carr-Purcell-Meiboom-Gill discussion of Section 2-3 and spin-locking in Section 8-3-**3**)

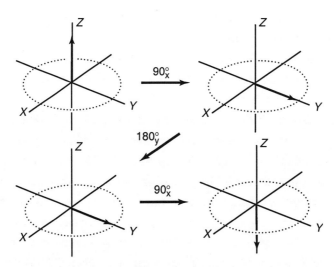

**Figure 8-16** A $90°_x - 180°_y - 90°_x$ pulse sequence for decoupling that rephases the chemical shifts of the decoupled nuclei with the aid of the $180°_y$ component

before being turned to the -Z axis. This sequence accumulates errors, as it is repeated, but these can be reduced by alternating the phases of the pulses as $90°_{-x} - 180°_{-y} - 90°_{-x}$. Sequences such as this are of the **MLEV** type and are often used in 2D NMR.

The **WALTZ** sequence is different $90°_x - 180°_{-x} - 270°_x$ (Figure 8-17). The purpose of the basic WALTZ is to reduce the sensitivity of the pulse sequence to phase errors in the excitations and this is done best when the sequence uses phase inversion in alternate cycles. Thus, WALTZ-4 is $(90°_x - 180°_{-x} - 270°_x) - (90°_{-x} - 180°_{-x} - 270°_x) - (90°_{-x} - 180°_x -$

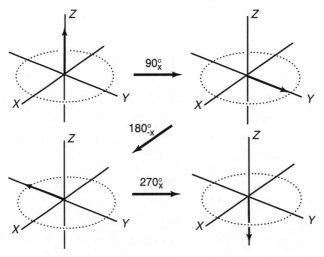

**Figure 8-17** The basic element of the WALTZ decoupling sequence $90°_x - 180°_{-x} - 270°_x$. This element is usually combined with others having phase shifts to reduce phase errors in the excitation

## 8-3 Some Techniques and Considerations for Multidimensional Spectra

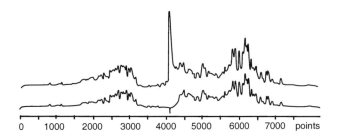

**Figure 8-18** Example of a sharply tuned digital filter to remove the water resonance from the proton NMR spectrum of the enzyme lysozyme A. For this example, the digital filtering was done "on the fly", being applied to each FID before it was added to the previously accumulated scans - a procedure that requires substantial computer resources. The NMR traces were kindly supplied by Bruker Instruments

$270^\circ_{-x}$) - ($90^\circ_{-x}$ - $180^\circ_{x}$ - $270^\circ_{-x}$). Such a sequence, with further permutations of pulses and phases can be extended to give WALTZ-8 and WALTZ-16.

These sequences are efficient and reduce the power needed for decoupling over what is needed for noise decoupling. Because the pulses increase by simple multiples of 90°, the basic sequence of 90°, 180°, 270° can be abbreviated as 1 x 90, 2 x 90, 3 x 90 and, through the whimsy of the experts the notation: $1,\bar{2},3; 1,\bar{2},3; \bar{1},2,\bar{3}; \bar{1},2,\bar{3}$ is named, even if derived by somewhat convoluted means, as WALTZ-4. The bars placed over the numbers in sequence such as $1,\bar{2},3$ indicate changes in phase.

**7. Suppression of Solvent Resonances.** Strong solvent resonances with solutes at low concentrations can give serious problems in time averaging by overrunning the data registers (Section 5-2-5). This is particularly important with water solutions because of its 55 M concentration. Many protein spectra, because of solubility or availability considerations may be taken at 1 mM or less concentrations. With the protons in water effectively at a 110 M concentration, it turns out that, for a single unique proton on a protein, the ratio of signal strengths would be 110:0.001 = 110,000. Use of deuterium oxide ($D_2O$), in place of water, can help enormously, but the exchangeable NH, OH and so on hydrogens of a protein will produce HOD, and that will again give a very substantial signal with respect to any particular single unique proton for which information may be desired.

A number of procedures can be used to suppress solvent signals. The simplest, but technically demanding to the spectrometer makers, is to use a sharp digital filter that will remove the solvent frequencies in each FID, **before** they are added to the others already accumulated. Digital filters of the requisite sharpness and computing capability of speeds needed to undertake the filtering essentially "on the fly" are now available. An application illustrating the results of their use is shown in Figure 8-18.

This procedure is appealing, because it can be used with any strong peak, although to give complete solvent suppression, the finite filter band width will necessarily enroach on some of the close-by peaks, as can be seen in Figure 8-18.

With water resonances, suppression can also take advantage of the long $T_1$ of water (on the order of 2 s) compared to those of the protons of many solutes. A somewhat greater than 90° pulse (either "hard" or "soft", see Section 7-9), can be followed by a field gradient pulse (Section 8-3-5) that will dephase all of the spins undergoing the 90° pulse. This will reduce the transverse magnetization to zero, but the solute protons with short $T_1$'s will recover their $Z$ magnetization more quickly than will the water protons. Then, at the point where the water protons are calculated to lose their excess $-Z$ magnetization and reach the null condition (see inversion recovery in Section 2-3) and, in hope that the solute protons have recovered sufficient $Z$ magnetization, a 90° pulse is applied to give a greatly reduced water signal. The weakness of the procedure derives from this last expression of hope. If the solute protons relax slowly, their resonances will be weak, if not missing. Further, any differences in the $T_1$ rates of relaxation of the solvent protons could make suppression of the solvent resonances difficult.

A quite different approach to water suppression than the above is to avoid exciting the water resonance. The need here is for the opposite of the soft pulses we discussed earlier for selective excitation. (Section 7-1). Here we want selective **nonexcitation**. It is hard to imagine giving a short pulse to a system of protons, including water as solvent, and not have the pulse exerted on the water protons as well. The trick is to excite the water resonance, but then turn the water magnetization back along the Z axis so that it has no transverse magnetization, and yet retain at least some measure of transverse magnetization of the solute protons.

This can be done far more simply than you might think. Suppose we have three sets of nuclei A, B, H$_2$O with $\nu_A > \nu_{H_2O} > \nu_B$ and we give them a $45^\circ_x$ pulse with the carrier frequency $\nu_0$ greater than $\nu_A$ (at one end of the spectrum), see Figure 8-19. Now, we wait a time $t_D$ that is precisely equal to $[2(\nu_0 - \nu_{H_2O})]^{-1}$ (or an appropriate multiple thereof). The result will be to bring the water magnetic vector into the $Y, Z$ plane, just 45° above the $Y$ axis (we are assuming the $T_1$ of water to be long compared to $[2(\nu_0 - \nu_{H_2O})]^{-1}$). Another $45^\circ_x$ pulse will achieve our objective and bring the water magnetization back along the Z axis, but this will not be true of $M_A$ and $M_B$. If we take, for purposes of illustration, the frequencies in the rotating frame to be $\nu_A = 1.5\nu_{H_2O}$ and $\nu_A = 0.75\nu_{H_2O}$ then, after $t_D$, $M_A$ will be above the $-X$ axis and $M_B$ will be 45° into the $X$, $-Y$ quadrant. After the second $45^\circ_x$ pulse, the transverse magnetization $M_{xy}$ for A will be 0.79 $M_A$, while $M_{xy}$ for B will be 0.37 $M_B$. Thus, the resulting first signal intensities

## 8-3 Some Techniques and Considerations for Multidimensional Spectra

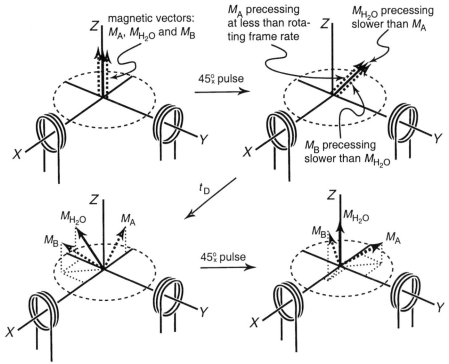

**Figure 8-19** Pulse sequence to null water resonances by turning the water magnetization away from the Z axis and then turning it back without complete loss of solute resonances that have chemical shifts different from those of water. The sequence here applies a $45°_x$ pulse and waits for a time $t_D$ that turns the water X, Y magnetic vector along the -Y axis. This is followed with another $45°_x$ pulse that returns the water magnetization to the Z axis. Two other resonances A and B are considered that, in the rotating frame, have precession frequencies such that, in the time interval $t_D$, wind up with their transverse components **not** along the -Y axis. The subsequent $45°_x$ pulse will be seen to leave the A and B nuclei with measurable transverse magnetizations

will be quite different from $M_A/M_B$. There will also be a phase problem, the second $45°_x$ pulse changes the phase of $M_B$ in one direction and $M_A$ in the opposite direction. What happens is that the peaks on one side of the suppressed water resonance in the final spectrum turn out to be 180° out-of-phase with those on the other side (one side is up, the other down).

This simple 1-1 sequence (where the pulses are multiples of 45°) can be improved by using 1-$\bar{2}$, but neither 1-1 or 1-$\bar{2}$ is as effective when the water lines are strong and broadened at the base as 1-2-1, 1-$\bar{2}$-1, 1-3-3-1 or 1-$\bar{3}$-3-$\bar{1}$. The 1-2-1 pulse sequence for the system of Figure 8-13 would be $22.5°_x$ - $t_D$ - $45°_x$ - $t_D$ - $22.5°_x$ and would have the same $t_D = [2(\nu_0 - \nu_{H_2O})]^{-1}$.

**8. Multiple-Quantum Coherences Revisited.** As I pointed out in Section 7-7, multiple-quantum coherences are real, not easy to visualize as physical entities, but

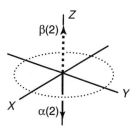

**Figure 8-20** Antiphase nuclear magnetic vectors produced by the INEPT pulse sequence described in Section 7-3

useful in many connections that are not obvious to those brought up on simple NMR spectroscopy as taught in elementary organic chemistry for structure elucidation, quantitative analysis or determination of reaction rates by analysis of changes in line shapes.

You may still have problems with the idea of coherence *vs.* noncoherence. An example may help. You might remember that in Section 7-3, we showed how we could generate an antiphase pair of $^{13}C$ magnetic moments along the Z axis as in Figure 8-20. (If you need to review, the procedure is described in Section 7-3.) This configuration, as shown in Figure 8-20 has no $^{13}C$ magnetization in the X, Y plane and no net $^{13}C$ magnetization along the Z axis, and so might seem to a casual observer to be identical with the sample at equilibrium in a **zero** field. However, we know there is a difference and we can expect that the system will evolve with time in accord with $T_1$ and, if subjected to a 90° pulse, be capable of giving an observable signal. This particular antiphase configuration is not itself a multiple-quantum coherence, but it shares the property of such coherences in being capable of evolution with time and, also, of not giving an observable magnetization until turned into the X, Y plane by a 90° pulse.

What more can we say about multiple-quantum coherences? First, we know that some of them correspond to magnetizations connected to transitions like $\alpha\beta \to \beta\alpha$ and $\alpha\alpha \to \beta\beta$ in Figure 8-11. Thus, we can imagine them as magnetizations formed by 180° pulses at those frequencies (but, of course, those are "forbidden transitions"). Multiple-quantum coherences only exist when *J* couplings are involved. With no *J* couplings, an AX system will behave like wholly separate A and X nuclei, assuming that they are not so close together in space that dipolar couplings would be important enough to give a nuclear Overhauser effect. However, although formation of multiple-quantum coherences requires *J* couplings, the coherences do not evolve in the same way as the antiphase vectors of a single-quantum state, such as on the right side of Figure 7-17. A simple way to understand this is shown by Figure 7-1, where you will see that both

8-3 Some Techniques and Considerations for Multidimensional Spectra            203

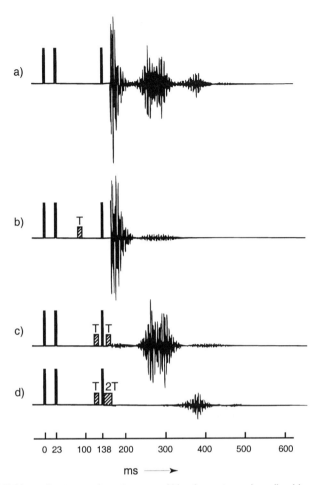

**Figure 8-21** Pulsed field-gradient experiments on an AX spin system, described in more detail in the text. Reproduced by the permission of *Chemical Physics Letters* and the authors[1]

states of the zero-quantum coherence are stabilized by $-J/4$, while both states of the double-quantum coherence are stabilized by $+J/4$, so the effect of $J$ will cancel out.

The really important things about multiple-quantum coherences are the role they play in magnetization transfer, and their sensitivity to phase changes. The latter behavior in a field gradient is particularly striking. A variety of coherences can be generated by the simple nonselective pulse sequence, $90°_x - t_1 - 90°_x$ and their presence detected by a subsequent $t'_1 - 90°_x - t_2$ (acquire) sequence. Here, the final 90° pulse brings the coherences back to the $Y$ axis, where they can be observed (see Section 7-7). A key property of multiple-quantum coherences is that they are defocused in a magnetic field gradient and the rate that they become defocused is directly proportional to the multiplicity order of the coherence; 0, 1, 2, 3 and so on. A wonderfully simple experiment that shows this uses the pulse sequence above, but applies gradient pulses in different ways during the $t'_1$ and $t_2$ periods.[1]

Figure 8-21 shows the 90° pulse and pulse field-gradient sequence used for a simple coupled two-proton AX system and what happens. First, look at the top of the figure **a)**, where there were 90° pulses, but no field gradient was applied. The observed signal was an FID arising from all of the coherences that were transformed in the X, Y plane by the final 90° pulse to observable single-quantum coherences.

The second trace from the top **b)**, used a field-gradient pulse (the shaded area marked T) of 5 ms in the $t'_1$ period before the final 90° pulse. It turns out that zero-quantum coherences are not affected by a field gradient (or by field inhomogeneities). However, all of the other coherences are sensitive to field gradients and thus lose phase coherence in the applied gradient. The final 90° pulse then turned the zero-order coherence into first-order coherence so it could give the observed FID.

In **c)**, equal-period field gradients were applied, one before and one after the last 90° pulse. These gradients had **opposite** slopes as a function of Z. The first gradient defocuses all but the zero-order coherences as in **b)** while the second gradient does two things. First, it defocuses and eliminates the X, Y magnetization from the zero-order quantum coherence seen in **b)** and then, acting oppositely from the first gradient it refocuses the first-order coherence so that it gives an observable signal.

The last trace **d)** shows what happens when a twice-as-long gradient pulse is employed after the last 90° pulse. This gradient destroys the Y magnetization from the zero- and single-quantum coherences that refocus and the double-quantum coherence is observed by its transformation as the result of the final 90° pulse to single-quantum magnetization in the X, Y plane. Why is it necessary to have twice as long a gradient pulse? First, we need to know that a double-quantum coherence defocuses at a rate twice that of single-quantum coherences in the first gradient pulse. Second, when the final 90° pulse transforms this double to single quantum coherence in the X, Y plane, being now single-quantum, it will then refocus only half as fast, so it requires **two** gradient pulses to get back into phase and be observed. Sounds impossible perhaps, but there is the evidence before you in Figure 8-21.

**9. Bloch-Siegert Shifts.** In our earlier treatment of the effect of a $B_1$ field on a system of magnetic nuclei (Section 2-1 and 4-2), we concentrated on the $B_1$ field produced along the X axis with no expectation that the strength of that oscillating field would have any influence on the shifts of the nuclei at resonance. However, Bloch and

---

[1] A. Bax, P.E. de Jong, A.F. Mehlkoff, and J. Smidt, *Chem. Phys. Lett.*, **69**, 567 (1980).

Siegert[2] in a rather turgid 1940 paper (quite a few years before Bloch demonstrated NMR for water protons and in contrast to his lucid 1946 paper[3]) derived an expression predicted changes in chemical shift as a function of $B_1$ field. The effects should be very small, a shift-correction factor of $(1 + B_1^2 / 16B_2^2)$ being predicted.

Larger shift changes, that are also called Bloch-Siegert shifts, can occur when a second rf field, call it $B_2$, is applied off-resonance to an AX system of nuclei as might be used in decoupling during an FID. The size of the shift depends on the frequency difference and on $B_2$. If the decoupling field strength is optimized in frequency and power for decoupling at frequency A, the change in shift at nucleus X is $B_2^2 / 2(\nu_A - \nu_X)$. The change in frequency is itself useful, because it can be used to determine the strength of the $B_2$ field. Bloch-Siegert shifts can cause substantial problems in 2D NMR when decoupling is involved and, also in 1D spectra, when it is desired to help untangle complex coupling patterns by comparing coupled and specifically decoupled spectra by subtracting one from the other to clarify the exact locations where spectral changes are taking place.[4] This type of Bloch-Siegert shifts does not occur in ordinary FT-NMR spectra ($90°_x$ - Acquire), because in ordinary FT spectra, no decoupling is applied during the FID. The effects of Bloch-Siegert shifts can be minimized by using the lowest possible $B_2$ power and employing techniques where decoupling is used only during evolution periods instead of during the acquisition periods.

## 8-4 The Homonuclear COSY Experiment

As mentioned earlier (see Section 8-1-1), the simple and original form of two-dimensional spectra called COSY (Correlation Spectroscopy) involves a very simple pulse sequence $90°_x$ - $t_1$ - $90°_x$ - $t_2$ and yet, as can be seen from Figure 8-2, COSY gives much more information than a 1D spectrum of the same substance. Exactly what happens in the overall process can't be very easily explained by the vector model we have used earlier because multiple-quantum coherences are involved that are only rigorously described by mathematical expressions, or less rigorously, in the semisymbolic product-operator formalism.[5]

As to homonuclear COSY, I will try now to give you a hand-waving flavor of

---

[2]F. Bloch and A. Siegert, *Phys. Rev.*, **57**, 522 (1940).
[3]F. Bloch, *Phys. Rev.* **70**, 474 (1946).
[4]J. K. M. Sanders and B. K. Hunter, *Modern NMR Spectroscopy*, Oxford University Press, N. Y., 2nd Ed., 1993, pp. 57-61.
[5]See R. Freeman, *Spin Choreography - Basic Steps in High-Resolution NMR*, Oxford University Press, New York, 1997, Chapter 4.

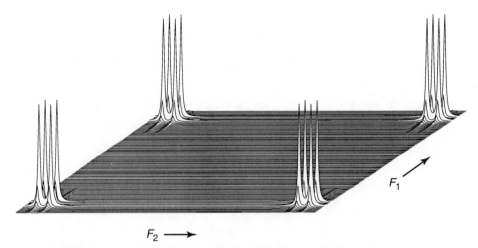

**Figure 8-22** COSY spectrum of an AX system displayed in the magnitude spectrum

what is involved without going into much detail. For in-depth discussions of practical details and problems, Derome[6] is excellent, while Freeman[5,7] gives about as simple a presentation of the product-operator formalism as you are likely to find.

We know from Chapter 7 that, if we give a short 90° pulse to a spin-spin coupled two-nucleus system with different chemical shifts, the magnetic vectors will evolve with time under the influence of the chemical-shift difference and $J$ coupling. Clearly, when we vary $t_1$ and supply a further 90° pulse, we will be observing the $t_1$ evolution of quite different magnetic states than in the $J$-modulated cases discussed in Section 8-2. A key characteristic of the sequence is that the second $90°_x$ pulse will turn all positive $M_y$ components of the spin systems to the -Z axis and so result in population inversions with some of the consequences of selective magnetization transfer, albeit modulated by the $J$ couplings. The evolution is complex and is expected by the product-operator formalism to involve a total of thirteen different product operators. Of these, four correspond to the double-quantum coherence, four correspond to antiphase orientations of vectors and one is magnetization along the Z axis.[7] All nine of these product operators correspond to unobservable magnetizations. The four remaining product operators can be matched to observable responses. Two of them correspond to chemical shifts (positioned along a diagonal like the one in Figure 8-2) and two others correspond to off-diagonal responses that involve mixing magnetization into the frequencies of one nucleus (including the $J$ coupling) with the corresponding frequencies of the other

---

[6] A.E. Derome, *Modern NMR Techniques for Chemical Research*, Pergamon Press, N.Y., 1987, Chapter 8.
[7] R. Freeman, *A Handbook of Nuclear Magnetic Resonance*, Addison Wesley Longman, Essex. Second Edition, 1997, pp. 185-194.

nucleus. To eliminate the problem represented by Figure 8-11, phase cycling is used extensively in the time averaging normally required to obtain COSY spectra.

A COSY spectrum of a homonuclear spin system is shown in Figure 8-22. You will see that it looks quite symmetrical and each group of peaks is an upright quartet. Remarkably simple, but there is a catch and that is that Figure 8-22 is a magnitude spectrum (see Section 5-2-**10** and Section 8-2). Consequently, Figure 8-22 does not convey any phase information and, furthermore, each peak has a broader base than it would have in a phase-sensitive COSY, a matter of some importance when high resolution is needed.

The problem with the phase-sensitive version is its complexity. Depending on which peaks you decide to have in the absorption mode the appearance will change. Because we are usually most interested in the cross peaks, the normal mode of presentation will have them phased to be absorption. Fair enough, but then the diagonal peaks (compare Figure 8-2) will be observed in the dispersion mode. Then, to make things more complicated, each of the off-diagonal quartets will have two upright and two downward absorption peaks. Clearly, the form of presentation of Figure 8-2 will be superior to using the magnitude representation of Figure 8-22, because then we can show the contours of all of the peaks whether above, or below, the plane of zero intensity.

Many of the other types of multidimensional NMR spectra provide similar information as COSY or use comparable styles of presentation.

## 8-5 Some Capabilities of 2D Programs

Besides J-modulated and homonuclear COSY, there are a plethora of 2D NMR procedures of very wide utility. A 1994 treatise[8] on two-dimensional NMR describes some 290 different pulse sequences! Many of these sequences have been worked out for those with special interests in large molecules, particularly protein structures and their conformations. In general, there will be differences between large and small molecules arising from differences in modes of relaxation. Nakanishi[9] provides an excellent survey of NMR spectral methods, both 1D and 2D for molecules of interest to organic chemists. I will list here a few 2D procedures that I believe to have special interest, but

---

[8] W.R. Croasman and R.M.K. Carlson (Editors), *Two-Dimensional NMR Spectroscopy*, VCH Publishers, N.Y., 1994, 2nd ed.
[9] K. Nakanishi (Editor), *One-dimensional and Two-dimensional NMR Spectra by Modern Pulse Techniques*, University Science Books, Sausalito, California, 1990.

this list must be regarded as both subjective and incomplete. The field changes with incredible rapidity and those who would try to cover everything, truly write on water.

**1. HETCOR.** Heteronuclear correlation is a form of COSY that is most often used with the proton spectrum along $F_2$ and the $^{13}$C spectrum along $F_1$. It can be carried on with, or without, decoupling.

**2. NOESY.** Nuclear Overhauser and exchange spectroscopy is a particularly important method for determination of three-dimensional structures. It is based on measuring the NOEs associated with dipole-dipole induced relaxation and the $r^{-6}$ dependence of dipole-dipole interactions. If we apply a selective 180° pulse to a particular nucleus and there is a second magnetic nucleus close enough so that $W_0$ and $W_2$ (Section 6-4) can act to equilibrate the spins, then we will expect a NOE when we apply the 90° pulse to the second nearby nucleus and acquire its FID. Clearly, this is a transient process different from the steady-state NOE discussed in Section 6-4. It will take a finite time to achieve substantial redistribution of the magnetization and, all the while, the respective $T_1$ relaxations will cause the NOE to disappear. This way of measuring NOEs is rather like selective magnetization transfer (Section 7-1) and also has some difficulty in that, because the redistribution of magnetization is slow, there is a problem in selecting the mixing time interval that corresponds to the maximum NOE. The need for selective pulses at particular frequencies can be avoided in somewhat the same way as used in INEPT.

In its 2D form, NOESY gives a COSY-like plot with off-diagonal elements representing NOE interactions. Normally, the NOE is not expected to be significant at distances of more than about 5 A. NOESY also can be used to detect chemical-exchange processes, such as of protons, between different sites in the molecules, because these provide ways of equilibrating disparities in magnetization.

**3. TOCSY.** Total correlation spectroscopy carries COSY a step further in identifying spins that have either smaller couplings or are part of a grouping so constituted that **multiple-quantum coherence** can be relayed from one nucleus to another then to a third and so on. TOCSY (or the closely related procedure HOHAHA, Homonuclear Hartmann-Hahn) can be used to delineate the spins belonging to particular amino-acid subunits in a peptide chain. TOCSY uses the Hartmann-Hahn condition (Section 8-3-3) and relatively long mixing times to achieve its total "correlations" of spins. A useful variation on TOCSY and HOHAHA has been described by Kupce and Freeman.[10]

**4. ROESY.** Rotating frame Overhauser enhancement spectroscopy also known

---

[10]E. Kupce and R. Freeman, *J. Am. Chem. Soc.*, **114**, 10671 (1993).

as CAMELSPIN fills an important gap in the measurement of nuclear Overhauser effects for large molecules, because of the interplay between the magnitude of the $B_0$ field and the molecular correlation time (see Section 6-3). ROESY, like TOCSY, uses the Hartmann-Hahn condition, but involves magnetization transfer rather than coherence transfer. The advantage for larger molecules is that the NOE is measured at lower field strengths with $T_{1P}$ (see Section 8-3-4) as the cross-relaxation time of interest.

5. **INADEQUATE.** Incredible natural-abundance double-quantum transfer experiment lives up to incredible in spades. The procedure provides a means of determining all of the one-bond $^{13}C$-$^{13}C$ coupling constants in a molecule. The natural-abundance of $^{13}C$ is only 1.1%, so the probability of adjacent $^{13}C$'s in the same molecule is on the order of 0.01%. This means that the sensitivity in the experiment is bound to be low (sample quantities can be critical) and, furthermore, there must be ultra-efficient filtering of the resonances of those $^{13}C$ carbons that are only connected to $^{12}C$ carbons. The magic of INADEQUATE is achieved through filtering double-quantum from single-quantum coherences by phase cycling.

The power of INADEQUATE lies in the way it shows which carbons are connected to which by C-C bonds. In principle, it can by itself establish a complex structure. Obviously, carbons such as methyl of a $CH_3$-O- group will show a $^{13}C$ peak, but give no cross peak in INADEQUATE.

6. **HMQC.** Heteronuclear Multiple Quantum Coherence spectroscopy has many variations, but is most useful as a procedure for detecting with enhanced sensitivity resonances of protons that are coupled to heteroatoms of low NMR sensitivity. For example, protons coupled to $^{15}N$, as in peptide linkages in proteins, can be detected separately from protons connected to $^{12}C$ or $^{13}C$ by first giving the protons a 90° pulse, waiting for the proton-$^{15}N$ vectors to evolve in a sequence like the one shown in Figure 7-17, then applying a 90° pulse to the $^{15}N$ to create antiphase proton magnetization, a $t_1$ waiting period that has in its center a 180° pulse and ends with a second 90° $^{15}N$ pulse has the result of filtering out all but the resonances of the protons attached to $^{15}N$. The $t_1$ period is varied to give a 2D plot with chemical-shift information.

7. **Proton Spectra without Spin-Spin Splittings.** The simplicity of the determination of chemical shifts in proton-decoupled $^{13}C$ spectra, such as shown in Figure 5-1, leads to possible suggestions of being able to do the same for proton spectra. In other words, can you decouple all of the protons from one another simultaneously? It almost sounds like a contradiction of terms because proton decoupling normally uses selective irradiation. But if you think about it and consider how selective proton transfer can be

achieved with hard pulses with INEPT (Section 7-3), you suspect that perhaps such spectra can be taken. R. Freeman, a Mozart of the NMR machine if there ever was one, has in fact shown how to do it.[11] Not by very simple means, but nonetheless with good linewidths and reasonable sensitivity. In the 2D spectra, when the stacked $F_2$ spectra are viewed straight on in the $F_1$ direction, one only sees the chemical shifts of completely decoupled proton peaks. In contrast, the 90° view along the $F_2$ direction shows all of the couplings.

## Exercises

**Exercise 8-1** Referring to Section 8-3-2, explain why a normally positive NOE becomes negative when dipolar relaxation dominates, but the viscosity of the medium is so high that $W_0 \gg W_2$.

**Exercise 8-2** Sketch out the features of the 2D COSY proton correlation that you would expect for the aldehyde with the structure shown below.

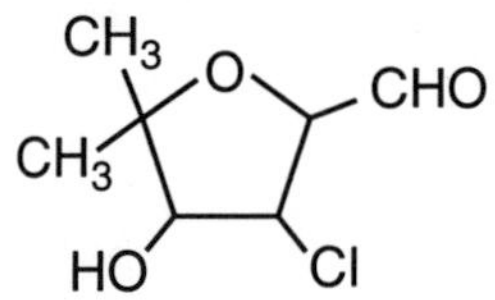

**Exercise 8-3** Explain how in a spin-lock experiment as described in Section 8-3-4 you would recool the proton reservoir.

**Exercise 8-4** Explain how for a $^{13}CH_3R$ system, you could spin lock the $^{13}C$ along the Y axis and, at essentially the same time, spin lock the protons along the X axis of their respective rotating frames.

**Exercise 8-5 a.** Devise a pulsed magnetic-field gradient experiment like the one shown in Figure 8-21 that would demonstrate the existence of the coherences possible for an AMX (all couplings small in comparison to the shift differences) proton spin system.
**b.** Explain qualitatively how you would expect the results of the same experiment would differ if, instead of AMX, one were to use an $A_2X$ system with energy levels as shown in Figure 7-18.

---

[11]P. Xu, Xi-Li and R. Freeman, *J. Am. Chem. Soc.*, **113**, 3596 (1991).

# Exercises

**Exercise 8-6** Draw out five isomeric structures of $C_4H_6O_2$ and explain how an INADEQUATE 2D spectrum might help you distinguish between them. Assume that the spectrum has $^{13}C$ shifts along the diagonal of a $F_1$, $F_2$ plot and $^{13}C$-$^{13}C$ couplings are off diagonal.

**Exercise 8-7** Devise suitable diagrams to show the information that you would expect to be obtained from a 3D proton-$^{13}C$-$^{15}N$ HETCOR spectrum using uniformly $^{13}C$ and $^{15}N$ labeled glycylglycine in aqueous solution with the indicated chemical shifts, in ppm with respect to the appropriate shift standards (assume rapid exchange of the $NH_3^+$ and NH protons).

$$\underset{349\quad 4.0}{\overset{42\quad 168}{H_3{}^{15}N{}^{13}CH_2{}^{13}C(O)}}\underset{260\quad 3.8}{\overset{45\quad 177}{{}^{15}NH{}^{13}CH_2{}^{13}CO_2^-}}$$

# 9
# Some Thoughts on Spin-Spin Splitting

An understanding of the quantum-mechanical basis for spin-spin splitting can be very helpful in structural analyses where spin-spin splittings are widely used to map out the arrangements of atoms in particular groups. In many cases, spin-spin splitting patterns conform to simple rules, as for example, the way that the protons of a -O-CH$_2$-CH$_3$ group appear in the NMR with the intensities of a 1:3:3:1 quartet and 3:6:3 triplet with equal line spacings. With most kinds of splitting complexities, it turns out that one or more chemical shifts are comparable in magnitude to one or more coupling constants. There are exceptions, as we shall see, but this is the common complicating factor, particularly in open-chain organic compounds.

In earlier times, when the usual magnetic field strengths were so low that protons came into resonance at 40 or 60 MHz, proton chemical shifts, measured in Hz, were very often comparable to proton-proton *J* values. And in those times, much effort was expended on spectral analysis, whereby elaborate procedures were developed to extract the chemical shifts and coupling constants, which often lead to very complex spectra. The effort expended was far from trivial; because, even a three-nucleus system can give spectra in which no simple measurement of a line spacing, or position of a line gives directly a coupling constant, or a chemical shift.

With the trend over recent years to utilization of higher-field, superconducting magnets, 10 to 15 times more intense than the earlier permanent or electromagnets, chemical shifts have increased greatly relative to coupling constants and much of the need has disappeared for detailed spectral analysis to obtain accurate couplings and chemical shifts. For this reason, you may wish to skip, or perhaps just skim, this chapter with the intent of returning later if you encounter systems that require more detailed understanding. However, in my opinion, much of the material is very intellectually interesting, in that it is possible to define the underlying parameters of a system of magnetic nuclei that shows literally hundreds of resonance lines so as to reproduce the positions and intensities within the experimental errors of measurement. We will consider here a number of aspects of how NMR spectra can become more complex than the simple rules suggest. At all times, we'll be dealing with nuclei of spin 1/2 (or zero). Extensions to nuclei of spin > 1/2 are certainly possible, but most cases are not very interesting, because quadrupole-induced relaxation (Section 6-4-1) is usually sufficiently rapid to cause the splittings to be washed out completely or to result in substantial line broadening.

## 9-1 What are our Specific Objectives?

The principal objective here is to introduce you to a qualitative quantum-mechanical analysis of a two-nucleus spin system, where the chemical-shift difference changes from zero to very large, relative to the coupling constant. If you persist in this, you will obtain a reasonable understanding (at least to the degree that quantum mechanics ever leads to real understanding) of many of the basics of spin-spin splitting.

With the two-spin system under our belts, we will next outline in greater detail than before the widely used alphabetical system for classifying spin systems in general. Then we will look into what is involved when you observe more lines than you might expect in situations where $\Delta v \sim J$. After that will come the concept and observable effects of negative $J$ values. Then, your intestinal fortitude will be tested by the fascinating cases where you get more lines than you expect, even when $\Delta v \ggg J$. If, after all of that, you are still on board, we will look at some situations that involve what are called **virtual couplings**, where you can make serious errors in structural analysis because line splittings may not correspond to couplings in ways that you would expect. Finally, we will take a quick look at how we can extract shift and coupling information from complex spectra. This is a heavy menu, but it does contain topics of general interest to those who plan to use NMR in a serious way.

## 9-2 The Two-Nucleus System

Figure 6-4 illustrates the energy levels and transitions that typify the simplest type of two-nucleus coupling system. There we made no note of potential complications that will arise when the coupling constant $J$ is comparable to the chemical-shift difference. Indeed, there was no need to, for the simple reason that $J$ was about 195 Hz, while the chemical-shift difference was about (600-150) = 450 MHz. So, in an intensity vs. frequency plot, we would expect a spectrum for $^{13}C$-labeled methanoate (formate anion), as shown in Figure 9-1.

This representation is somewhat inaccurate in one important respect, namely that the two $^1H$ resonances should be about 60 times more intense than the two $^{13}C$ resonances, because of the differences in magnitudes of the magnetic moments of the

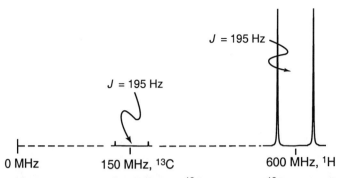

**Figure 9-1** Schematic representation of the proton and $^{13}C$ spectrum of $^{13}C$-labeled methanoate anion at 600 MHz. The intensity of the $^{13}C$ doublet is expected to be about 1/60th of the intensity of the proton doublet because of the smaller magnetic moment of $^{13}C$

two kinds of nuclei. Irrespective of that, the components of each of the individual doublets will have equal intensities to great precision.

However, when the chemical-shift difference $\Delta\nu$ between the nuclei becomes comparable to $J$, the intensities begin to deviate from simple expectations and, finally, when $\Delta\nu \ll J$, no splitting is observed (see Figure 9-2). This kind of result leads to the simple catechism that nuclei having the same, or very nearly the same, chemical shifts

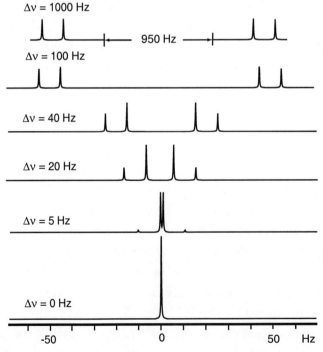

**Figure 9-2** Line positions and relative intensities for the resonances of a two-nucleus system (A and B) with $J = 10$ Hz and a selection of chemical-shift differences $\Delta\nu$ ranging from 1000 to 0 Hz

## 9-2 The Two-Nucleus System

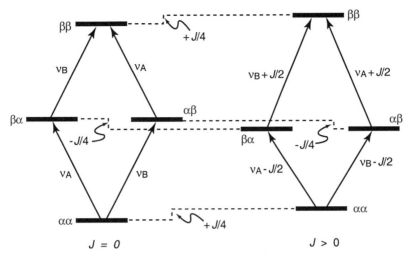

**Figure 9-3** Energy diagram that might be constructed to determine the transition energies for a system of two nuclei (A and B) with the same chemical shifts, i.e., $v_A = v_B$ and $\Delta v = 0$. On the left, $J$ is zero and, on the right, $J$ is greater than zero. The states $\alpha\alpha$ and $\beta\beta$ are raised in energy by $J/4$, while the states $\alpha\beta$ and $\beta\alpha$, have their energies lowered by $J/4$. The transitions shown are all single-quantum transitions, where $v_A$ and $v_B$ represent energy changes that should correspond to the frequencies of the different nuclei, respectively. The energy diagram suggests that, when $J$ is different from zero, a two-line spectrum will result with two coincident transitions at $v_A - J/2$ and $v_B - J/2$, along with two coincident transitions at $v_A + J/2$ and $v_B + J/2$. The inference then is contrary to experiment, in which it is found that two nuclei with the same chemical shift show no splitting of their resonances

do not split each other's resonance. However, it turns out that even this simple apple-pie "rule" has serious exceptions in more complex systems, but we shall leave these fascinating cases for later. If you are not familiar with the quantum-mechanical treatment of spin-spin splitting, you will have difficulties with the line spacings of Figure 9-2, especially if your eyes were sharp enough to notice that, when $J \sim \Delta v$, there were two line spacings that correspond directly to $J$, but **nothing** corresponds in a direct way to the chemical-shift difference.

Figure 6-5 gives no hint of why the states should change their properties so much when v approaches 0. Indeed, if we modify Figure 6-5 to have the chemical shifts be equal, we might come up with the energy levels and transitions shown on the right side of Figure 9-3, which suggests that there should be two transitions ($\alpha\alpha \rightarrow \alpha\beta$ and $\alpha\alpha \rightarrow \beta\alpha$) of equal energy, separated by $J$ from two other equal energy transitions ($\beta\alpha \rightarrow \beta\beta$ and $\alpha\beta \rightarrow \beta\beta$) (see the left side of Figure 9-4).

What is wrong with this diagram? Why is only one line observed? The answer is rooted in quantum mechanics and comes about because quantum mechanics does not allow $\alpha\beta$ and $\beta\alpha$ to have the same energy and, at the same time, exist as separate and

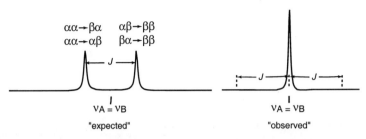

**Figure 9-4** The left side shows the transitions and energies expected for two, spin 1/2, nuclei with the same chemical shift and a coupling constant J on the basis of Figure 9-3. The right side shows the spectrum actually observed with vertical dashed lines at the positions that would be predicted for the outside lines (if they could be observed) by extrapolation of the spectral changes shown in Figure 9-2

distinct states. What this means is that αβ and βα can exist only as mixed states and, to a greater or lesser degree, this will be true of all αβ and βα states, depending on how large the chemical-shift difference Δν is with respect to the coupling constant J.

At this point, we need to go beyond αβ and βα as simple descriptions of nuclear spin orientations of particular states. We need to regard them instead as **quantum-mechanical wave functions**. In this role, αβ and its brethren can become explicit statements of the quantum-mechanical ψ functions (such as Equation 9-1), which are most interesting to us in the formulation that leads to calculation of the energy of a particular state E where H is a Hamiltonian operator and x, y and z are cartesian coordinates.

$$E = \int_{-\infty}^{\infty}\int_{-\infty}^{\infty}\int_{-\infty}^{\infty} \Psi \mathcal{H} \Psi^* \, dxdydz \, / \, \int_{-\infty}^{\infty}\int_{-\infty}^{\infty}\int_{-\infty}^{\infty} \Psi \Psi^* \, dxdydz \qquad \text{Eqn. 9-1}$$

This formidable expression takes on a much more benign appearance in our application, even if not really changed in content. For us, the complex conjugate ψ* equals ψ and the triple integrals are usually abbreviated so that, for the state αβ when Δν >> J, we write Equation 9-2, provided that <αβ|αβ> = 1. If you look back at Figures 6-1 and 6-2,

$$E = <\alpha\beta|\mathcal{H}|\alpha\beta>/<\alpha\beta|\beta\alpha> = -v_A/2 + v_B/2 - J/4 \qquad \text{Eqn. 9-2}$$

you will be able to see that we arrived at the same energy for αβ in a more qualitative way. The key elements in the equation for E is the **Hamiltonian operator** H and how it operates on the wave function αβ. It is not as complicated as it looks, but we will not go into these matters here (but see Appendix 4), but instead emphasize the results.[1]

Matters become more complicated when the states have to be mixed, as when Δν

---

[1] A discussion of the evaluation of the integrals is provided by Appendix 4 and by J. D. Roberts, "An Introduction to the Analysis of Spin-Spin Splitting in High-Resolution Nuclear Magnetic Resonance Spectra," W.A. Benjamin, Inc., New York, 1961; a book only slightly longer than its title.

## 9-2 The Two-Nucleus System

~ $J$. We will consider first the extreme case, where $\Delta v = 0$ and the quantum-mechanical treatment predicts observation of a single-frequency spectrum. The wave function for mixing $\alpha\beta$ and $\beta\alpha$ is $c_1\alpha\beta + c_2\beta\alpha$, where $c_1$ and $c_2$ are the coefficients that express the degree to which each starting wave function contributes. Only states of the same total quantum number mix and just as many mixed states will emerge as there were states that provided wave functions to be mixed. A restraint on the possible values of $c_1$ and $c_2$ is the requirement that $c_1^2 + c_2^2 = 1$ (this is called the **normalization** condition). When $\Delta v = 0$ and the energy of $\alpha\beta$ is equal to $\beta\alpha$, it is not unreasonable to expect that the absolute value of $c_1$ will equal that of $c_2$ after mixing of the two functions. This turns out to be the case and the separate mixed wave functions are:[1]

$$(1/\sqrt{2})\alpha\beta + (1/\sqrt{2})\beta\alpha \text{ or } 1/\sqrt{2}(\alpha\beta + \beta\alpha) \qquad \text{Eqn. 9-3}$$

$$(1/\sqrt{2})\alpha\beta - (1/\sqrt{2})\beta\alpha \text{ or } 1/\sqrt{2}(\alpha\beta - \beta\alpha) \qquad \text{Eqn. 9-4}$$

where for Equation 9-3, $c_1 = c_2 = 1/\sqrt{2}$. For Equation 9-4, $c_1 = 1/\sqrt{2}$ and $c_2 = -1/\sqrt{2}$. You will see that the requirement $c_1^2 + c_2^2 = 1$ is satisfied for both cases, because $(1/\sqrt{2})^2 + (1/\sqrt{2})^2 = 1$ and $(1/\sqrt{2})^2 + (-1/\sqrt{2})^2 = 1$.

The real importance of these two mixed wave functions is the difference in algebraic sign. Significant information about the character of these, and other, wave functions comes from interchanging the numbering of the nuclei. Clearly, changing $\alpha(1)\alpha(2)$ to $\alpha(2)\alpha(1)$ does not change anything and the same will be true for $\beta\beta$. Also, with $1/\sqrt{2}(\alpha\beta + \beta\alpha)$, the interchange gives $1/\sqrt{2}[\alpha(2)\beta(1) + \beta(2)\alpha(1)]$, which when rearranged gives $1/\sqrt{2}[\beta(1)\alpha(2) + \alpha(1)\beta(2)]$. This will be seen to be equivalent to the original function. Whenever the interchange of nucleus numbering does not change the sign of the function, it is a **symmetric wave function**. Thus, $\alpha\alpha$, $\beta\beta$ and $1/\sqrt{2}(\alpha\beta + \beta\alpha)$ are all symmetric functions. The situation with the remaining function is different. Interchanging the numbering of $1/\sqrt{2}(\alpha\beta - \beta\alpha)$ gives $1/\sqrt{2}[\alpha(2)\beta(1) - \beta(2)\alpha(1)]$ and this rearranges to $1/\sqrt{2}[\beta(1)\alpha(2) - \alpha(1)\beta(2)]$, which has the overall result of producing a change in sign of the function. This result leads to classification of $1/\sqrt{2}(\alpha\beta - \beta\alpha)$ as an **antisymmetric wave function**. An important conclusion provided by quantum mechanics is that transitions between states with symmetric wave functions to states with antisymmetric wave functions are strictly forbidden. This means that a system of two equivalent coupled nuclei have a state to, and from, which no NMR transitions at all are allowed! Further meat in this coconut comes from the fact that the symmetric state $1/\sqrt{2}(\alpha\beta + \beta\alpha)$ is destabilized by $J/4$, while the antisymmetric state $1/\sqrt{2}(\alpha\beta - \beta\alpha)$ is stabilized by $3J/4$. The resulting sequence of states, their energies and the possible

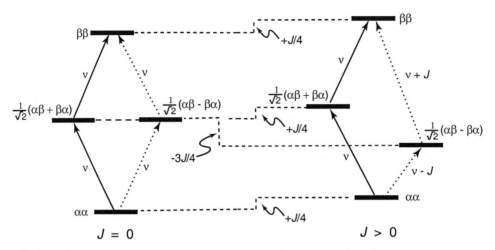

**Figure 9-5** Energy diagram for two-nucleus system with $\Delta\nu = 0$ that takes into account the quantum-mechanical mixing of the $\alpha\beta$ and $\beta\alpha$ states. On the left, $J$ is zero and, on the right, $J$ is greater than zero. If $J = 0$, two new states emerge with the same energy. One is the **symmetric** state $1/\sqrt{2}(\alpha\beta + \beta\alpha)$ and the other is the **antisymmetric** state $1/\sqrt{2}(\alpha\beta - \beta\alpha)$. If $J > 0$, the symmetric state $1/\sqrt{2}(\alpha\beta + \beta\alpha)$ is destabilized by $J/4$, just as are the $\alpha\alpha$ and $\beta\beta$ states. Therefore, the one-quantum transitions between the states have the energy $\nu$. When $J > 0$, the antisymmetric state $1/\sqrt{2}(\alpha\beta - \beta\alpha)$ is stabilized by $3J/4$. This might be expected to lead to two transitions, one at $\nu + J$ and the other at $\nu - J$. However, transitions from the symmetric states to, and from, the antisymmetric state are forbidden so that the transitions at $\nu + J$ and $\nu - J$ do not show up in the actual spectrum. Only the two equal, single-quantum symmetric-to-symmetric transitions are observed. These have just the intensities to disguise the fact that the transitions to one magnetic state of the system are not possible. The quantum-mechanical treatment is necessary to explain why two nuclei that have $\Delta\nu = 0$ and $J > 0$ show only a single resonance line

i transitions are shown in Figure 9-5. From this figure, you will see that there are just two allowed transitions with the frequency $\nu$. Interestingly, if the transitions from symmetric to antisymmetric were allowed, there would be two more transitions separated by $J$ from $\nu$.

With just two symmetric-to-symmetric transitions allowed, instead of the usual four, you might think that the intensity of NMR absorption would only be one half of the total intensity observed for the four-line spectra when $\Delta\nu \gg J$. Surprisingly, this is not the case. The transition probability of the symmetric-to-symmetric transitions increases as $\Delta\nu$ approaches zero and, when $\Delta\nu = 0$, each of the allowed transitions has twice the normal intensity. The result is that the existence of the antisymmetric state, to and from which no transitions are allowed, is wholly disguised. Perhaps it is really not there? However, remember that, even in the presence of high magnetic fields, there is very little difference in the populations of the various possible magnetic states, so we have to predict that, at equilibrium, there should be just 25% of the antisymmetric state

## 9-2 The Two-Nucleus System 219

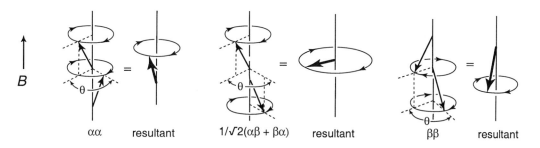

**Figure 9-6** Representation of the symmetric states of a system of two equivalent nuclei. The key idea here is that, because the nuclei have the same precession frequency, their precessional phase angles will remain constant. For each state, there is a representation of the separate precessions of the nuclei and the resultant sum (not to scale) for some phase angle θ. If θ is small, it should be easy to see how the resultant can be large and interact very effectively with an oscillator with the same frequency

and 75% of the three symmetric states. The simplest possible case of two equivalent nuclei is dihydrogen $H_2$. This substance can be separated into two components, one called *ortho*-hydrogen and the other called *para*-hydrogen. They occur in the ratio 75:25, respectively and it can be shown that *ortho*, but not *para*, gives an NMR signal. The *para* form can be converted to *ortho* by shuffling the nuclei around, as with a metal catalyst. In effect, you can think of this process as breaking the H-H bond and recombining the atoms in a new array of magnetic states. The eventual result will be equilibration of the states to the normal 75:25 ratio of symmetric states to antisymmetric states.

How can it be that a system of magnetic nuclei does not give an NMR signal? A very qualitative, but perhaps useful, explanation takes the crux of the matter to be the phase relationships between the precessing nuclei and recognizes that, if $\Delta \nu$ is precisely zero, these phase relationships will be constant. Let us represent the states of a pair of precessing nuclei with $\Delta \nu = 0$ by the kind of diagrams we have used before. First, the symmetric states represented in Figure 9-6. Because the two nuclei involved here will have the same precession frequency, they will precess without changing their phase relationship to one another. For the symmetric states, the nuclei behave as if there is a constant phase angle θ between their precessions and the magnitude of θ is such as to make the destabilizing interaction between the nuclei equal to $J/4$.

The value of these diagrams is to show that there is a straightforward expectation that an rf oscillator with a frequency at, or near to, ν can "get a grip" on the resultant magnetic moments of the symmetric states that are precessing at the frequency ν. Furthermore, the picture provides a qualitative rationale for the increased intensity of the allowed transitions because the oscillator acts on a constant **resultant** of two nuclear moments, which interaction should be more effective than an interaction involving a pair

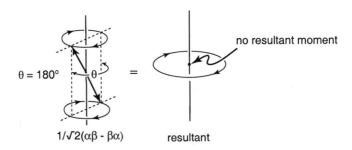

**Figure 9-7** Representation of the antisymmetric state of a system of two equivalent nuclei. Again, as in Figure 9-6, the key idea is that, because the states have the same precession frequency, their precessional phase angle will remain constant. Here, the nuclear moments are represented as having a constant phase angle θ of 180°. In this situation, the moments will cancel one another and there will be no resultant magnetization

of moments that, with a chemical-shift difference, are going in- and out-of-phase with one another.

What about the antisymmetric state? Because it gives no NMR signal, the inference is that it has **no resultant** magnetic moment. This can be readily envisioned, if the two nuclear moments precess at a constant phase angle of 180° so that their moments cancel one another, as shown in Figure 9-7. It is not unreasonable that, at a constant phase angle of 180°, the magnetic interaction between the nuclei would be stabilizing and at $-3J/4$ be larger than the usual $-J/4$ for αβ states with a large chemical-shift difference.

So, with the aid of the concepts of symmetric and antisymmetric nuclear magnetic states, as is dictated by the quantum-mechanical treatment, we can rationalize how it might be that there no observed splitting when $\Delta v = 0$, as well as the extra intensity of those transitions that are observed. The same general kind of treatment applies to electrons, so a system of two equivalent electrons has three symmetric electronic (and magnetic) states called **triplet states** along with a nonmagnetic **singlet state**.

What happens when $\Delta v$ is non-zero and large enough to be comparable to $J$? Here, the important idea is that, unlike pregnancy, it is possible for states to be just a little symmetric or just a little antisymmetric. This might be clearer, if you look back at the diagrams for the symmetric and antisymmetric states with the wave functions $1/\sqrt{2}(\alpha\beta + \beta\alpha)$ and $1/\sqrt{2}(\alpha\beta - \beta\alpha)$. If $\Delta v$ is different from zero, the phase angles between the precessing nuclei will not be constant. If $\Delta v$ is small, say 1 Hz, then once every second, we might expect the phase angle to be 180° and the state will be effectively antisymmetric and, at other times, when the phase angle is not 180°, symmetric. However, the situation is not that simple. What we find is that the states αβ and βα, when

$\Delta v$ is not equal to zero, "mix" to give composite wave functions of the types shown in Equations 9-5 and 9-6

$$\Psi_s = c_1\alpha\beta + c_2\beta\alpha \quad \text{symmetric - like} \qquad \text{Eqn. 9-5}$$

$$\Psi_A = c_2\alpha\beta - c_1\beta\alpha \quad \text{antisymmetric - like} \qquad \text{Eqn. 9-6}$$

Here, as before, $c_1^2 + c_2^2 = 1$. Clearly, if $c_1$ approaches zero, and this is the situation when $\Delta v \gg J$, the states then become essentially $\beta\alpha$ and $\alpha\beta$, as in our earlier formulations.

The line positions and relative intensities for the two-nucleus case with $\Delta v$ more or less comparable to $J$ can be calculated by Equations 9-7 to 9-9, where the conventions are $\Delta v = v_B - v_A$ and $v_{AB} = (v_A + v_B)/2$ and the line positions are as shown:

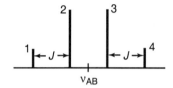

$$v_3 - v_1 = v_4 - v_2 = [(\Delta v)^2 + J^2]^{1/2} \qquad \text{Eqn. 9-7}$$

$$Q = J / \{\Delta v + [(\Delta v)^2 + J^2]^{1/2}\} \qquad \text{Eqn. 9-8}$$

$$\text{Intensity}(2)/\text{Intensity}(1) = I_2/I_1 = I_3/I_4 = [(1+Q)/(1-Q)]^2 \qquad \text{Eqn. 9-9}$$

You will see for the two-nucleus system that, wherever $J$ appears in the above equations, it appears as $J^2$. One critical significance of this is that changing the sign of $J$ does not change anything observable, when the spectra are taken in the usual way.

## 9-3 Classification of NMR Spin Systems

There is much variety in the structural units that give NMR spectra. However, many of these represent arrangements of spins that have common elements and can be analyzed in common ways. An important element in discussions of these entities is to have a useful classification system. The key part of the most widely used system is to provide for differentiation between systems where $\Delta v \gg J$, ones where $\Delta v \sim J$ and ones where $\Delta v = 0$. For the first type, where $\Delta v \gg J$, a two-nucleus system would be designated as AX, that is with letters far apart in the alphabet. Then, when $\Delta v \sim J$, we use AB (letters close together in the alphabet) and finally when $\Delta v = 0$, the designation would be $A_2$. Extensions are fairly obvious: ABC will be three nuclei with similar, but not identical shifts; AMX represents three nuclei with dissimilar shifts; ABX is two nuclei with similar shifts and one with a very different shift. For these cases, we can expect the various couplings, $J_{AB}$, $J_{AC}$, $J_{BC}$ for ABC, $J_{AM}$, $J_{AX}$, $J_{MX}$ for AMX and $J_{AB}$, $J_{AX}$, $J_{BX}$ for ABX.

An added level of complexity is possible with three or more nuclei, when two nuclei have the same shift, but are also coupled differently to another nucleus. This is not a common situation with three nuclei, but we can contrive a case in the form of 1,2,3,4-tetrachlorobenzene-1-$^{13}$C.

Here, except for a very small chemical-shift difference between H5 and H6, (because H5 is flanked by two $^{12}$C atoms, while H6 is flanked by a $^{12}$C and a $^{13}$C), the system might be taken as $A_2X$, with X = $^{13}$C. However, $^2J_{H6^{13}C}$ will certainly be different than $^3J_{H5^{13}C}$ because one is a two-bond coupling while the other is a three-bond coupling. Therefore we designate this system to a high level of approximation as AA'X, where A and A' have the same shift, but are coupled differently to X.

Two geometrically different $A_2X_2$ arrangements need to be distinguished in the classification scheme. 1,1-Difluoro-1,2-propadiene is pure $A_2X_2$, because the $J_{HF}$ couplings are all equal.

In contrast, 1,1-difluoroethene has two equal cis-$J_{HF}$ and two equal trans-$J_{HF}$ couplings. There is no reason to expect that cis-$J_{HF}$ should necessarily be equal to trans-$J_{HF}$. The proper designation for a compound with this kind of symmetry is therefore AA'XX'.

We shall return to the consequences for spin-spin splitting attendant to these systems shortly.

## 9-4 More Lines than Expected

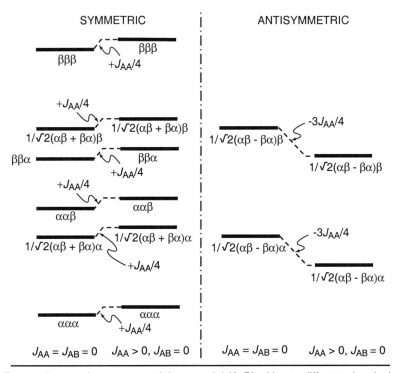

**Figure 9-8** Energy diagram for a system of three nuclei ($A_2B$) with two different chemical shifts, *i.e.* $\nu_A$ and $\nu_B$. On the far left, are the symmetric states with $J_{AA} = J_{AB} = 0$ and, next to them to their right, is shown what happens when $J_{AA} > 0$ and $J_{AB} = 0$. Here, **each symmetric state** is raised in energy by $+J_{AA}/4$. In contrast, on the right side of the figure, we see that the **antisymmetric states** are each lowered in energy by $-3J_{AA}/4$. The importance of this figure is that it shows that there is no differential effect of $J_{AA}$ on the energies of the symmetric states. This is also true for the antisymmetric states. The result is that we can take $J_{AA}$ as a constant factor and omit it from our later figures dealing with $A_2B$ systems. However, it would be a mistake to believe that all couplings analogous to $J_{AA}$ are unimportant in NMR spectra. We will consider cases where they are important later

### 9-4 More Lines than Expected

The simplest case in which more lines can be observed than anticipated from simple counting on your fingers is $A_2B$. This case provides a worthwhile extension of the energy diagram of the possible magnetic states arrived at earlier for two nuclei, and it will show you the general manner of how one goes about treating more complex systems. For $A_2B$, there will be two chemical shifts $\nu_A$ and $\nu_B$ along with two couplings $J_{AA}$ and $J_{AB}$ ($J_{BA}$ is here by definition equal to $J_{AB}$). Let us assume $\nu_A < \nu_B$.

The eight possible states are now $\alpha\alpha\alpha$, $1/\sqrt{2}(\alpha\beta + \beta\alpha)\alpha$, $1/\sqrt{2}(\alpha\beta - \beta\alpha)\alpha$, $\alpha\alpha\beta$, $\beta\beta\alpha$, $1/\sqrt{2}(\alpha\beta + \beta\alpha)\beta$, $1/\sqrt{2}(\alpha\beta - \beta\alpha)\beta$ and $\beta\beta\beta$. The left-side of Figure 9-8 shows the arrangement of the resulting magnetic states when the two possible couplings $J_{AA}$ and $J_{AB}$ are both zero. First, let us consider the situation when $J_{AA}$ is greater than zero,

**Table 9-1**

Transitions for the $A_2B$ System

| Beginning state | Final state | Energy change |
|---|---|---|
| $1/\sqrt{2}\,(\alpha\beta-\beta\alpha)\alpha$ | $\rightarrow$ $1/\sqrt{2}\,(\alpha\beta-\beta\alpha)\beta$ | $\nu_B$ |
| $\alpha\alpha\beta$ | $\rightarrow$ $\beta\beta\alpha$ | $2\nu_A-\nu_B$ |
| $\alpha\alpha\alpha$ | $\rightarrow$ $1/\sqrt{2}\,(\alpha\beta+\beta\alpha)\alpha$ | $\nu_A$ |
| $\alpha\alpha\alpha$ | $\rightarrow$ $\alpha\alpha\beta$ | $\nu_B$ |
| $1/\sqrt{2}\,(\alpha\beta+\beta\alpha)\alpha$ | $\rightarrow$ $\beta\beta\alpha$ | $\nu_A$ |
| $1/\sqrt{2}\,(\alpha\beta+\beta\alpha)\alpha$ | $\rightarrow$ $1/\sqrt{2}\,(\alpha\beta+\beta\alpha)\beta$ | $\nu_B$ |
| $\alpha\alpha\beta$ | $\rightarrow$ $1/\sqrt{2}\,(\alpha\beta+\beta\alpha)\beta$ | $\nu_A$ |
| $\beta\beta\alpha$ | $\rightarrow$ $\beta\beta\beta$ | $\nu_B$ |
| $1/\sqrt{2}\,(\alpha\beta+\beta\alpha)\beta$ | $\rightarrow$ $\beta\beta\beta$ | $\nu_A$ |

but $J_{AB}$ is equal to zero. The result is also shown in Figure 9-8. Here, you will see that each symmetric state is raised in energy by $J_{AA}/4$, while each antisymmetric state is lowered by $3J_{AA}/4$. We will be concerned about mixing of states farther along, but we can be sure that the symmetric and antisymmetric states will not mix. For this reason, we can disregard $J_{AA}$ because it has a constant effect on the symmetric states, as well as a constant, but different, effect on the antisymmetric states. So we now can construct Figure 9-9 in which we omit consideration of $J_{AA}$.

If you have followed us through Figure 9-3, you will see that there is not much new in the energy sequences of the symmetric states on the far left side of Figure 9-9, where $J_{AB} = 0$. The exceptions are a possible one-quantum **combination** transition from $\alpha\alpha\beta \rightarrow \beta\beta\alpha$ and a clearly allowed antisymmetric-to-antisymmetric transition from $1/\sqrt{2}(\alpha\beta - \beta\alpha)\alpha \rightarrow 1/\sqrt{2}(\alpha\beta - \beta\alpha)\beta$ that will have the frequency $\nu_B$ and will be wholly unaffected by mixing. There will be the seven other symmetric-to-symmetric one-quantum transitions listed in Table 9-1 and you should verify these to become familiar with the system.

All but the combination transition are allowed and, when $\nu_A \ll \nu_B$ with $J_{AB} = 0$, we expect to have four A transitions at $\nu_A$ and four B transitions at $\nu_B$. Because transitions of the type $\alpha \rightarrow \beta$ will have half the intensity of the type $\alpha\alpha \rightarrow 1/\sqrt{2}(\alpha\beta + \beta\alpha)\beta$ (Section 9-2), the total intensities at $\nu_A$ and $\nu_B$ will be in the ratio 2:1, see Figure 9-10.

## 9-4 More Lines than Expected

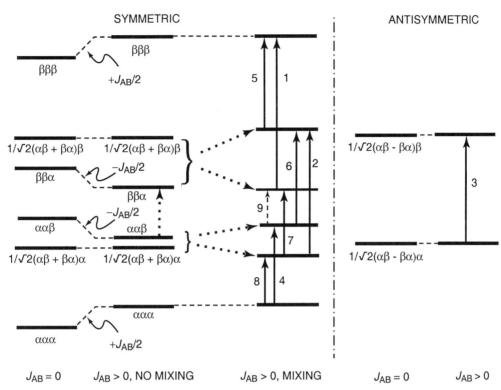

**Figure 9-9** Energy diagram for a system of three nuclei ($A_2B$) with two different chemical shifts, *i.e.* $v_A$ and $v_B$. On the far left, are the symmetric states with $J_{AB} = 0$ (we ignore $J_{AA}$ in this diagram) and, next to them to their right, is shown the result of the changes when $J_{AB} > 0$. The consequence is that $\alpha\alpha\beta$ and $\beta\beta\alpha$ are both decreased in energy by $-J_{AB}/2$, while $1/\sqrt{2}(\alpha\beta + \beta\alpha)\alpha$ and $[1/\sqrt{2}(\alpha\beta + \beta\alpha)\beta$ are unchanged. This increases the extent of mixing for one pair of states, but decreases it for the other. The four new mixed states that emerge are involved in **all of the symmetric-to-symmetric transitions.** Consequently, none of the energies of these transitions are predictable from simple theory. In contrast, on the right side of the figure, we see that the **antisymmetric states** are unchanged by $J_{AB} > 0$, and the one **antisymmetric-to-antisymmetric** transition is simply $v_B$. The other important thing is that mixing makes transition **9** allowed, at least when $J_{AB}$ is comparable to $v_A$ and $v_B$, although it is always very weak, as can be seen in Figure 9-10

When $J_{AB} > 0$ (this corresponds to the second sequence of energies from the left in Figure 9-9), then we see $\alpha\alpha\alpha$ and $\beta\beta\beta$ raised in energy by $J_{AB}/2$ and this is reasonable because there are two interactions between nucleus (3) and the equivalent nuclei (1) and (2). The states $\alpha\alpha\beta$ and $\beta\beta\alpha$ are both reduced in energy by $J_{AB}/2$ and this also fits with earlier expectations. The situation with states like $1/\sqrt{2}(\alpha\beta +\beta\alpha)\alpha$ is more complex. It turns out that $J_{AB}$ does not affect their energies and the simple answer as to why this is so, is that equivalent nuclei as $\alpha\beta$, or $\beta\alpha$, interacting with a third nucleus $\alpha$, will have one positive and one negative interaction. These interactions will be equal and cancel each other because each interaction involves $J_{AB}$.

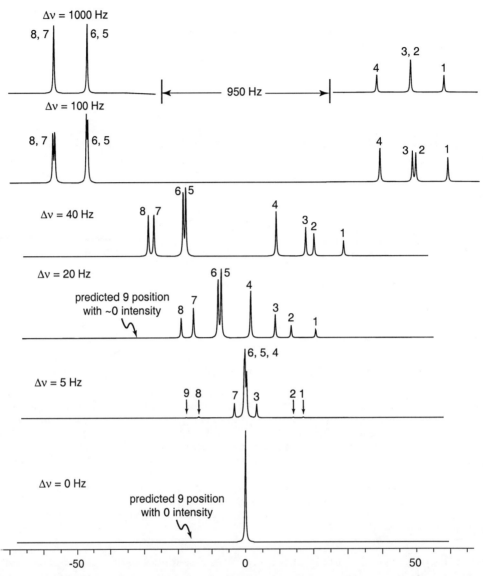

**Figure 9-10** Line positions and relative intensities for the resonances of $A_2B$ spin systems with $J = 10$ Hz and a selection of chemical-shift differences $\Delta\nu$ ranging from 1000 to 0 Hz. The numbering of the transitions corresponds to that shown in Figure 9-9. Note how the combination transition is not observed when the shift difference is either very large or very small

Now, if there is no mixing of states, but $J_{AB}$ is greater than zero, we can work out the transitions as shown at the top of Figure 9-10 for the case of $J_{AB} = 10$ Hz and $\Delta\nu = \nu_B - \nu_A = 1000$ Hz. There are two transitions with energies of $\nu_A - J_{AB}/2$ (**7** and **8**) and two with $\nu_A + J_{AB}/2$ (**5** and **6**). Then there are two transitions at $\nu_B$ (**2** and **3**), and one at

$v_B + J$ (**1**) and one at $v_B + J$ (**4**). The result is the simple 1:1 doublet and the 1:2:1 triplet expected from counting on your fingers.

Mixing produces interesting changes. First, you must recognize that if $v_A = v_B$, which is the $A_3$ system, then the states $1/\sqrt{2}(\alpha\beta + \beta\alpha)\alpha$ and $\alpha\alpha\beta$ disappear and we have then to derive mixed states from $\alpha\alpha\beta$, $\alpha\beta\alpha$ and $\beta\alpha\alpha$. These turn out to be of the form $(c_1\alpha\alpha\beta \pm c_2\alpha\beta\alpha \pm c_3\beta\alpha\alpha)$. Because all of the nuclei will be equivalent, there will be no $J_{AB}$, only $J_{AA}$. Irrespective of the magnitude of $J_{AA}$, the system will give a single line at $v_A$. The message of the $A_3$ system is, if we start with an $A_2B$ system with $v_A < v_B$ and reduce $v_B$ until $v_A - v_B \sim J_{AB}$, mixing of the states $1/\sqrt{2}(\alpha\beta + \beta\alpha)\alpha$ and $\alpha\alpha\beta$, and also of $1/\sqrt{2}(\alpha\beta + \beta\alpha)\beta$ with $\beta\beta\alpha$, will become important. This may seem odd, but remember that the shifts of all three nuclei are coming closer together and approximating the $A_3$ case.

Now, if you look at Figure 9-9, you will see that $J_{AB}$ brings the members of one pair of states closer together and pushes the members of the other pair farther apart. The rule is that, the mixing coefficients being the same, the closer the states are together in energy, the more mixing there will be. Thus, the states that emerge, when two states of nearly equal energy are mixed, will be separated much more by mixing than the original states, as compared to mixing of two states of quite different energies. This fact is reflected in the center column of energy states in Figure 9-9. One interesting result of mixing is that all of the transition energies, except for transition **3**, are changed by mixing. The most significant change for us here, caused by mixing, relates to the combination transition **9**, which is forbidden in the absence of mixing and does not exist in the $A_3$ extreme, but although always very weak, should be observable when $\Delta v \sim J_{AB}$, see Figure 9-10. Here then, we have more lines than expected from simple theory, as a consequence of mixing of states. This is not at all an unusual occurrence in more complex spin systems and combination transitions can be expected whenever chemical-shift differences are comparable to the couplings.

## 9-5 Positive versus Negative J Values

We have already shown at the end of Section 9-2 that the sign of $J_{AB}$ does not affect the line positions or intensities of AB, no matter how large, or how small, the chemical-shift difference. This is somewhat less obvious for the $A_2B$ system. However, if you look carefully at Figure 9-9, you will see that reversing the sign of $J_{AB}$ will cause a symmetrical set of changes. Thus, the transitions **6** and **5** trade their energy changes with **7** and **8**, respectively. The same is true of **1** and **4**, but **3**, **2** and **9** will not change. The labeling of the lines will be different in Figure 9-10, but the positions and intensities will not change. So both AB and $A_2B$ are insensitive to whether $J$ is positive or negative.

To have different transition energies for nuclear spin systems, because of changing the sign of one of more coupling constants, requires that there be at least two couplings that, unlike $J_{AB}$ in the $A_2B$ system, can cause the spectrum to change. The simplest system in which this happens is ABX and it involves the two couplings $J_{AX}$ and $J_{BX}$. The special role of $J_{AB}$ in this kind of system will be discussed a little later on. The states for ABX are like those for $A_2B$, except that there will be no fully symmetric, and no fully antisymmetric, states as long as $\Delta \nu_{AB} > 0$, as well as $J_{AX}$ and $J_{BX} \neq 0$. We will use as our starting place the energy levels on the left-hand side of Figure 9-11, where $J_{AX} = J_{BX} = 0$.

Now, if we let $J_{AX} \sim J_{BX}$, and both are positive, we get the left-of-center sequence of energy states shown in Figure 9-11. This sequence was generated using the concept that, for the $\alpha\alpha\beta$ state, simple spin-spin interactions decrease the energy by $J_{AX}/4$ and $J_{BX}/4$. This is because both the $\alpha(1)\ldots\beta(3)$ and the $\alpha(2)\ldots\beta(3)$ interactions are expected to be stabilizing. Similarly for $\alpha\alpha\alpha$, both the $\alpha(1)\ldots\alpha(3)$ and $\alpha(2)\ldots\alpha(3)$ interactions should be destabilizing. For $\beta\alpha\beta$, $\beta(1)\ldots\beta(3)$ will be destabilizing (+ $J_{AX}/4$) and $\alpha(2)\ldots\beta(3)$ stabilizing (- $J_{BX}/4$). You should satisfy yourself that these considerations were correctly applied in Figure 9-11. Now, $\alpha\beta\alpha$ will mix with $\beta\alpha\alpha$ and $\alpha\beta\beta$ will mix with $\beta\alpha\beta$. There will be no significant mixing of $\alpha\alpha\beta$ and $\beta\beta\alpha$, because these states have different total magnetic quantum numbers (+1/2 and -1/2).

At this point, let us see what happens when the sign of $J_{AX}$ but not $J_{BX}$, is reversed. The right-of-center sequence of Figure 9-11 was derived on the basis that, if $J_{BX}$ is negative, the $\alpha(2)\ldots\beta(3)$ interaction will be destabilizing and $\alpha(2)\ldots\alpha(3)$ will be stabilizing. The contrast with $J_{AX}$ and $J_{BX}$ both being positive is striking. Clearly, $\alpha\beta\alpha$ and $\beta\alpha\alpha$ will mix far less if $J_{BX}$ is negative, while $\alpha\beta\beta$ and $\beta\alpha\beta$ should mix only slightly less. The message is clear, the change in sign will drastically change the transition energies and this is also illustrated in Figure 9-11. The spectral differences for a typical set of parameters is shown in Figure 9-12. The question of what happens when both $J_{AX}$ and $J_{BX}$ are negative is interesting and I leave it to you to investigate that possibility. Also, could you tell whether it is the sign of $J_{AX}$ or the sign of $J_{BX}$ that is reversed?

A very interesting situation can arise with ABX spectra because there are combinations of coupling constants where both $J_{AX}$ and $J_{BX}$ are positive and where one is positive and the other negative, but can give **exactly the same line positions**. The conditions for this happening are that the values of $J_{AB}$, $|J_{AX} + J_{BX}|$ and of $|(J_{AX} - J_{BX})/2| + |\nu_A - \nu_B|$ be the same for each combination of parameters. Figure 9-13 shows

## 9-5 Positive versus Negative J Values

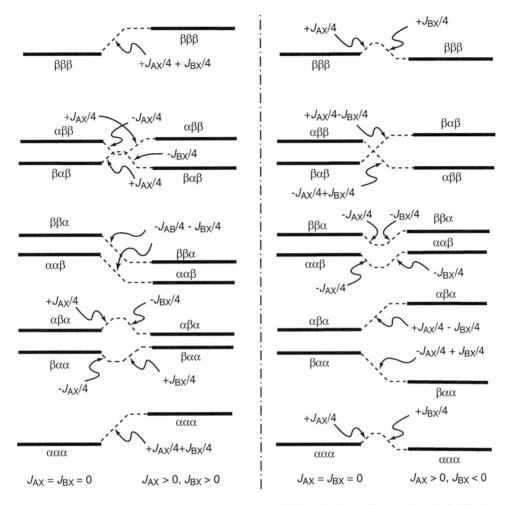

**Figure 9-11** Energy diagram for a system of three nuclei (ABX) with three different chemical shifts, *i.e.* $\nu_A$, $\nu_B$ and $\nu_X$. The point of this figure is to show that changing the signs, but not the absolute values, of $J_{AX}$ and $J_{BX}$, can cause drastic changes in the spectrum. On the far left, is shown the sequence of states when $J_{AX} = J_{BX} = 0$. We are cheating a bit here by neglecting $J_{AB}$; because, unlike the $A_2B$ case, $J_{AB}$ can be very important in determining the transition energies, as we shall see shortly. However, if $J_{AB}$ is relatively small compared to $J_{AX}$ and $J_{BX}$, the effects will not be large and our diagram will be much simpler. Moving to the right, we see the effect on the energies of the states when $J_{AX} > 0$ and $J_{BX} > 0$ ($J_{AX}$ is taken to be slightly smaller than $J_{BX}$). There will be mixing of states $\beta\alpha\alpha$ and $\alpha\beta\alpha$, as well as of $\beta\alpha\beta$ and $\alpha\beta\beta$. Now, on the far right, you can see that, when $J_{AX} > 0$ and $J_{BX} < 0$, without change in the absolute magnitudes, the energies of $\beta\alpha\alpha$ and $\alpha\beta\alpha$, as well as $\beta\alpha\beta$ and $\alpha\beta\beta$, become quite different from the situation where both couplings have the same sign. Consequently, the spectra corresponding to these differences will be quite different, see Figure 9-12. (Be sure that you understand that, a negative $J_{BX}$, $+ J_{BX}/4$ corresponds to **stabilization** of the state involved)

an example and you will see that, while the line positions and intensities are identical in both the A and X parts of the spectrum, in the X part, the intensities are different.

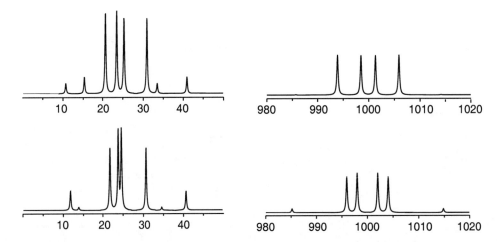

**Figure 9-12** Comparison of two ABX spectra that only differ in the signs of the AX and BX coupling constants. The result is in accord with the predictions of Figure 9-11 that quite different line positions would be expected. The spectral parameters that give these two spectra are for the top spectrum: $v_A = 20$ Hz, $v_B = 30$ Hz, $v_X = 1000$ Hz, $J_{AB} = 10$ Hz, $J_{AX} = 4$ Hz and $J_{BX} = 8$ Hz; and, for the lower spectrum, are the same, except that $J_{AX} = 4$ Hz and $J_{BX} = -8$ Hz

Indeed, with couplings of opposite sign, two X combination transitions emerge. So, this is a case where we can tell whether $J_{AX}$ and $J_{BX}$ have the same, or opposite, signs by analysis of the pattern of intensities of the X lines.

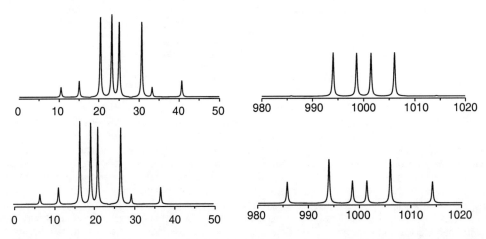

**Figure 9-13** Examples of ABX spectra that have exactly the same line positions but have very different line intensities in the X part of the spectrum. The spectral parameters that give these two spectra are for the top spectrum: $v_A = 20$ Hz, $v_B = 30$ Hz, $v_X = 1000$ Hz, $J_{AB} = 10$ Hz, $J_{AX} = 4$ Hz, $J_{BX} = 8$ Hz; and for the lower spectrum: $v_A = 20$ Hz, $v_B = 22$ Hz, $v_X = 1000$ Hz, $J_{AB} = 10$ Hz, $J_{AX} = 4$ Hz, $J_{BX} = -16$ Hz. The criterion for this kind of situation is that $J_{AB}$, $|J_{AX} + J_{BX}|$ and $|(J_{AX} - J_{BX})/2| + |v_A - v_B|$ be the same for both sets. Thus, for one set, we have $|(4 - 8)/2| + |20 - 30| = 12$ and, for the other, $|4 - (-16))/2| + |20 - 22| = 12$

## 9-5 Positive versus Negative J Values

While we have said much about the way that negative coupling constants can influence NMR spectra, we have said nothing about the molecular situations in which they occur. First, we might ask how is it that a coupling can be negative. This is not so easy to understand. You may remember how, Section 6-4-3, we drew an analogy between the interaction of two magnetic nuclei and two simple bar magnets. Thus,

$$\uparrow \uparrow \sim \begin{array}{|c|c|} \hline N & N \\ \hline S & S \\ \hline \end{array}$$

represented extra destabilization of the $\alpha\alpha$ state, while

$$\uparrow \downarrow \sim \begin{array}{|c|c|} \hline N & S \\ \hline S & N \\ \hline \end{array}$$

represented extra stabilization of the $\alpha\beta$ state. These interactions we took as the result of positive couplings.

Clearly, if we were able to insert between our nuclear magnets some suitable polarized magnetic substance, it might change the orientations that correspond to stabilization, or destabilization, of the states.

$$\uparrow \begin{array}{|c|c|} \hline S & S \\ \hline N & N \\ \hline \end{array} \uparrow \sim \begin{array}{|c|c|c|c|} \hline N & S & S & N \\ \hline S & N & N & S \\ \hline \end{array}$$

bonding electrons

At the molecular level, the polarizable magnetic medium is constituted of the pairs of bonding electrons. If we assume that the electrons, which themselves have large magnetic moments but, although normally paired in chemical bonds, can interact with magnetic nuclei in such a way to have some mixing of states with pairing of the electron moments with the nuclear moments (but of unpairing of electrons with each other), we can see that, in a two-electron bond, the coupling might have a positive sign.

With an intervening nonmagnetic nucleus, such as $^{12}C$ in the structural unit $CH_2$, it is not intuitively obvious how mixing could occur in a way that would tell you which nuclear state is stabilized and which is destabilized; but, in most cases, the $J$ values across two bonds do have signs opposite to the $J$ values for one-bond couplings. Most three-bond $J$ values have the same signs as one-bond couplings.

Some strange anomalies occur with respect to signs of couplings. For example, the two-bond $^1H$-$^{13}C$ couplings between the $^{13}C$ and the nonequivalent $CH_2$ protons in $H_2C=^{13}CHCl$ have opposite signs.[2]

---

[2] F.J. Weigert and J.D. Roberts, *J. Phys. Chem.*, **73**, 449 (1969)

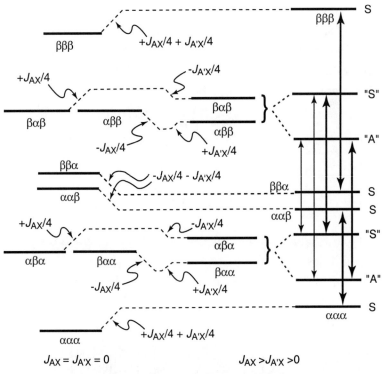

**Figure 9-14.** Energy diagram for a system of three nuclei (AA'X) with a single AA' chemical shift, *i.e.* $v_A = v_{A'}$ and two different coupling constants $J_{AX} > J_{A'X} > 0$. On the far left are the energies expected for $J_{AX} = J_{A'X} = 0$ and then, moving to the right, the results of taking into account $J_{AX} > J_{A'X} > 0$. As with Figure 9-11, we are cheating a bit by neglecting $J_{AA'}$ because we shall see that this coupling can be important in determining the transition energies. On the far right, the states resulting from mixing of βαα and αβα as well as of βαβ and αββ, are shown. The X transitions that are between like states (*i.e.*, being more or less symmetric or more or less antisymmetric) are shown with heavy lines and those between unlike states with light lines. Clearly, for this case, we can expect more transitions than for the A₂X kind of spin system, and, most important, the message is that the mixing of states is here independent of the $v_A$, $v_X$ chemical-shift difference

## 9-6 Unexpected Splittings by Ostensibly Equivalent Nuclei

The conventional simplest prototype of cases, where ostensibly equivalent nuclei give more splittings than expected, even when $\Delta v \gg J$ (a condition not true for A₂B), is AA'XX'. However, this case involves a rather large number of states and it is easier to illustrate the principles involved with the AA'X case, mentioned in Section 9-3. Let us suppose that $J_{AX} > J_{A'X}$. The energy diagram, which is intermediate between A₂B and ABX is shown in Figure 9-14.

Here, we see that the states αβα and βαα with $J_{AX} > J_{A'X}$ will mix and give two new states, one being symmetric-like, marked "S", and the other being antisymmetric-like, and marked "A". The same will be true for αββ and βαβ. Assuming that $J_{AX}$ and $J_{A'X}$ are both positive, it will be the upper of the mixed states that will be the more symmetric. To simplify Figure 9-14, only the transitions are shown that involve the X nucleus. Two X transitions are pure S → S and are delineated with heavy lines. The other X transitions all involve mixed states and two of these are "S" → "S" and "A" → "A", respectively, and are also delineated with heavy lines. The less-favored X transitions are "S" → "A" and "A" → "S", respectively, and are shown with lighter lines.

The message here is that, while $A_2X$ with $J_{AX} > 0$ would give a 1:1 $A_2$ doublet and a 1:2:1 X triplet, AA'X with $J_{AX} \neq J_{A'X}$ can show as many as six different X resonances, even if two can be quite weak. Details of how to deal with AA'XX' cases are available.[1]

## 9-7 When What You See is Not Necessarily What is There

In Section 9-5 dealing with the ABX case, we glossed over the role of $J_{AB}$. We had shown for $A_2B$ that $J_{AA}$ does not affect the spectrum and you might have been lulled into thinking that $J_{AB}$ behaves similarly. This is not the case. The special role of $J_{AB}$ in cases like this is that it is the **mixing coefficient** for the functions such as α(1)β(2)α(3) and β(1)α(2)α(3) and also of α(1)β(2)β(3) and β(1)α(2)β(3) (see Figure 9-11). Other things being the same, the larger $J_{AB}$ is relative to Δν, $J_{AX}$ and $J_{BX}$, the greater the mixing and the more the resulting upper state becomes symmetric-like and the more the lower state becomes antisymmetric-like. To add to the complexity, as mixing becomes strong, the observed couplings between AB and X approach $(|J_{AX}| + |J_{BX}|)/2$. Thus, if $J_{AB}$ is large and Δν is small, the relative sizes of $J_{AX}$ and $J_{BX}$ are immaterial and the spacings are the average of $J_{AX}$ and $J_{BX}$. Even if one of these couplings is zero! See Figure 9-15 for an example. This kind of situation, where a nucleus can appear to be coupled to another nucleus, even though its actual coupling constant is zero, is called **virtual coupling**.

What is unsettling about the ABX case with $J_{AB}$ large and Δν, $J_{AX}$ and $J_{BX}$ small, is that it is possible to have a quartet of lines for X, such as would seem to allow for direct measurement of $J_{AX}$ and $J_{BX}$, but, as can be seen in Figure 9-15, may actually represent a situation where one of these couplings is zero! This is particularly pertinent in the analysis of stereochemical relationships in structures such as:

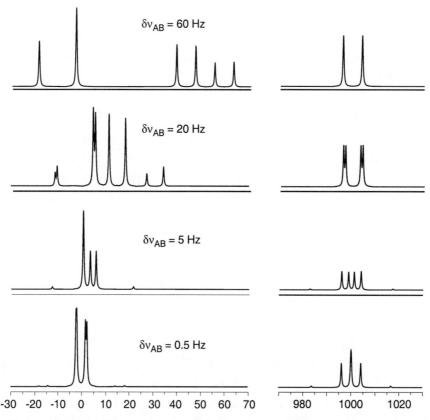

**Figure 9-15** An example of how a spin-spin splitting pattern can appear as though coupling constants are changing, when in fact it is the result of chemical-shift changes. In this example of an ABX spectrum, the values of $J_{AB}$, $J_{AX}$ and $J_{BX}$ are 16, 0 and 8 Hz, respectively. In the topmost spectrum, where the chemical shift is large with respect to the couplings, the X part is a simple 8-Hz doublet as expected for $J_{AX} = 0$ Hz and $J_{BX} = 8$ Hz. However, as the chemical shift decreases, and if $J_{AB}$ is reasonably large with respect to both the other couplings and the chemical-shift difference, the X part tends to become more complex with multiplicities quite unlike what one would expect on a simple basis. Finally at very small shift differences, the spectrum becomes a triplet with a spacing that is the average of $J_{AX}$ and $J_{BX}$, even if one of those couplings is actually zero

Here, $H_A$ and $H_B$ may have a small chemical-shift difference so that the $J_{AX}$ and $J_{BX}$ couplings observed for $H_X$ may not reflect the actual stereochemical relationships in the molecule. The situation is exacerbated if $H_A$ and $H_B$ have chemical shifts that blend into others in the molecule and the clearest transitions are those of $H_X$. Beware!

One way to avoid interpretational errors in dealing with this kind of situation should be evident from Figure 9-15, where we see that the spacings of the splittings in the X part of the spectrum are a function of the $v_A - v_B$ chemical-shift difference. What this means for a given system is that the **observed splittings will violate all simple**

9-8  How Can We Analyze Spin-Spin Splitting Patterns?

**expectations and depend on the magnetic-field strength!** Therefore, if we wish to draw stereochemical implications from observed couplings in the X part of an ABX system, where the AB resonances are masked by other resonances, we should take spectra at different field strengths. If the spacings of the X part do not change with the field, then we can be sure that they represent the real couplings. On the other hand, if the X spacings do change with a change in field, it will be necessary to do a more complete spectral analysis before drawing any stereochemical conclusions.

## 9-8  How Can We Analyze Spin-Spin Splitting Patterns?

It turns out to be much easier to take a set of shifts and couplings and calculate corresponding line positions and intensities than it is to take line positions and intensities and calculate shifts and couplings. One problem is that a given experimental spectrum may not correspond to a unique set of shift and coupling parameters.

Knowing that we can calculate a spectrum, one approach we could take in the analysis of a spectrum would be to guess an initial set of parameters, then see how closely the corresponding calculated line positions and intensities fit the experimental spectrum. After that, we could try to make systematic changes in the parameters until the calculated and experimental agree. For even three nuclei, this can be an exasperatingly slow process, a little like trying to push a grand piano into an exact spot on a skating rink. This is true even when you are quite sure that your three shifts and three couplings are in the right ball park. The procedure here has the same defects as extracting a spectrum from an FID by the method described in Section 5-3.

There are a number of ways to carry out spectral analyses utilizing computer programs, the one illustrated here is called LAOCOON that basically does the job of fitting the parameters as described in the preceding paragraph. This program has gone through a number of versions and is the brainchild of A.A. Bothner-By.[3] Originally designed to be used with punched cards on IBM mainframes of the distant past, it has been converted for use on personal computers.

The procedure for using LAOCOON is, in principle, straightforward but, in practice, somewhat messy. It is best illustrated by going through a specific example. Consider the experimental spectrum and its integral shown in Figure 9-16. This spectrum is of the ABC variety and there are no peak positions in it that clearly

---

[3]A.A. Bothner-By and S.M. Castellano in D.F. DeTar, Editor, *Computer Programs For Chemistry*, Vol. I, W.A. Benjamin, Inc. New York, N.Y. 1968.

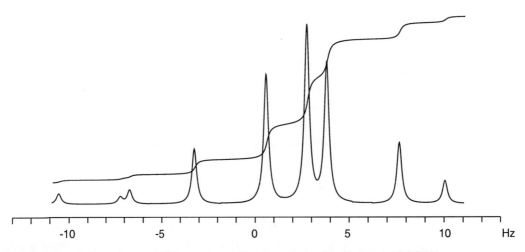

**Figure 9-16** An experimental ABC spectrum with integral to be fitted by the LAOCOON program

correspond to a chemical shift or to a coupling constant. Our first task is to guess at a set of possible parameters and calculate a trial spectrum that will provide a basis for an iterative fit by LAOCOON. However, this may not be easy, if the spectrum does not have well-defined first-order characteristics, which is to say that mixing of energy levels is not important. If our choice of trial parameters is a poor one, the calculated spectrum will usually look wholly different from the experimental spectrum and you will be hard pressed to know how to proceed further.

The first rule is to use all of the information that you can glean from the experimental spectrum to assist in the choice of the initial trial parameters. One possible aid to analysis for relatively simple spectra is the large collection of line-position and line-intensity data complied by Wiberg and Nist.[4] You may be able to get some clues as to suitable parameters by comparing the appearances of your experimental spectra with their calculated spectra. If the Wiberg and Nist reference is not available, you may have to proceed on foot, perhaps something along the lines of the following analysis of the spectrum in Figure 9-16.

As I have said, there are no peak positions in Figure 9-16 that clearly correspond to chemical shifts or couplings. However, if you look at the integral, it is not hard to guess that at least two of the nuclei have shifts greater than zero, because there is more intensity above, than below, the zero-frequency point of the spectrum. The integral also suggests that there might be one nucleus with a shift at perhaps -1, another at perhaps +3 and a third at perhaps +8 Hz. Further, we would probably be justified by the

---

[4]K.B. Wiberg and B.J. Nist, *The Interpretation of NMR Spectra*, W.A. Benjamin, Inc., New York, N.Y. 1962

## 9-8 How Can We Analyze Spin-Spin Splitting Patterns?

substantial spread of the peaks in the spectrum to conclude that one or more of the $J$ values is likely to be comparable to, if not larger than, the chemical shifts. This is not a lot of help, when we know that there are six separate parameters, as well as the line widths, that define the peak positions, relative intensities and general appearance of the spectrum. But if we make a list of the peak positions we find that there are several peak separations of 6 ± 1 Hz and 6 Hz could correspond to one or more of the $J$ values. This is enough information to make a trial run and we might use as initial parameters, shifts of -1, 3, 8 Hz and couplings of 6, 6, 6 Hz. But we need still more information to actually proceed to calculate the appearance of the corresponding spectrum.

An important feature of LAOCOON is the excellent way that it handles cases, like ABX, without the necessity of entering large shift differences between A, B and X. At one point in the data entry, you are asked to classify the system and for an ABX case, you simply enter 112 or some similar combination of integers to show that the shifts of A and B are comparable, but that of X is, in effect, infinitely different. The result is that, no matter what shift you assign to X relative to A and B, there is no mixing of states such as $\alpha\beta\alpha$ and $\alpha\alpha\beta$. For our example of an ABC system, we simply enter 111, which means that all of the chemical shifts are comparable.

Another input is the level of peak intensity to ignore. This parameter is mostly important for complex spin systems, where the calculations may indicate that there could be a hundred or more peaks with intensities so low that they would never be seen in an ordinary spectrum. To winnow the wheat from this kind of chaff, you can specify that all peaks with relative intensities less than say 0.05 or 0.001 be rejected from consideration. We can try 0.001, recognizing that it is easy to change, if necessary.

When we make a plot of the calculated spectrum, a line width needs to be selected for the spectral peaks (it is only possible to input one value) and inspection of the peaks of Figure 9-16 shows that their widths at half height are approximately 0.3-0.5 Hz wide. We can try a line width of 0.4 Hz as a starting average value and if this turns out to be unsatisfactory, it can also be easily changed.

The result of the first-try calculation (Figure 9-17) might well be characterized as encouragingly discouraging. As you can see, there are some sixish Hz spacings, and the general appearance is indeed rather like the experimental spectrum of Figure 9-16. At this stage, we could make more parameter changes to see if we could get a better fit, or else try to find out whether the iterative procedure can do something with the calculated line positions and intensities in Figure 9-17.

If we take the second course of action, we will need to make the best match that we can between the experimental and calculated frequencies that correspond to our trial

```
CASE #1
First_try
NUMBER OF NUCLEI = 3
FREQUENCY RANGE TO BE COVERED IS -11 TO 11 HZ
MINIMUM LINE INTENSITY OF INTEREST IS .001
NUMBER OF ITERATIONS 0
```

| NUCLEUS # | NUCLEUS TYPE | SHIFT |
|---|---|---|
| 1 | 1 | − 1.000 |
| 2 | 1 | 3.000 |
| 3 | 1 | 8.000 |

| NUCLEUS # | NUCLEUS # | COUPLING CONSTANT |
|---|---|---|
| 1 | 2 | 6.000 |
| 2 | 3 | 6.000 |
| 3 | 3 | 8.000 |

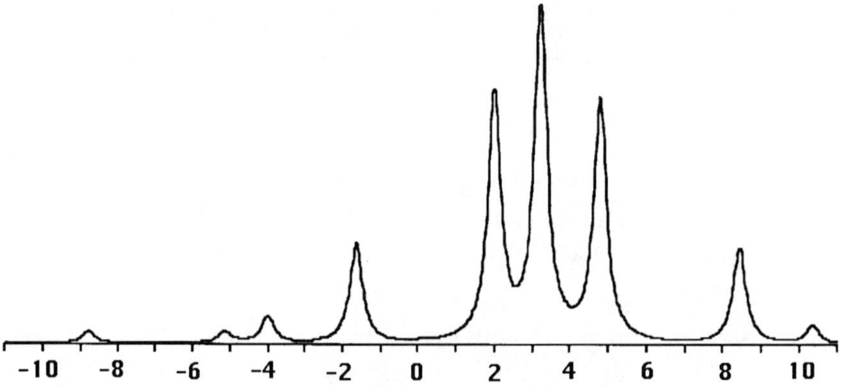

ORDERED LINES CASE #1

| LINE | EXP FREQ | CALC FREQ | ITEN | ERROR |
|---|---|---|---|---|
| 10 | - | −10.428 | 0.011 | |
| 15 | **−10.6** | −8.782 | 0.135 | |
| 7 | **−6.8** | −5.137 | 0.134 | |
| 14 | **−7.3** | −4.000 | 0.286 | |
| 11 | **−3.3** | −1.646 | 1.075 | |
| 1 | **0.5** | 2.000 | 2.632 | |
| 12 | - | 3.137 | 0.088 | |
| 4 | **2.7** | 3.218 | 3.433 | |
| 8 | - | 3.646 | 0.023 | |
| 13 | **3.7** | 4.782 | 2.579 | |
| 9 | **7.6** | 8.428 | 1.037 | |
| 2 | **10.0** | 10.354 | 0.194 | |

**Figure 9-17** Printout of the results of a first-try simulation of the experimental spectrum of Figure 9-16. The top table lists the input parameters. The plot of the calculated spectrum employed a line width of 0.4 Hz. In the lower part of the figure, the positions of the calculated lines and their intensities are tabulated. A possible set of assignments that can be derived by inspection of Figure 9-16 is written in under the heading EXP FREQ

## 9-8 How Can We Analyze Spin-Spin Splitting Patterns?

parameters. In making this matchup, we need to pay close attention to the intensities for help in deciding what corresponds to what. At the bottom of Figure 9-17, eight first-try assignments have been written in under the column headed EXP FREQ. You will see that no assignments have been made for the lines with very small calculated intensities and that some of the discrepancies between EXP FREQ and CALC FREQ are quite large. Thus, LINE 14 is assigned to an EXP FREQ that is 2.8 Hz greater than the corresponding CALC FREQ. In making the subsequent data input into the program, the LINE numbers and EXP FREQ values must be entered correctly. Nothing throws the fitting process off as well as a gross entry error in just one of either the LINE numbers, or the EXP FREQ values. Usually, it is better in the early iterative rounds to assign only those EXP FREQ line positions of which you are reasonably sure. In later rounds, you will want to use as much of the available information as possible.

LAOCOON accepts input of your trial shifts and couplings along with the experimental line positions keyed to the particular transition numbers. Another input is how many iterations you want it to try; nine is standard, but the program usually converges before that and stops. Or alternatively, blows up and gives ridiculous values. You can also specify which parameters you want to have varied in the iterative fit. This is an excellent feature for those parameters that do not affect the spectrum, such as $J_{AA}$ in the $A_2B$ case, or for parameters that have values you are quite sure of as the result of inspection of the spectrum. For our ABC case, we would enter the shift-set designators (not the shift values!) 1, 2 and 3 as SHIFT SET #1, SHIFT SET #2 and SHIFT SET #3, respectively. The coupling sets will be 12, 13 and 23 as COUPLING SET #1, COUPLING SET #2 and COUPLING SET #3, respectively. These entries ensure that all of the shifts and all of the couplings will be varied individually in the iteration.

The printout of the results of the iterative fit are shown in Figure 9-18 and the corresponding plotted spectrum is displayed at the top of Figure 9-19. You will see that the agreement is greatly improved and the calculated spectrum is quite close in appearance to the experimental spectrum of Figure 9-16. However, experience indicates that the agreement is not as good as it might be. For one thing, the probable errors of the parameter sets are substantial, compared to the accuracy of measurement, and note also that LINE 7 has a rather sizable error and especially that the two resonances in Figure 9-16, at around -7 Hz, have their intensities in reverse order from those of the top spectrum in Figure 9-19. This fact suggests that a try should be made with the assignments of LINE 7 and LINE 14 reversed. The printout of the second iteration, in which the assignments of LINE 7 and LINE 14 are reversed and the other parameters are those found to give the best fit in the first iteration, is shown in Figure 9-20. Now, it

```
BEST VALUES FOR CASE #1
CASE #1
First_try
NUMBER OF NUCLEI = 3
FREQUENCY RANGE TO BE COVERED IS -11 TO 11 HZ

        NUCLEUS #           NUCLEUS TYPE            SHIFT
           1                     1                - 3.176
           2                     1                  2.899
           3                     1                  7.102

        NUCLEUS #           NUCLEUS #         COUPLING CONSTANT
           1                     2                  6.188
           2                     3                  5.962
           3                     3                  8.400

CHEMICAL SHIFT PARAMETER SETS TO BE VARIED    COUPLING CONSTANT SETS TO BE VARIED
SHIFT SET #1                                  COUPLING SET #1
MEMBERS ARE:        1                         MEMBERS ARE            12
SHIFT SET #2                                  COUPLING SET #2
MEMBERS ARE:        2                         MEMBERS ARE:           13
SHIFT SET #3                                  COUPLING SET #3
MEMBERS ARE:        3                         MEMBERS ARE:           23

       SUMMARY OF RMS ERRORS OF CALCD AND ASSIGNED LINE POSITIONS.

       FOR ITERATION       1,   THE RMS ERROR =        1.601
       FOR ITERATION       2,   THE RMS ERROR =         .198
       FOR ITERATION       3,   THE RMS ERROR =         .132
       FOR ITERATION       4,   THE RMS ERROR =         .157

   ERROR VECTORS AND STANDARD ERRORS      PROBABLE ERRORS OF PARAMETER SETS
   STANDARD ERROR = .13668                     1            .1674
   STANDARD ERROR = .66706                     1            .3459
   STANDARD ERROR = .17703                     1            .2519
   STANDARD ERROR = .212715                    1            .4117
   STANDARD ERROR = 1.28396                    1            .7399
   STANDARD ERROR = .325636                    1            .3406

                         ORDERED LINES CASE #1

            LINE     EXP FREQ     CALC FREQ      ITEN       ERROR
             15      -10.600       -10.850       0.195       0.250
              7       -7.300        -7.005       0.265      -0.295
             14       -6.800        -6.693       0.168      -0.107
             11       -3.300        -3.456       1.068       0.156
              1        0.500         0.389       2.472       0.111
             12                      0.702       0.011
              4        2.700         2.618       3.367      -0.118
              8                      3.506       0.002
             13        3.700         3.818       2.637      -0.118
              9        7.600         7.663       1.177      -0.063
              2       10.000        10.012       0.444      -0.012
```

**Figure 9-18** Printout of an iterative fit to the ABC spectrum of Figure 9-16 that used shifts for $\nu_A, \nu_B, \nu_C$ of -1, 8 and 16 Hz with all of the couplings $J_{12}$, $J_{13}$ and $J_{23}$ equal to 6 Hz and an intensity rejection level of less than 0.001.

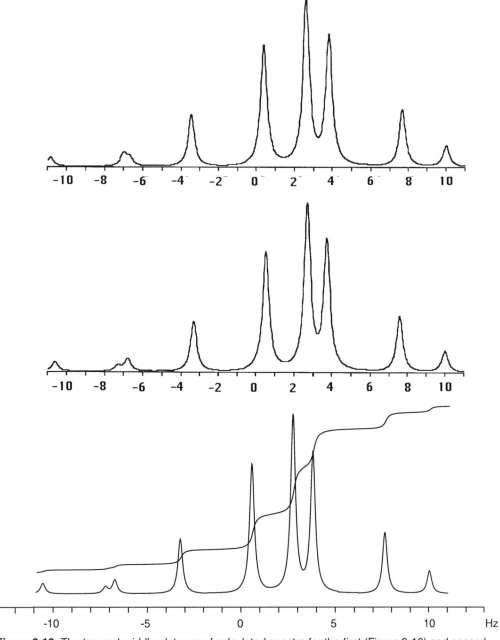

**Figure 9-19** The top and middle plots are of calculated spectra for the first (Figure 9-18) and second (Figure 9-20) iterations with LAOCOON in attempts to fit the experimental spectrum, bottom, which is duplicated from Figure 9-16. Note that the second iteration produces the correct order of intensities for the two resonance peaks in the vicinity of -7 Hz. It also gives a better line-position fit to the resonance at about -10.5 Hz than does the first iteration. The spectral parameters are tabulated in Figure 9-19 for the top calculated spectrum and in Figure 9-20 for the middle calculated spectrum

```
BEST VALUES FOR CASE #1
CASE #1
Second_try
NUMBER OF NUCLEI = 3
FREQUENCY RANGE TO BE COVERED IS -11 TO 11 HZ

         NUCLEUS #        NUCLEUS TYPE           SHIFT
             1                 1                - 3.023
             2                 1                  3.017
             3                 1                  6.956

         NUCLEUS #         NUCLEUS #        COUPLING CONSTANT
             1                 2                  5.886
             2                 3                  6.220
             3                 3                  9.027

CHEMICAL SHIFT PARAMETER SETS TO BE VARIED    COUPLING CONSTANT SETS TO BE VARIED
SHIFT SET #1                                  COUPLING SET #1
MEMBERS ARE:        1                         MEMBERS ARE       12
SHIFT SET #2                                  COUPLING SET #2
MEMBERS ARE:        2                         MEMBERS ARE:      13
SHIFT SET #3                                  COUPLING SET #3
MEMBERS ARE:        3                         MEMBERS ARE:      23

        SUMMARY OF RMS ERRORS OF CALCD AND ASSIGNED LINE POSITIONS.

        FOR ITERATION     1,   THE RMS ERROR =       1.640
        FOR ITERATION     2,   THE RMS ERROR =        .123
        FOR ITERATION     3,   THE RMS ERROR =        .019
        FOR ITERATION     4,   THE RMS ERROR =        .019

     ERROR VECTORS AND STANDARD ERRORS      PROBABLE ERRORS OF PARAMETER SETS
     STANDARD ERROR = .13668                    1              .0207
     STANDARD ERROR = .66706                    1              .0471
     STANDARD ERROR = .17703                    1              .0359
     STANDARD ERROR = .212715                   1              .0522
     STANDARD ERROR = 1.28396                   1              .0873
     STANDARD ERROR = .325636                   1              .0386

                        ORDERED LINES CASE #1

         LINE    EXP FREQ    CALC FREQ    ITEN     ERROR
          15     -10.600     -10.617      0.203    0.017
          14      -7.300      -7.300      0.121   -0.000
           7      -6.800      -6.783      0.266   -0.017
          11      -3.300      -3.317      1.048    0.017
          12                   0.000      0.017
           1       0.500       0.517      2.486   -0.017
           4       2.700       2.700      3.405   -0.000
          13       3.700       3.733      2.676   -0.033
           9       7.600       7.567      1.177    0.033
           2      10.000      10.000      0.444   -0.000
```

**Figure 9-20** Printout of the second iterative fit to the ABC spectrum of Figure 9-16. This iteration used the best-fit shifts and couplings from Figure 9-18 and an intensity rejection level of less than 0.001. The assignments of EXP FREQ for lines 7 and 14 were interchanged relative to Figure 9-18. The corresponding calculated spectrum with a line width of 0.4 Hz is shown in the middle of Figure 9-19

will be seen that the RMS error is much smaller and the probable errors are less than 0.1 Hz, which is less than the experimental error. The spectrum corresponding to the best-fit parameters of the second iteration is plotted with a line width of 0.3 Hz in the bottom half of Figure 9-19.

The general program structure and FORTRAN source code for LAOCOON have been described in considerable detail.[3] The punched-card procedure for data entry was particularly suitable for the early versions of this program, because the experimental frequencies were entered one to a card and, if any of the line assignments were unsatisfactory, only particular card(s) needed to be changed before entering the data again in the computer. With present-day computers, this part of the operation is not so convenient and different means of changing the input data files are necessary. Methods for making such changes have been written into the programs, but to avoid not having to type a list of perhaps 200 transitions each time you run the program, just because one or two transitions have to be reassigned, requires more complex source code. However, the newer versions do provide much greater flexibility in plotting and/or archiving the results.

### Exercises

**Exercise 9-1** Explain how you can rationalize that, for two equivalent spin 1/2 nuclei, the symmetric state $1/\sqrt{2}(\alpha\beta + \beta\alpha)$ is destabilized by $J/4$, just as are $\alpha\alpha$ and $\beta\beta$, but the antisymmetric state is stabilized by $-3J/4$.

**Exercise 9-2** Use Equations 9-7 to 9-9 and the data of Figure 9-1 to calculate the difference in intensity of the two resonance lines of the CH proton of $^{13}$C-labeled methanoate for the 600-MHz proton signal where $^1J_{13CH}$ is 195 Hz.

**Exercise 9-3 a.** Suppose you observed a 1:3:3:1 quartet with equal line spacings, could this be a simple AB splitting? Explain.
**b.** Explain how you could prove that a 1:3:3:1 quartet resulting from spin coupling involved a -$CH_3$ or -$CF_3$ group rather than a fortuitous combination of shifts and couplings?

**Exercise 9-4** Suppose one had a sample of pure *para* hydrogen that gave no proton NMR signal (Section 9-2) and exposed it to a nickel surface, whereby the sample was slowly converted to the equilibrium mixture of *ortho* and *para* hydrogen. Does the interconversion surely prove that H-H bonds are broken and reformed on the metal surface or can you devise a different mechanism to convert the antisymmetric hydrogen to symmetric hydrogen?

**Exercise 9-5** Suppose you could only observe half of the lines of an AB quartet in a complex NMR spectrum. One of the lines has 0.64 of the intensity of the other and the lines are 12.3 Hz apart. Show how you can calculate the chemical-shift difference?

**Exercise 9-6** Calculate the difference in chemical shift $(v_A - v_B)$ that, with a $J_{AB}$ of 10 Hz, would give an intensity of the outside lines of an AB quartet equal to 98% of the inside lines.

**Exercise 9-7** Explain how you can tell (or not tell) from the spectra of Figure 9-13 whether it is $J_{AX}$ or $J_{BX}$ that has a negative sign.

**Exercise 9-8** If you have access to LAOCOON or a similar spectral analysis program, use it to extract the shifts and coupling constants of the three-nuclei spectra of **a.** Figure 9-21A, and **b.** Figure 9-21B (AB),C (X) of an ABX system.

**Exercise 9-9** Classify each of the following spin systems according to the procedures of Section 9-3. Include $^{13}C$ only if specifically labeled.
a. chloroethene (count Cl as nonmagnetic)
b. $^{13}CH_3F$
c. cyclobutene
d. 1,1-difluoro-1,2,3-butatriene
e. 1,2-dichlorethane-1-$^{13}C$
f. 1,4-cyclohexadiene
g. benzene

**Exercise 9-11** Identify the transitions of the A nuclei in Figure 9-14 and reason which of them can be expected to give rise to strong resonances and which weak resonances. Show you reasoning.

**Exercise 9-12** In this exercise you are to consider the AA'X type of nuclear spin systems described in Section 9-6. The first example is H-C≡$^{13}C$-H that has the potential of allowing direct observation of $^3J_{HH}$ in ethyne, while the second example is of a different AA'X system afforded by 1,2,3,4-tetrachlorobenzene-1-$^{13}C$. For the $^{13}C$-labeled

# Exercises

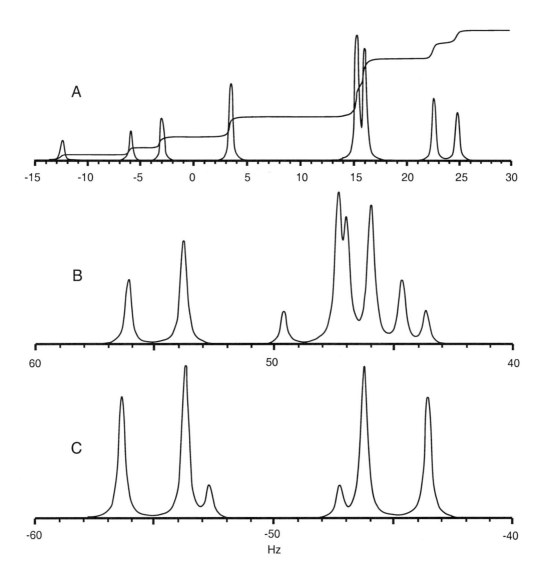

**Figure 9-21** Spectra for Exercise 9-8

ethyne $^1J_{^{13}CH}$ is 250 Hz, $^2J_{^{13}CH}$ is 50 Hz and assume that $^3J_{HH}$ is 15 Hz. For the $^{13}C$-labeled tetrachlorobenzene, $^2J_{^{13}CH}$ is about -4 Hz, $^3J_{^{13}CH}$ is +12 Hz and $^3J_{HH}$ is +8 Hz. (For both systems, assume that the AA' shift difference is zero, *i.e.*, the $^{13}C$ isotope effect on the shifts is negligible.)

**a.** Make up energy diagrams for each of these cases (graph paper will be helpful), starting with all *J*'s equal to zero, correcting each energy level for the *J* values given above as appropriate and neglecting mixing.

**b.** Make a bar graph of the proton and $^{13}C$ spectra expected for each case on the basis of your energy diagrams in **a.**

**c.** Make up another energy diagram assuming that $J_{AA'} = 10$ Hz and $J_{AX} = J_{A'X} = 20$ Hz. Write the $\psi$ functions for each of the states. Will the magnitude or sign of $J_{AA'}$ affect the appearance of the spectrum predicted for this system? What about the relative signs of $J_{AX}$ and $J_{A'X}$? Explain.

**d.** Return to your earlier energy diagrams in **a.** and indicate which of the states will mix and whether the degree of mixing will be different for the two $^{13}C$-labeled compounds. Make a qualitative assessment of mixing for each case and prepare a bar graph of the expected proton and $^{13}C$ spectra. As part of that graph, also assess the expected line intensities, recognizing that, if mixing is important, the more antisymmetric states will be of lower energy than the more symmetric-like states.

**e.** Go back to **d.** and determine how a change in the magnetic field of the spectrometer, say from 300 to 600 MHz, would affect the appearance of the spectrum. Now, do a similar analysis for the sign of $J_{AA'}$ and then for the relative signs of $J_{AX}$ and $J_{A'X}$.

**f.** Return again to **d.** and show how you can use Equations 9-7 to 9-9 to calculate precisely the positions and intensities of the proton lines for each of the $^{13}C$-labeled compounds. (Hint: look at group transitions when the $^{13}C$ is $\alpha$ or when it is $\beta$).

# 10

# Some Thoughts about Chemical Shifts

It is not my intention in this Chapter to provide a treatise on chemical shifts in NMR spectra. Virtually all books on the basis of NMR spectroscopy provide excellent and usually comprehensive discussions of almost all aspects of chemical shifts. Less commonly, and indeed, the important topic in this chapter, is rationalization of the very large ranges of chemical shifts of most magnetic nuclei compared to the limited range of proton chemical shifts. Application of the principles involved is particularly useful in $^{15}$N NMR.

## 10-1 Why is the Range of Proton Chemical Shifts So Small?

Much work has been done on theoretical approaches and rationalizing of proton chemical-shift differences. Stereochemical, ring-current, electronegativity, steric and other effects have been documented *ad nauseam*. However, the question of why the range of proton shifts is so limited compared to those for carbon, nitrogen, fluorine and phosphorous is much less frequently given serious attention.

What is the problem? The normal range of proton shifts is only about 15 ppm, whilst the ranges for some other elements are much larger; about 300 ppm for carbon, 900 ppm for nitrogen, 600 ppm for fluorine and about 500 ppm for phosphorous. Among the less-common nuclei, the shifts of lithium and other alkali metals are like protons, in falling in rather narrow ranges. What is the reason for this difference? One important element clearly lies in the kind of orbitals used in chemical bonding. The elements with small shift-ranges form bonds primarily using s orbitals, while those with the larger shift-ranges use p orbitals.

Matters are made more complex by the fact that some, possibly cherished, simple rationalizations of chemical shifts do not work well, when viewed in different contexts from those for which they were originally formulated. For example, shifts are often rationalized by consideration of simple diamagnetic shieldings. This usually boils down to considering the relative magnitudes of electron density surrounding the nuclei under consideration. For example, when the different proton shifts of ethanol are compared, it is usual to consider each of the $CH_3$ protons as being connected to a relatively weakly electron-attracting carbon, thus these protons are regarded as being surrounded by a greater density of electrons than the $CH_2$ protons, whose carbon, although connected to another carbon, is expected to have its electron density at least somewhat depleted by being also connected to a strongly electron-attracting oxygen. Further, when we consider the shift of the ethanol OH proton, the usual argument is that, because this proton is directly connected to the electron-attracting oxygen, its nucleus is expected to have a still lower density of surrounding electrons than either the protons of the $CH_2$ or of the $CH_3$ groups. "Equating" diamagnetic shielding to electron density for ethanol, leads to the simple, satisfying prediction that the ethanol protons, at constant frequency, should require higher magnet fields to achieve resonance in the order, OH < $CH_2$ < $CH_3$, in accord with common experience. The validity of this kind of analysis is buttressed by the finding that the differences in proton shifts between the $CH_2$ and $CH_3$ protons of compounds of the type $XCH_2CH_3$, depend essentially linearly on the electronegativity of X.

However, there is a major problem with these simple and straightforward ideas, because it turns out that the shifts of the ethanol OH protons depend greatly on the ethanol concentration in aprotic solvents such as $CCl_4$. Indeed, if the OH proton shift is extrapolated to zero concentration, the shift of what is expected to be non-hydrogen-bonded OH shows that such protons are **more** shielded than the $CH_3$ protons of ethanol. Similar large shieldings are found for the protons of ammonia, hydrogen fluoride and hydrogen iodide when hydrogen bonding is minimized. Interesting and related shift effects are noted in $^{15}N$ NMR spectra. Proton addition to ammonia does cause the $^{15}N$ resonance of the nitrogen to shift to lower magnetic fields of about 20-30 ppm, in accord with the notion that a positive nitrogen will have less, close-in, electronic shielding than a neutral nitrogen. However, protonation of aminobenzene (aniline) produces an **upfield** $^{15}N$ shift of about 9 ppm. The epitome of this kind of effect is observed in the protonation of azabenzene (pyridine), where the result is **greater** $^{15}N$ shielding by about 100 ppm - the exact value depending on the solvent and the nature of the counterion. There is a direct counterpart in $^{13}C$ spectra, where conversion of benzene through replacement of a hydrogen by sodium or lithium, causes the anionic carbon to be **less shielded** rather than more.

As we have mentioned earlier, the shifts of many nuclei such as $^{13}C$, $^{15}N$, $^{19}F$ and $^{31}P$, cover much larger ranges than protons. What is surprising with $^{15}N$ is the substantial parallelism between the $^{15}N$ shifts and the color of the compound containing the nitrogen. Here, it turns out that absorption at **longer** wavelengths corresponds generally to **less** shielding. An extreme is nitrosobenzene, which is green and its nitroso nitrogen nucleus is 900 ppm less shielded than the nitrogen of ammonia. The most useful spectroscopic transitions for these correlations seem to be the $n \rightarrow \sigma^*$ and $n \rightarrow \pi^*$, where the $n$ electrons are unshared electron pairs on nitrogen (or on oxygen, which shows rather similar behavior). The smaller the transition energy, the less shielded is the nucleus with the unshared pair.

What happens on protonation? Binding of a proton to a nitrogen unshared pair is certainly expected to increase the energy required to raise one of the electrons of the unshared pair to a higher electronic orbital and thus should tend to decrease the deshielding effect that correlates with the optical transition. This effect will usually compete very effectively with the diamagnetic effect expected from increasing the degree of positivity of the atom by protonation. Hydrogen bonding to an unshared pair of nitrogen or oxygen is expected to have a similar, but smaller, effect and this is also found experimentally. Thus, the $^{15}$N resonance of azabenzene (pyridine) taken in cyclohexane moves upfield by 6 ppm, when taken in $HCCl_3$, as the result of weak hydrogen bonding to the solvent, 16 ppm with stronger hydrogen bonding to methanol, and 29 ppm with still strong hydrogen bonding to $HOCH_2CF_3$. The very large changes of shift with protonation of, or hydrogen bonding to, nitrogen are characteristic of the =N:- and ≡N: groupings and have many important chemical and biochemical applications.

With this glimpse of the basic outlines of the panoply of very large shift effects expected for nuclei of atoms that form bonds using p orbitals, the question is still remains as to **why**. Diamagnetic influences, including ring currents, are much too small and, as we have just seen, often operate in quite the wrong direction.

## 10-2 The Elements of the Second-Order Paramagnetic Effect

The following explanation of how to understand the role of electronic excitation in influencing the chemical shifts of nuclei that form bonds with p orbitals is due to the late W. Moffitt, a person who had a remarkable ability to convert the abstruse to the understandable. The effect involved is called the **second-order paramagnetic effect**. Let us see how it works.

Consider a single electron in an isolated p orbital of a magnetic nucleus that has its orbital axis directed along the Z axis of the coordinate system; i.e., a $p_z$ orbital. If a magnetic field is also along the Z axis, you can expect a relatively small stabilization arising from a diamagnetic circulation around the field axis, Figure 10-1. A similar situation would hold for an s orbital. However, the electron can also circulate around the field axis in one of two combinations of $p_x$ and $p_y$ orbitals. One combination,

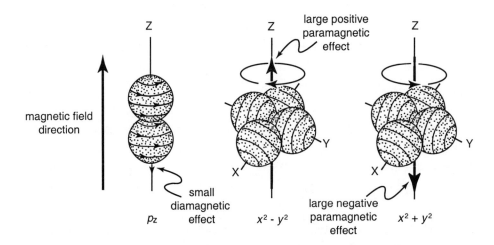

**Figure 10-1** Qualitative representation of possible modes of circulation of an electron in p orbitals in a magnetic field. At the left is a $p_z$ orbital in which the electron would circulate about the Z axis and have a small diamagnetic stabilization. In the middle, and on the right, the electron would circulate about the Z axis in combinations of $p_x$ and $p_y$ orbitals. These circulations would produce sizable magnetic dipoles, which would be oriented along the +Z or -Z axis and would represent the states of rather different energy, as shown in Figure 10-2.

symbolized by $(x^2 - y^2)$, has the electron circulating in one direction; while the other, $(x^2 + y^2)$, has it circulating in the other direction. Each of these states has angular momentum with respect to the Z axis and each has a sizable magnetic moment directed along, or opposite to, the magnetic field direction. The state that has its magnetic moment aligned with the magnetic field is the more stable state (see Figure 10-2) and the degree of stabilization will be proportional to the strength of the magnetic field.

Of course, chemists are primarily interested in bonded states and we can provide a crude simulation of such situations by locating two positive charges along the + and - X axes of the coordinate system for our p electron. In the absence of a magnetic field, the electrostatic attractions of the positive changes for the electron are expected to be sufficiently powerful to "quench" the orbital angular momentum and make the $p_x$ orbital the most favorable location for the electron, see Figure 10-3. Now, when a magnetic field is applied, at least some measure of stabilization is possible by having even a small degree of circulation of the electron in the favorable direction. The result of this electron circulation will be to generate a component of magnetization directed along the field axis and centered at the nucleus. This component will augment the field at the nucleus and will thus constitute a **paramagnetic effect** on the chemical shift of the nucleus. Because the circulation of the electron and the associated degree of extra

## 10-2 The Elements of the Second-Order Paramagnetic Effect

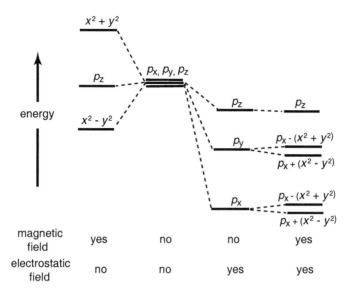

**Figure 10-2** A very qualitative representation of possible energy levels for an electron in different $p$ orbitals; on the left, in the presence of, or the absence of, a magnetic field and, toward the right, in the presence of an electrostatic field with, or without, a magnetic field. The electrostatic field, shown in Figure 10-3 along the $X$ axis, is expected to reduce the energy of all of the orbital arrangements, with the largest stabilization for the $p_x$ orbital. With this arrangement, in the presence of a magnetic field, there will be a small extra stabilization of system as the result of the $p_x$ electron circulation around the $Z$ axis, that will produce a paramagnetic chemical-shift effect at the nucleus. This stabilization can be regarded as the result of some degree of mixing of the $p_x$ and the $(x^2 - y^2)$ orbitals in the ground-state wave function

stabilization will be directly proportional to the strength of the applied magnetic field, there will be a chemical-shift effect on the resonance frequency of the nucleus involved. This then is the second-order paramagnetic effect - a field-dependent circulation of p electrons around the magnetic-field axis. The representations of Figures 10-1 to 10-3 are not very sophisticated, but they do provide a useful rationale of why atoms that use p orbitals in bonding have much larger average chemical-shift differences than atoms that primarily use s orbitals in their bonds.

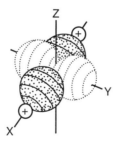

**Figure 10-3** Representation of an electron in a $p_x$ orbital with positive charges on opposite sides of the $X$ axis to quench the electron's orbital angular momentum and simulate, in a very qualitative way, the effect of bonding on the circulation of the electron around the $Z$ axis

## 10-3 The Second-Order Paramagnetic Effect and Nitrogen Shifts

Application of the simple representations to real atoms in real molecules requires further amplification. Consider a nitrogen-15 atom in ethanonitrile, $CH_3C\equiv{}^{15}N$: (acetonitrile). The nitrogen uses p orbitals in its bonds to carbon, but there is a difference now because the nitrogen 2s and 2p orbitals are filled with the normal eight electrons and would thus seem to allow only for diamagnetic circulation of these electrons. The solution to the dilemma involves a somewhat unorthodox idea, the mixing into the usual ground-state wave functions of the molecule, additional wave functions corresponding to the excitation of one of the electrons on nitrogen to a higher energy state, such as would then allow circulation of electrons in p orbitals in the manner described earlier. The degree to which such states mix in will depend on the degree of magnetic stabilization that they can supply and hence will depend on the applied magnetic field. Mixing in will, of course, also depend on the difference in energy between the ground and the relevant excited states. The smaller the energy difference, the more the circulating excited states can contribute to the paramagnetic chemical shift of the nuclei around which circulation occurs. Because the energies of the relevant excited states can often be correlated with the transition energies in ultraviolet and visible spectroscopy, it should not come as a surprise that, in many cases, the chemical shifts of nuclei, for which the second-order paramagnetic effect is important, show a rough parallelism between color and chemical shift. Thus, as mentioned earlier, nitrosobenzene (bright green) has large downfield $^{15}N$ chemical shifts, about 520 ppm compared to nitrobenzene (faint yellow) and about 900 ppm compared to ammonia (colorless).

An important excited state for nitrogens carrying unshared electron pairs and involved in multiple-bond formation often corresponds to a $n\to\pi^*$ electronic transition, where $n$ is an unshared pair and $\pi^*$ is a higher energy, normally unoccupied, $\pi$ orbital of the multiple bond. Here, ethanonitrile, $CH_3C\equiv N$:, provides an excellent example.

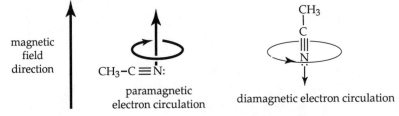

An interesting point with respect to ethanonitrile, is that the second-order paramagnetic effect on the nitrogen nucleus operates maximally when the principal molecular axis is perpendicular to the magnetic field, while the smaller diamagnetic effect takes over when the principal axis is parallel to the magnetic-field direction.

An important corollary to the application of the second-order paramagnetic effect to substances with the general structural features of $CH_3C\equiv N:$ is the effect of forming an additional bond to the unshared pair on nitrogen, as for example by protonation.

$$CH_3C\equiv N: + H^{\oplus} \rightarrow CH_3C\equiv \overset{\oplus}{N}:H$$

Making a bond to the unshared pair converts the $n \rightarrow \pi^*$ transition to a $\sigma \rightarrow \pi^*$ transition, because now the unshared pair is involved in sigma bonding to the added proton. It is reasonable to expect that excitation of one of these electrons to the $\pi^*$ excited state will have a much larger transition energy than for the unshared pair only, because of bonding to the proton. An unequivocal prediction would be that the second-order paramagnetic effect should be substantially diminished. This would result in a larger shielding of the nitrogen molecule. The effect can be large, and, in the case of ethanonitrile, protonation results in an increase in nitrogen shielding of about 100 ppm!

Such shift changes have many chemical applications especially in the studies of biomolecules,[1] because not only does protonation cause large shifts, but hydrogen bonding is also very effective. It is not surprising then that the $^{15}N$ resonance of azabenzene (pyridine) is more shielded by almost 40 ppm on going from the gas phase to solution in the strongly hydrogen-bonding solvent, trifluoroethanol.

## 10-4 The Second-Order Paramagnetic Effect and Proton Shifts

The foregoing discussion should help throw some light on the much larger chemical-shift ranges of the magnetic nuclei whose atoms use p orbitals in forming chemical bonds compared to hydrogen, that normally uses s orbitals. This by itself, does not tell us why proton shifts are so strongly influenced by hydrogen bonding. To reiterate what was discussed in Section 10-1, a decrease in hydrogen bonding, as of an OH function, causes an increase in the magnetic shielding of the proton involved.

---

[1] For some specific applications to biochemically interesting systems see K. Kanamori and J.D. Roberts, *Accts. Chem. Res.*, **16**, 35 (1983) and W. Phillipsborn and R. Müller, *Angew. Chem.* (Intl Ed.), **25**, 383 (1986).

We can rationalize this change in shielding in two different ways, each deriving from the second-order paramagnetic effect. Either, or both, could be important contributors. One way to influence the shielding of a proton by hydrogen bonding would involve changes in the second-order paramagnetic effect of the atom to which the proton is attached. Consider the hydroxyl proton of an alcohol. If the oxygen is not hydrogen bonded, the paramagnetic effect at the oxygen will be large, but the paramagnetic circulation will result in diamagnetic shielding at the hydrogen, in a manner analogous to the way that the diamagnetic circulation of electrons in the π orbitals of benzene are postulated to cause paramagnetic shifts of the resonances of the benzene protons.

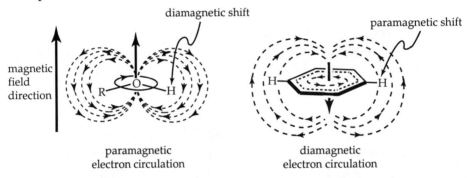

Certainly, if hydrogen bonding is important to the unshared electron pairs on oxygen then mixing in a higher energy electronic state into the ground state will be less favorable and the diamagnetic shielding of the proton will diminish. For an aliphatic alcohol, a higher energy state that would allow for a paramagnetic electron circulation might correspond to a $n \rightarrow \sigma^*$ electronic transition, where $\sigma^*$ is an antibonding $\sigma$ orbital. This mode of influencing proton shifts that arises from the degree of hydrogen bonding to unshared pairs might be expected, for an aliphatic alcohol such as ethanol, to act to also influence the $^{13}C$ chemical shift of the C1 carbon. The effect should be much smaller because the C-O bond is longer (1.42 A) compared to the O-H bond (1.09 A), and the through-space component of the shielding should depend on $r^{-3}$. For ethanol, the actual C1 $^{13}C$ shift change on dilution in $CCl_4$ is almost negligibly small.

An alternative way for accounting for the influence of hydrogen bonding on proton chemical shifts would be to have the electron-pair donating atoms perturb the normal hydrogen s orbital to allow mixing in some p character along the line of centers between the atoms. Mixing in some p character to the 1s orbital, along with excited-state functions that allow for paramagnetic circulation around the hydrogen, could cause a direct second-order paramagnetic effect at the hydrogen.

## 10-4 The Second-Order Paramagnetic Effect and Proton Shifts

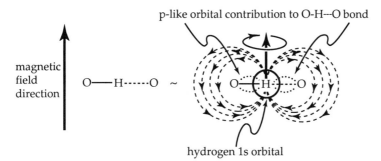

This mechanism for causing paramagnetic shifts of proton resonances, as the result of hydrogen bonding, seems particularly appropriate for intramolecular hydrogen bonds such as in 2, 4-pentadione where intermolecular hydrogen bonding is small.

2,4-pentanedione

The shift of the enolic proton in 2, 4-pentanedione is 15.2 ppm, relative to tetramethylsilane, while that of the ethanol proton in liquid ethanol is 5.2 ppm.

The commonly observed sequence of proton shieldings for C-H bonds is

$$-\overset{|}{\underset{|}{C}}-H \quad > \quad \equiv C-H \quad > \quad \overset{\backslash}{\underset{/}{C}}-H$$

Clearly, either the acetylenic or the ethylenic protons can be regarded as anomalous in the sequence, because the chemical characteristics of these groupings suggest that the sequence should be

$$-\overset{|}{\underset{|}{C}}-H \quad > \quad \overset{\backslash}{\underset{/}{C}}-H \quad > \quad \equiv C-H$$

The usual explanation for the observed order of C-H proton shifts postulates a diamagnetic circulation of the triple-bond electrons around the long axis of the triple bond.

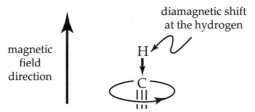

The effect will, of course, be greatest when the axis of the triple bond is parallel to the applied magnetic field and minimal when the orientation is perpendicular to the field.

This mechanism is simple and attractive, but there are theoretical arguments that it is actually not large enough to account for the observed shift differences. An alternative explanation is the second-order paramagnetic effect, which will be maximal when the triple bond is perpendicular to the magnetic-field direction. The second-order paramagnetic effect is expected to be more effective in changing the shift of a hydrogen attached to a triple bond than one attached to a double bond, because the promotion energy of an electron to a $\pi^*$ state should be less for a triple bond than a double bond.

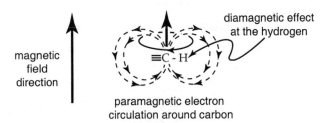

For the case of the triple bond, both kinds of electron circulation should increase the shielding of the proton, but are expected to act in opposition in determining the overall shielding of the triple-bonded carbon. The $^{13}$C shifts of ethane (6.5 ppm), ethene (123.5 ppm) and ethyne (71.5 ppm) are in the same order as their C-H proton shifts. This order of $^{13}$C shifts would seem to be contrary to what would be expected for the second-order paramagnetic effect, but it must be remembered that ethene has more axes for paramagnetic electron circulation than does ethyne and circulation about these axes need not affect the proton shifts in the same way. In any case, the greater magnitudes of the $^{13}$C-shift differences compared to proton-shift differences clearly suggests that second-order paramagnetic effects must be important in determining the $^{13}$C shifts.

### Exercises

**Exercise 10-1** Would you expect the second-order paramagnetic effect to be important, or not, in contributing to $T_1$ relaxation that occurs by the chemical-shift anisotropy mechanism? Explain.

**Exercise 10-2** Explain how you would design an experiment to determine whether changes in hydrogen bonding to 2-propanone (acetone) would affect the chemical shift (in ppm) of the carbonyl oxygen's $^{17}$O resonance. Would you expect the changes to be large (tens of ppm) or small (less than 2-4 ppm)? Give your reasoning.

# Exercises

**Exercise 10-3 a.** How would you expect the $^{15}N$ chemical shift of 1-aza-2,4-cyclopentadiene (pyrrole) to change if the NH proton were to be removed by a strong base? Give your reasoning.

**b.** The $^{15}N$ chemical shifts of the two nitrogens of diazocyclopentadiene, with $^{15}N$-nitric acid as zero, come at 106 and -9 ppm and those of benzenediazonium ion at 150 and 57 ppm. (On the $^{15}N$- nitric acid scale, ammonium ion comes at +355 ppm.) Reason as to which of the two nitrogens in each of these pairs of compounds will be the least shielded. How could you determine by experiment whether your predictions were correct or not? Interestingly, 4-hydroxybenzenediazonium ion shows $^{15}N$ shifts at 147 and 51 ppm in acid solution and at 117 and 10 ppm in basic solution. Explain what happens when the pH is changed.

**c.** When azobenzene ($C_6H_5N=NC_6H_5$, red) is dissolved in strong sulfuric acid, its $^{15}N$ resonance moves 150 ppm in the direction of **increased shielding**. Account for this behavior.

**d.** The $^{15}N$ shifts of the following structures cover more than 160 ppm. Arrange them in a predicted order of $^{15}N$ shifts with those having the least-shielded nitrogens coming first. Show your reasoning.

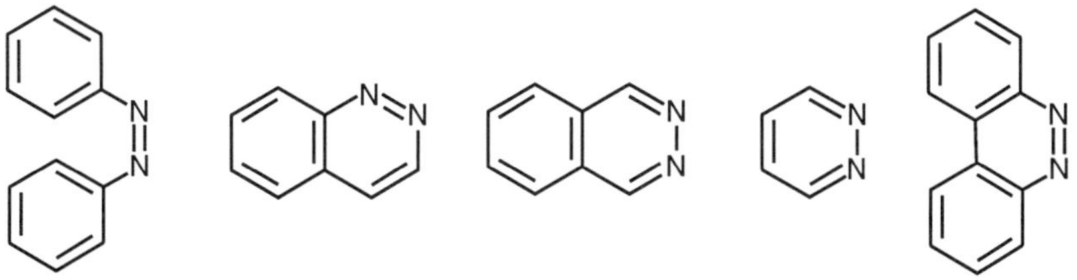

**Exercise 10-4** Explain how would you expect the proton shift of the hydrogen in an O-H··O bond to change with the O-H-O bond angle between 90° and 180°.

**Exercise 10-5** The cyclic polyenes, 18-annulene and 16-annulene (see below) show quite different patterns of proton chemical shifts at low enough temperatures so that they are conformationally immobile. Under these conditions, the inner protons of 18-annulene are at -4.2 ppm and the outer protons at 10.8 ppm relative to tetramethylsilane. For 16-annulene, the inner protons come at 10.3 ppm and the outer protons at 5.3 ppm. Explain how one might account for these striking shift differences. At room temperature, both 18-annulene and 16-annulene are conformationally mobile and give single averaged proton resonances. Calculate what you expect the averaged shifts to be. Show your method.

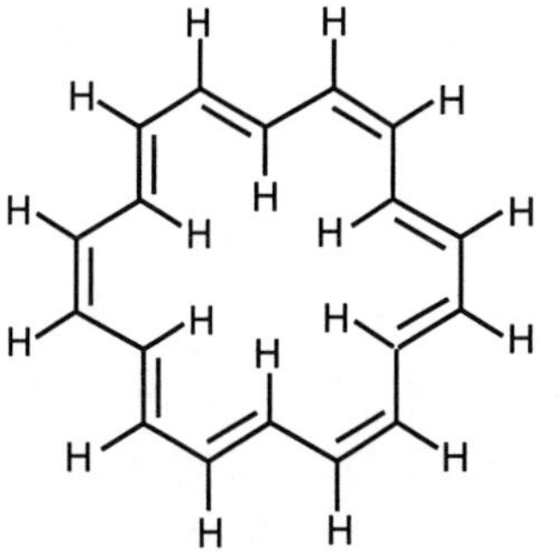

18-annulene

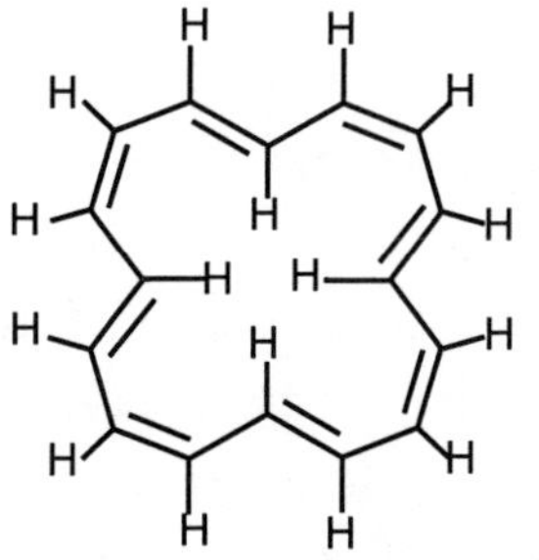

16-annulene

**Exercise 10-6** Here, you are to interpret experimental results that show the power of $^{15}N$ NMR to measure tautomeric equilibrium and subtle effects of hydrogen bonding to nitrogen. The players and some of their $^{15}N$ shifts in water or acidic solutions follow:

| | | $^{15}N$ shifts, ppm relative to $HNO_3$ |
|---|---|---|
| imidazole | 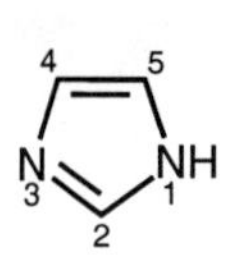 (ring with N3, C2, N1-H, C5, C4) | 171.0 (one peak) |
| imidazolium ion | 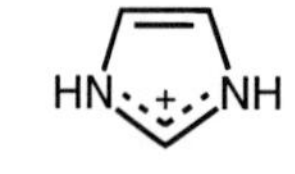 HN⋯+⋯NH | 202.0 |
| N-imidazole |  N≡︎N-CH₃ | 211.5, 128.5 |
| N-imidazolium ion |  HN⋯+⋯N-CH₃ | 204.1, 203.6 |
| 4-methylimidazole | 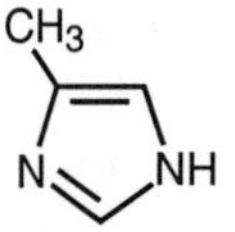 CH₃ on ring, N≡︎NH | 172.8, 164.4 |
| 4-methylimidazolium ion | 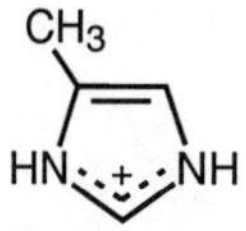 CH₃ on ring, HN⋯+⋯NH | 202.6, 198.6 |

**a.** Interpret the $^{15}N$ chemical shifts listed above, assigning each shift to its appropriate nitrogen and recognizing that, when $^{15}N$ shifts are close (within 1-4 ppm or so), the differences are not usually very significant. Show your reasoning and remember that positive shifts from nitric acid represent increased shieldings (see Exercise 10-3b).

**b.** When N-methylimidazole is dissolved in benzene, the $^{15}N$ chemical shifts become 215.7 and 111.4 ppm relative to $^{15}N$-$HNO_3$. Assign these shifts to the appropriate nitrogens and explain why these shifts occur.

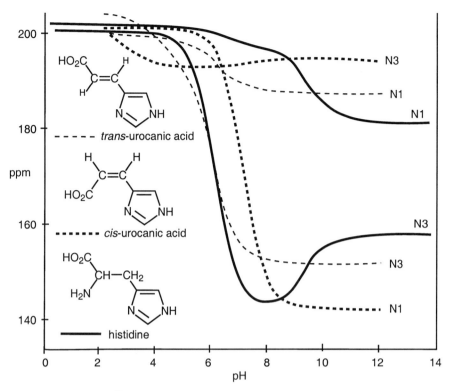

**Figure 10-4** Plots of pH vs. $^{15}$N for Exercise 10-6c

**c.** Figure 10-4 shows the changes in $^{15}$N-shift observed for the **imidazole** nitrogens of *trans*-urocanic acid, *cis*-urocanic acid and histidine as a function of pH. You are to interpret the general features of these spectra and it is suggested that you start by writing out the possible equilibria and influences on the $^{15}$N-shift changes for *trans*-urocanic acid, then proceed to *cis*-urocanic acid (take note of which nitrogens give which pH curve) and finally explain the changes observed for histidine. Be sure you consider why the ionization of *cis*-urocanic acid around pH 7-8 seems to be less favorable than those of the other two acids. You should recognize that both tautomeric equilibrations and hydrogen bondings will influence the shifts. Also, that imidazole rings are protonated and carboxyl groups unionized at pH < 3. Further, that carboxylic acids are stronger acids than imidazolium cations by about 2-3 p$K_A$ units and that ammonium groups are somewhat weaker acids than imidazolium cations.

# 11

# Measurement of Rates by NMR

Nuclear magnetic resonance has extraordinary possibilities for determination of reaction rates. The simplest of them is to make conventional measurements of concentration vs. time by simply following the changes in peak height (or better the integral) of a reactant, a product or both. This procedure is widely applicable and can often be carried out using an NMR tube as the reaction vessel and the NMR spectrometer as the venue for the reaction, with heating or cooling of the sample as necessary to attain a convenient reaction rate. However, more subtle and far-reaching procedures are possible for the determination of reaction-rate constants by NMR that are applicable to many, very rapid, reactions and draw extensively on the concepts of nuclear magnetization and transfer of such magnetization that we have considered earlier.

## 11-1 An Example of the Determination of Rates by NMR

We indicated as far back as Chapter 1 that the line widths of the transitions between the nuclear spin states depend on the lifetimes of the nuclei in those states. Long lifetimes are associated with narrow lines and short lifetimes with broad lines (see Section 1-2). One way for a lifetime to be short is for a process to occur that changes the nature of the state and this can be as simple as a change in chemical shift or coupling constant. Such a chemical-shift change can take place by a change in shielding of one or more of the component nuclei in a spin state. The classic example of such behavior is observed with *N,N*-dimethylmethanamide-1-*d*, (*N,N*-dimethylformamide-1-*d*), a substance that is usually formulated as having substantial double-bond character in its C(O)-N bonds as the result of electron delocalization of the nitrogen unshared electron pair into the double-bond orbitals of the C=O bond.[1] Such delocalization is ascribed to contributions of resonance structures with a carbon-nitrogen double bond and charges on nitrogen and oxygen. We use the deuterio compound as our example here to avoid complications in the proton NMR spectrum of the amide as the result of small $^4J_{HH}$ couplings between the HC(O) and $(CH_3)_2$ protons that occur in the all-protio compound.

---

[1] However, see K.B. Wiberg, and K.E. Laidig, *J. Am. Chem. Soc.*, **109**, 5935, (1987) for an alternative structural formulation supported by quantum-mechanical calculations.

## 11-1 An Example of the Determination of Rates by NMR

$$\underset{\underset{CH_3}{|}}{\overset{\overset{O}{\|}}{D-C}}{\overset{}{\diagdown}}_{N-CH_3} \quad \longleftrightarrow \quad \underset{\underset{CH_3}{|}\atop +}{\overset{\overset{O^-}{|}}{D-C}}{\overset{}{\diagup\!\!\!\diagdown}}_{N-CH_3}$$

Because of the expected partial double-bond character of the C(O)-N bond in the hybrid, it is expected that the molecule will be most stable in the planar configuration and there will be a barrier to rotation about this bond that will be greater than for an ordinary C-N single bond, but less than for a full C=N double bond. The proton NMR spectrum of the pure liquid at room temperature shows two equal-intensity resonances separated by 0.09 ppm. These resonances are just as expected, with signals arising from the $CH_3$ protons that are respectively *cis* and *trans* to the carbonyl oxygen, with, of course, rapid rotation about the N-$CH_3$ bonds.

The interesting thing about this spectrum is that, on heating, the two $CH_3$ resonances begin to broaden, move closer together, then later coalesce to give a single broad peak centered on the average of the two chemical shifts and finally produce a single sharp line. An idealized sequence of the changes of lineshapes of the $CH_3$ resonances is shown in Figure 11-1. The reproducibility of the shapes of the resonances is, to a large degree, subject to vagaries of spectrometer performance, but the temperature at which the separate methyl resonances coalescence occurs is usually rather easy to determine, although there are often serious problems with calibration of the actual internal temperatures in NMR sample tubes. The temperature at which the separate lines coalescence depends on the strength of the magnetic field of the spectrometer. Spectra taken in a strong magnetic field, as corresponds to 500-MHz proton spectra, will have higher coalescence temperatures than those taken at a low field such as 40-90 MHz proton frequencies and, at 500 MHz, the coalescence point for *N,N*-dimethylmethanamide-1-*d*, may be above the normal temperature operating range of your NMR probe.

What is happening here is that the rate of rotation about the C(O)-N bond increases with temperature and, as the lifetimes of the protons of the $CH_3$- groups in the *cis* or *trans* locations become short relative to the reciprocal of the line width (in rad/sec) of their resonances in each location, the lines begin to broaden. As the lifetimes of the states decrease with further increases in temperature, the lines continue to broaden and begin to move toward each other and approach coalescence.

The temperature at the coalescence point is expected to depend on the chemical-shift difference by the following argument. Suppose the rate of rotation is essentially

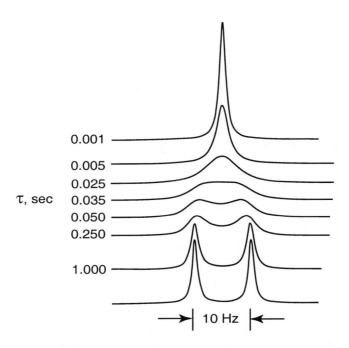

**Figure 11-1** Changes in NMR lineshapes for exchanges involving an equally populated two-site system as a function of $\tau$, the mean lifetime of the nuclei at each site, with a chemical-shift difference of 10 Hz and no spin-spin coupling. The curves were calculated by the procedure described in Section 11-2, specifically by Equations 11-32 to 11-36

infinitely fast. Then, if we take a CW spectrum, or accumulate and transform a FID, the individual resonance frequencies of the $CH_3$ protons in each location will not affect the spectrum, all that can be observed is their average resonance frequency. The observed line width of the average line will then depend on the uncertainty in the lifetime of the average of the states involved. The greater the chemical-shift difference between the locations, the faster the rotation will have to be to effectively average the resonance frequencies to a single line. Because the chemical shift depends on the applied field, and the rate of rotation depends on the temperature, we expect the coalescence point to depend on the strength of the applied field. If we have a system in which there are exchanges of nuclei between two sites and spin-spin coupling is not evident, so that the spectral changes are like those shown in Figure 11-1; then, provided that the equilibrium constant is unity and the line widths in the slow, or in the fast, exchange limits are substantially less than the chemical-shift difference, the exchange rate at the coalescence point is given by Equation 11-1, where $(\nu_A - \nu_B)$ is the chemical-shift difference in Hz.

$$k = \pi(v_A - v_B)/\sqrt{2} \qquad \text{Eqn. 11-1}$$

Applied to the system of Figure 11-1, which shows lineshapes for a system with a 10-Hz shift difference, the $\tau$ (= $1/k$) calculated value for coalescence is 0.045 s. This is a reasonable fit with Figure 11-1, because when $\tau$ is 0.035 s, the individual peaks can not be seen but, when $\tau$ is 0.050, the spectra are clearly not quite to coalescence.

## 11-2 Derivations of Lineshapes in the Absence of Spin-Spin Splitting

A more quantitative treatment of the changes of NMR spectrum with changes in reaction rate, for exchange between two sites with different chemical shifts, such as rotation about a bond, or a simple intermolecular process like an exchange between an OH proton of an alcohol and solvent water, can be derived using the Bloch equations (Chapter 4), **provided that there are no complications from spin-spin splitting**. The derivation can be simplified by using the slow-passage equations developed for CW operation and assuming that the $B_1$ power level is small enough so that $M_z$ can be approximated by $M_0$ which is the case when $\gamma^2 B_1^2 T_1 T_2 \ll 1$ (see Equations 4-22, 4-23 and 4-24 in Section 4-3). The changes in spectra calculated in this way are directly applicable to pulse-Fourier transform operation, as long as time-averaging saturation effects are absent, as would be true when $T_1$ relaxation is essentially complete before the pulse is repeated. The point of this Section is to show how the Bloch equations can be applied to the lineshapes of exchange processes, but if you are not interested in that you should immediately skip to Section 11-4 that emphasizes applications.

If you start with Equations 4-22 to 4-24, take $\gamma^2 B_1^2 T_1 T_2 \ll 1$ and neglect exchange, you can write:

$$M_z = M_o \qquad \text{Eqn. 11-2}$$

$$v = \frac{-\gamma B_1 M_0 T_2}{1 + (\omega_0 - \omega)^2 T_2^2} \qquad \text{Eqn. 11-3}$$

$$u = \frac{\gamma B_1 (\omega_0 - \omega) T_2^2 M_0}{1 + (\omega_0 - \omega)^2 T_2^2} \qquad \text{Eqn. 11-4}$$

The magnetization, call it G, in the X, Y plane will be $u + iv$ for one nucleus, or a set of equivalent nuclei. So, from Equations 11-3 and 11-4, we have:

$$G = \frac{\gamma B_1 (\omega_0 - \omega) T_2^2 M_0}{1 + (\omega_0 - \omega)^2 T_2^2} + i \frac{\gamma B_1 M_0 T_2}{1 + (\omega_0 - \omega)^2 T_2^2} \qquad \text{Eqn. 11-5}$$

If we look at the rate problem in somewhat more general terms than before, we can consider the effect of intramolecular interchange of nuclei at site A with those at site B.[2] The reaction is reversible and we can write:

$$A \underset{k_B}{\overset{k_A}{\rightleftarrows}} B \quad \text{for which} \quad K = \frac{[B]}{[A]} \qquad \text{Eqn. 11-6}$$

where $k_A$ and $k_B$ are the respective rate constants, the equilibrium constant is $K$ and the concentrations of A and B are $[A]$ and $[B]$. For the rate processes we are considering, the concentrations will be those at equilibrium, but the equilibrium constant will not be required to be unity. Assuming that the magnetization, going into and coming out of each site, is proportional to the chemical concentration at that site, we have:

$$\frac{dG_A}{dt} = -k_A G_A + k_B G_B \qquad \text{Eqn. 11-7}$$

What this equation tells us is that the magnetization at site A is decreased at the rate $k_A G_A$ and increased at the rate $k_B G_B$. A similar equation can be written for $G_B$:

$$\frac{dG_B}{dt} = -k_B G_B + k_A G_A \qquad \text{Eqn. 11-8}$$

To make further progress, we need to rewrite the equations for $G_A$ and $G_B$ in terms of dispersion $u$ and absorption $v$. For $G_A$, we have:

$$G_A = u_A + i v_A \qquad \text{Eqn. 11-9}$$

$$\frac{dG_A}{dt} = \frac{du_A}{dt} + i \frac{dv_A}{dt} \qquad \text{Eqn. 11-10}$$

Using the derivatives $du/dt$ and $dv/dt$ of the Bloch equations (Section 4-2), we can write:

$$\frac{du_A}{dt} = -v_A(\omega_{0A} - \omega) - \frac{u_A}{T_{2A}} \qquad \text{Eqn. 11-11}$$

$$\frac{dv_A}{dt} = +u(\omega_{0A} - \omega) - \frac{v_A}{T_{2A}} - \gamma B_1 M_{0A} \qquad \text{Eqn. 11-12}$$

Combining Equations 11-10 to 11-12 gives Equation 11-13 and this, after some algebra, settles down as Equation 11-14.

---

[2] H.M. McConnell, *J. Chem. Phys.*, **28**, 430 (1958).

## 11-2 Derivations of Lineshapes in the Absence of Spin-Spin Splitting

$$\frac{dG_A}{dt} = -v_A(\omega_{0A} - \omega) - \frac{u_A}{T_{2A}} + iu_A(\omega_{0A} - \omega) - i\frac{u_A}{T_{2A}} - i\gamma B_1 M_{0A} \qquad \text{Eqn. 11-13}$$

$$= -(u_A + iv_A)\frac{1}{T_{2A}} - (v_A - iu_A)(\omega_{0A} - \omega) - i\gamma B_1 M_{0A}$$

$$= -G_A \frac{1}{T_{2A}} + i(u_A + iv_A)(\omega_{0A} - \omega) - i\gamma B_1 M_{0A}$$

$$= -G_A \frac{1}{T_{2A}} + iG_A(\omega_{0A} - \omega) - i\gamma B_1 M_{0A}$$

$$= -G_A \left[ \frac{1}{T_{2A}} + i(\omega_{0A} - \omega) \right] - i\gamma B_1 M_{0A} \qquad \text{Eqn. 11-14}$$

and for $G_B$, we can write Equation 11-15.

$$\frac{dG_B}{dt} = -G_B \left[ \frac{1}{T_{2B}} - i(\omega_{0B} - \omega) \right] - i\gamma B_1 M_{0B} \qquad \text{Eqn. 11-15}$$

To get the complete story on the change of $G_A$ and $G_B$ with time, we combine the Bloch components with the rate components. The resulting equations, when further combined with appropriate equations for the decay of the Z magnetization at site A and site B, could be evaluated by numerical integration. However, the experimental conditions usually can be arranged so that this is not necessary. Combining Equation 11-8 with Equation 11-15 and introducing $p_A$ as the fraction of A and $p_B$ as the fraction of B, $p_A + p_B = 1$, $M_{0A} = p_A M_0$ and $M_{0B} = p_B M_0$, provides us with Equation 11-16 that has general applicability. Of course, there will also be a corresponding equation for $dG_A/dt$.

$$\frac{dG_B}{dt} = -k_B G_B + k_A G_A - G_B \left[ \frac{1}{T_{2B}} - i(\omega_{0B} - \omega) \right] - i\gamma B_1 p_B M_0 \qquad \text{Eqn. 11-16}$$

Under steady-state conditions of slow passage, the $B_1$ power input, effects of the reaction rates, relaxation and so on balance out, and equilibrium is set up so that:

$$\frac{dG_A}{dt} = \frac{dG_B}{dt} = 0 \qquad \text{Eqn. 11-17}$$

and the total X, Y magnetization will be $G = G_A + G_B$. Now, to make the equations simpler, we define the quantities $D_A$, $D_B$, and $C$ as:

$$D_A = \frac{1}{T_{2A}} - i(\omega_{0A} - \omega) \qquad \text{Eqn. 11-18}$$

$$D_B = \frac{1}{T_{2B}} - i(\omega_{0B} - \omega) \qquad \text{Eqn. 11-19}$$

$$C = \gamma B_1 M_0 \qquad \text{Eqn. 11-20}$$

Then, substituting Equations 11-18 and 11-20 into the counterpart of Equation 11-16 that deals with $dG_A/dt$, we can write for the steady state:

$$G_A(k_A + D_A) = k_B G_B - iCp_A \qquad \text{Eqn. 11-20}$$

$$G_A = \frac{k_B G_B - iCp_A}{k_A + D_A} \qquad \text{Eqn. 11-21}$$

If we substitute $G_A$ of Equation 11-21 into the steady-state version of Equation 11-16, where $dG_A/dt = 0$, we get:

$$0 = -k_B G_B + k_A \left( \frac{k_B G_B - iCp_A}{k_A + D_A} \right) - G_B D_B - iCp_B \qquad \text{Eqn. 11-22}$$

Algebraic manipulations of Equation 11-22 lead us finally through Equation 11-23 to Equation 11-24 and its counterpart for $G_A$, Equation 11-25:

$$G_B \left( k_B - \frac{k_A k_B}{k_A + D_A} + D_B \right) = -\frac{ik_A Cp_A}{k_A + D_A} - iCp_B \qquad \text{Eqn. 11-23}$$

$$\frac{G_B(k_A k_B + k_B D_A - k_A k_B + k_A D_B + D_A D_B)}{k_A + D_A} = iC \frac{(k_A p_A + p_B k_A + D_A p_B)}{k_A + D_A}$$

$$G_B(k_B D_A + k_A D_B + D_A D_B) = iC[k_A(p_A + p_B) + D_A p_B]$$

$$G_B = -\frac{iC(k_A + D_A p_B)}{(k_A D_B + k_B D_A + D_A D_B)} \qquad \text{Eqn. 11-24}$$

$$G_A = -\frac{iC(k_B + D_B p_A)}{(k_A D_B + k_B D_A + D_A D_B)} \qquad \text{Eqn. 11-25}$$

## 11-2 Derivations of Lineshapes in the Absence of Spin-Spin Splitting

These last two equations can now be combined to give a general equation for G under slow-passage conditions with small $B_1$, (the lineshapes so defined will be those observed in usual CW or pulse operations).

$$G = G_A + G_B = -iC\left(\frac{k_A + k_B + D_A p_B + D_B p_A}{k_A D_B + k_B D_A + D_A D_B}\right)$$ Eqn. 11-26

To make further progress, we need to substitute for $D_A$ and $D_B$ as defined by Equations 11-18 and 11-19, then separate out the imaginary terms (remembering that $i^2 = -1$). This is straightforward, but tedious. For now, let us proceed by simplifying the system.

For rotation about the bonds of N,N-dimethylmethanamide-1-d, $p_A$ will necessarily be equal to $p_B$ and hence $k_B = k_B = k$. Now, if we assume that $T_2 = T_{2A} = T_{2B}$ (unlikely to be exactly true), the equations become simpler, but are still tedious to work through. An often-used simplifying assumption is to suppose that, in the most interesting part of the changes in lineshape, the observed line widths will be much greater than $2/T_2$, so that in effect $1/T_2 = 0$. With this assumption, the mathematics further simplify and the appropriate substitution in Equation 11-26 for $D_A$ and $D_B$ (Equation 11-18 and 19), along with the simplifying assumption gives:

$$G = iC\frac{2k - i(\omega_{0A} + \omega_{0B} - 2\omega)/2}{-ik(\omega_{0A} + \omega_{0B} - 2\omega) - (\omega_{0A} - \omega)(\omega_{0B} - \omega)}$$ Eqn. 11-27

Here, it is convenient to define the quantity $\Delta\omega$ that is the difference between the average of the shifts of nuclei A and B and the observing frequency. Thus, $\Delta\omega = (\omega_{0A} + \omega_{0B})/2 - \omega$. Substituting $\Delta\omega$ into Equation 11-27 and further tedious algebra followed by separating out the imaginary $v$ term gives Equation 11-28 for the absorption mode signal.

$$G = -iC\left(\frac{2k - i\Delta\omega}{-2ik\Delta\omega - (\omega_{0A} - \omega)(\omega_{0B} - \omega)}\right)\frac{[(2ik\Delta\omega - (\omega_{0A} - \omega)(\omega_{0B} - \omega)]}{[(2ik\Delta\omega - (\omega_{0A} - \omega)(\omega_{0B} - \omega)]}$$

$$= -iC\left(\frac{4ik^2\Delta\omega + 2k\Delta\omega^2 - (\omega_{0A} - \omega)(\omega_{0B} - \omega)(2k - i\Delta\omega)}{4k^2\Delta\omega^2 + (\omega_{0A} - \omega)^2(\omega_{0B} - \omega)^2}\right)$$

$$G = -C\left(\frac{4k^2\Delta\omega + 2ik\Delta\omega^2 - (2ik + \Delta\omega)(\omega_{0A} - \omega)(\omega_{0B} - \omega)}{4k^2\Delta\omega^2 + (\omega_{0A} - \omega)^2(\omega_{0B} - \omega)^2}\right)$$

$$v = -C\left(\frac{2k[\Delta\omega^2 - (\omega_{0A} - \omega)(\omega_{0B} - \omega)]}{4k^2\Delta\omega^2 + (\omega_{0A} - \omega)^2(\omega_{0B} - \omega)^2}\right) \qquad \text{Eqn. 11-28}$$

In much of the literature concerning NMR and reaction rates, a time $\tau$ is defined for the general case of a two-site system as:

$$\tau = \frac{\tau_A \tau_B}{\tau_A + \tau_B} \qquad \text{Eqn. 11-29}$$

The rate constants for the separate sites are $k_A = 1/\tau_A$ and $k_B = 1/\tau_B$. So, for the simple case, where $k_A = k_B = k$, $\tau$ is equal to $1/(2k)$. There is sometimes confusion as to whether any given author has equated $1/\tau$ to the rate constant $k$ of the reaction, or has used $\tau = 1/(2k)$, as it is defined above. If the magnitudes of reported rate constants obtained by NMR are important, it is necessary to be certain which convention is employed.

If you square $\Delta\omega = (\omega_{0A} + \omega_{0B})/2 - \omega$ and substitute for $\Delta\omega^2$ in Equation 11-28, you find that $\Delta\omega^2 - (\omega_{0A} - \omega)(\omega_{0B} - \omega) = (\omega_A - \omega_B)^2/4$. Then, letting $k = 1/(2\tau)$, we can write:

$$v = -C\frac{\frac{1}{4\tau}(\omega_{0A} - \omega_{0B})^2}{\frac{1}{\tau^2}\Delta\omega^2 + (\omega_{0A} - \omega)^2(\omega_{0B} - \omega)^2} \qquad \text{Eqn. 11-30}$$

multiplying through by $\tau$ and including $4^{-1}$ in $C$ gives:

$$v = -C\frac{\tau(\omega_{0A} - \omega_{0B})^2}{\Delta\omega^2 + \tau^2(\omega_{0A} - \omega)^2(\omega_{0B} - \omega)^2} \qquad \text{Eqn. 11-31}$$

The lineshape, where $C$ is to be regarded simply as a proportionality constant, is easily calculated from Equation 11-31 as a function of $\omega$ for specified values of $\omega_A$, $\omega_B$ and $\tau$. The shapes will usually be quite accurate for the intermediate values of $\tau$, where the rate process produces substantial line broadening, but will be inaccurate when $\tau$ is very long or very short, because $1/T_2$ was taken to be zero and that assumption corresponds to infinitely sharp lines when exchange is either very fast or very slow.

Solution of the steady-state equation for the imaginary part of $G$ when $p_A \neq p_B$ and $1/T_{2A} \neq 1/T_{2B} \neq 0$ is straightforward, but tedious and leads to Equation 11-32.

$$G = -C\frac{\left\{P\left[1 + \tau\left(\frac{p_B}{T_{2A}} + \frac{p_A}{T_{2B}}\right)\right] + QR\right\}}{(P^2 + R^2)} \qquad \text{Eqn. 11-32}$$

## 11-4 Lineshape Analysis to Determine Rates of Intramolecular Processes

where $P$, $Q$, $R$ and $\tau$ are defined by Equations 11-33 to 11-36, respectively. Equation 11-36 may not look like Equation 11-29, but it is an equivalent definition of $\tau$.

$$P = \tau\left[\frac{(\omega_A - \omega_B)^2}{4} - \Delta\omega^2 + \frac{1}{T_{2A}T_{2B}}\right] + \frac{p_A}{T_{2A}} + \frac{p_B}{T_{2B}} \qquad \text{Eqn. 11-33}$$

$$Q = \tau\left[\Delta\omega - \frac{(\omega_A - \omega_B)}{2}(p_A - p_B)\right] \qquad \text{Eqn. 11-34}$$

$$R = \Delta\omega\left[1 + \tau\left(\frac{1}{T_{2A}} + \frac{1}{T_{2B}}\right)\right] + \frac{\tau\Delta\omega}{2}\left(\frac{1}{T_{2B}} - \frac{1}{T_{2A}}\right) + \frac{\omega_A - \omega_B}{2}(p_A - p_A)\right] \qquad \text{Eqn. 11-35}$$

$$\tau = \tau_A p_B = \tau_B p_A \qquad \text{Eqn. 11-36}$$

Equations 11-32 to 11-36 can be used for intermolecular exchange processes where only two chemical shifts and no spin-spin couplings are involved as for exchange of the hydroxyl protons of water and an alcohol like 2-methyl-2-propanol (chosen to avoid the complication of spin-spin coupling that would be expected in the slow exchange extreme for an alcohol with protons on the α carbon, see Exercise 1-1). For this kind of rate measurement, it is necessary to use appropriate values of $p_A$ and $p_B$ that represent the respective fractions of the nuclei having the respective chemical shifts at slow exchange. As before though, it may be necessary to provide appropriate corrections for changes of shift with temperature before the lines are substantially overlapped. In the fast-exchange regime, the observed chemical shift will be a weighted average of the separate shifts. An example is shown in Figure 11-2, where the chemical-shift difference is as in Figure 11-1 equal to 10 Hz, but now the ratio $p_A/p_B$ is equal to two.

### 11-3 Multi-Site Intermolecular Exchange Processes

The lineshape changes associated with multi-site exchange processes where spin-spin couplings are not involved can be treated mathematically very much in the same way as was done for the two-site situation in Section 11-2. The key point is that the overall lineshape function $G$ will be the sum of the contributions from the exchanging nuclei at each site. Thus, for a three-site exchanging system, we can modify Equation 11-16 and write the separate $G_n$ functions as follows:

$$\frac{dG_A}{dt} = -\tau_{AB}^{-1}G_A - \tau_{AC}^{-1}G_A + \tau_{BA}^{-1}G_B + \tau_{CA}^{-1}G_C - G_A\left[\frac{1}{T_{2A}} - i(\omega_{0A} - \omega)\right] - i\gamma B_1 p_A M_0 \qquad \text{Eqn. 11-37}$$

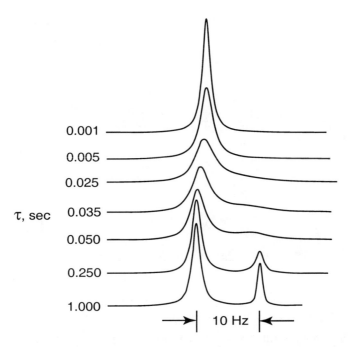

**Figure 11-2** Changes in NMR lineshapes for exchanges involving an unequally populated two-site system as a function of the exchange time $\tau = \tau_A p_B = \tau_B p_A$. The chemical shift-difference is 10 Hz, there is no spin-spin coupling and $T_{2A} = T_{2B} = 0.5$ Hz. The curves were calculated by the procedure described in Section 11-2, specifically by Equations, 11-32 to 11-36

$$\frac{dG_B}{dt} = -\tau_{BA}^{-1}G_B - \tau_{BC}^{-1}G_B + \tau_{AB}^{-1}G_A + \tau_{CB}^{-1}G_C - G_B\left[\frac{1}{T_{2B}} - i(\omega_{0B} - \omega)\right] - i\gamma B_1 p_B M_0 \quad \text{Eqn. 11-38}$$

$$\frac{dG_C}{dt} = -\tau_{CA}^{-1}G_C - \tau_{CB}^{-1}G_C + \tau_{AC}^{-1}G_A + \tau_{BC}^{-1}G_B - G_C\left[\frac{1}{T_{2C}} - i(\omega_{0C} - \omega)\right] - i\gamma B_1 p_C M_0 \quad \text{Eqn. 11-39}$$

You will see in Equations 11-37 to 11-39 the rates at which the magnetization comes and goes at each of the sites along with the $T_2$ relaxation and $B_1$ driving terms. To make further progress on this system of differential equations. we use the steady-state approximation that $dG_A/dt = dG_B/dt = dG_C/dt = 0$. However, as you can see from the steps taken after Equation 11-16 in Section 11-2, working out the final equations for the imaginary component of $G_n$ for a multi-site case, even with only three sites, will require some very heavy algebra. Sandström[3] gives an excellent account of the difficulties and possible solutions. In some multi-site systems, such as the exchange of the

---

[3] J. Sandström, *Dynamic NMR Spectroscopy*, Academic Press, New York, 1982, pp. 18-25.

## 11-4 Lineshape Analysis to Determine Rates of Intramolecular Processes

hydroxyl protons of two different alcohols and water it would be expected that the protons on any given site could go to any other site at somewhat comparable rate constants. In other systems, exchange between, say Site A and Site C, might only take occur through Site B as an intermediate. The lineshapes in the two kinds of systems will be different in general and thus provide a test for the exchange mechanism.

### 11-4 Lineshape Analysis to Determine Rates of Intramolecular Processes

Equations 11-32 to 11-36 may look formidable, but in fact they are quite easy to program to calculate useful lineshapes with any respectable PC. For example, lineshapes involving 300 points can be calculated with those equations using True BASIC with a not-very-fast Macintosh in less than 2 seconds. Programming the algorithm is trivial, what is tedious is programming the data input, display and output so that one can easily match calculated lineshapes to experimental lineshapes. In the ideal case, there should be sufficient control over the plotting parameters to be able to actually superimpose the calculated lineshapes on the experimental spectra. Another approach is to use digitized experimental lineshapes from the spectrometer, then make all-in-the-computer fits by procedures that systematically vary the parameters to attain some specified degree of precision for fits between experimental and calculated lineshapes.

The results of such an analysis of an exchanging two-site system is shown for rotation about the C-N bonds of thiourea in Figure 11-3. You can write resonance structures for thiourea that are quite analogous to those written for dimethylmethanamide (Section 11-1) and these predict some degree of double-bond character and resistance to rotation about the C-N bonds.

With thiourea, the sulfur can be taken to be involved in resonance with two nitrogens rather than one as for the oxygen of dimethylmethanamide, so it is not surprising that the free energies of activation for rotation about the C-N bonds of thiourea and urea are respectively about 14 kcal/mol and 11 kcal/mol and thus considerably smaller than the free energy of activation of about 20 kcal/mol for rotation about the C-N bonds of dimethylmethanamide and most other normal amides.

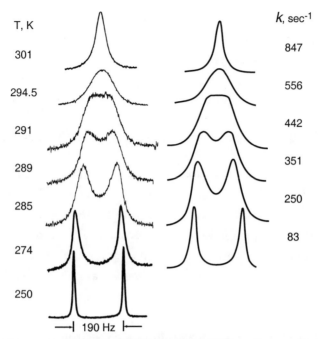

**Figure 11-3** Comparison of experimental and calculated NMR resonance lineshapes for thiourea dissolved in dimethylmethanamide/deuteriodimethyl sulfoxide solution as a function of temperature. The rate constants are $1/\tau$ s.[4] The system is simpler than might otherwise be expected because the $^2J_{HH}$ couplings are close enough to zero so they can be neglected. Spectra taken by K.A. Haushalter

A typical case where one would need to use the complete capabilities of Equations 11-32 to 11-36 would be in measuring the rates of inversion of N-methyl-2-methylaziridine by following the change of the NMR resonances of the N-methyl group.

N-methyl-2-methylaziridine

The equilibrium constant between the two isomeric forms of N-methyl-2-methylaziridine is not expected to be unity because the forms will surely differ in energy. Likewise, the chemical shift of the N-methyl group is not expected to be the same when it is on the same side of the ring as the 2-methyl group as when it is on the opposite side. Here then, we have to take account of the values of $p_A$, $p_B$, $\omega_{0A}$, and $\omega_{0B}$, as well as $T_{2A}$ and $T_{2B}$ for the best accuracy. We will expect the experimental and calculated

---

[4] K.A. Haushalter, J. Lau and J.D. Roberts, *J. Am. Chem. Soc.*, **118**, 8891 (1996).

## 11-4 Lineshape Analysis to Determine Rates of Intramolecular Processes

spectra to have lineshapes very much like those in Figure 11-2.

The important thing to remember for lineshape calculations is that each requires use of parameters appropriate to the experimental conditions of the NMR spectrum that you hope to duplicate. If the rate measurements involve changes in temperature, we cannot necessarily expect that $p_A$ and $p_B$, as well as $\omega_{0A}$ and $\omega_{0B}$, $T_{2A}$ and $T_{2B}$ will remain constant, wholly independent of temperature. Consequently, for best results, it is desirable to check the manner in which each of these parameters changes with temperature, preferably by extrapolation from the slow-exchange region where it is easiest to determine the individual parameters. In many systems, it is possible to make quite substantial errors by assuming that $\omega_{0A} - \omega_{0B}$ is constant over the range of rates being investigated. This is particularly true when, rather than a complete lineshape analysis, one of several possible relationships is used between chemical shifts, peak positions and reaction rates without correcting for changes in chemical shifts associated with temperature or other experimental variables. In general, rate processes themselves do not cause significant changes in peak position until the individual peaks broaden enough to overlap substantially. When there is evidence that the chemical shifts are dependent on other factors than changes in rate, then it is important to measure and extrapolate the shifts into the experimental conditions of interest and use the extrapolated shifts in the lineshape calculations.

Determination of proper $T_{2A}$ and $T_{2B}$ values to use in lineshape calculations can be difficult. Fortunately, as we explained in the derivations of Equations 11-27 to 11-28, the values of $T_{2A}$ and $T_{2B}$ are usually not very important when the resonances are substantially broadened and overlap as the result of rate processes. If it is not already clear, it is necessary to understand that $T_{2A}$ and $T_{2B}$ are not directly related to the line widths observed in the intermediate rate situation. Instead, they are related to the line widths that would be observed if everything else were the same, including field inhomogeneities, but the rate process itself did not occur. It is easy to check whether $T_{2A}$ or $T_{2B}$ is important in determining a given calculated lineshape by changing one or the other value and seeing whether the exchange lineshape changes. If it does, then it will be important to determine the line widths (and from them the $T_2$ values, see Section 4-2) in the fast- and slow-exchange domains as a function of the experimental conditions and extrapolate these into the intermediate ranges to calculate more accurate lineshapes.

In determining reaction rates, as a function of a particular experimental variable, such as temperature, it is also very important to determine a large number of experimental NMR lineshapes in the region where the shapes are most sensitive to the variable. Errors can be very large, especially entropies of activation, when only two or

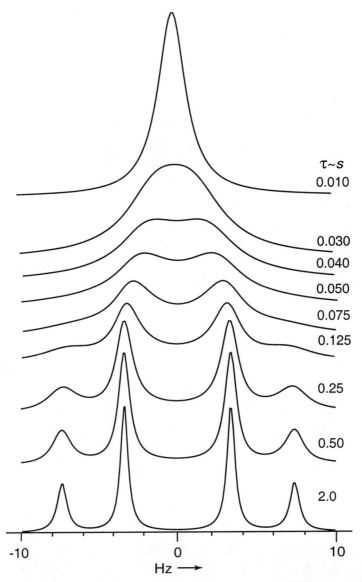

**Figure 11-4** Calculated lineshapes as a function of $\tau \sim s$ for an exchanging AB system made with the aid of Equations 11-39 to 11-43. The chemical-shift difference $v_A - v_B$ was equal to 10 Hz, the $J_{AB}$ coupling was 4 Hz and the line width in the extreme slow-exchange regime was 0.5 Hz. The progression of lineshapes with exchange rate demonstrates for this kind of system how sensitive matching lineshapes is for determination of exchange rates. In general, the rate constants should not be expected to be more accurate than 5%

three lineshapes are determined and fitted in the most sensitive region, while other curves are used that are too close to the limits of detectability of differences between experimental and calculated lineshapes by being near to the fast- or slow-exchange extremes. For this reason, it is very common in the literature to report only the free energy of activation $\Delta G^\ddagger$ obtained from measurement of the rate found at the coalescence point where the accuracy is greatest.

## 11-5 Lineshape Analysis for J-Coupled Systems

As we have seen earlier, when $J$ couplings get to be comparable to the chemical-shift differences, the Bloch equations are no longer directly applicable and derivations of the equations for NMR lineshapes are difficult to derive. The usual procedure in such circumstances is to employ the density-matrix formalism and, with it, it is possible, but not at all easy, to deal with quite complex spin systems. Even the simple AB case of two exchanging nuclei with chemical shifts comparable to the magnitude of the $J$ coupling leads to formidable algebraic equations, although these are usually relatively easy to program. As one example, the following equations[5] allow calculation of the imaginary component of the magnetization $v(v)$ for two mutually exchanging nuclei of an AB system where the sites have equal probability and it is assumed that both nuclei have the same $T_2$. Here, $k$ is the rate constant in s$^{-1}$ and $v$ is the frequency in Hz at the desired along the lineshape.

$$v(v) = const\left[\frac{r_+b_+ - sa_+}{a_+^2 + b_+^2} + \frac{r_-b_- - sa_-}{a_-^2 + b_-^2}\right] \qquad \text{Eqn. 11-39}$$

$$a_\pm = 4\pi^2(v_0 - v \pm J/2)^2 - (k + 1/T_2)^2 - \pi^2(v_A - v_B)^2 - \pi^2 J^2 + k^2 \qquad \text{Eqn. 11-40}$$

$$b_\pm = 4\pi(v_0 - v \pm J/2)(k + 1/T_2) \mp 2\pi Jk \qquad \text{Eqn. 11-41}$$

$$r_\pm = 2\pi(v_0 - v \pm J) \qquad \text{Eqn. 11-42}$$

$$s = 2k + 1/T_2 \qquad \text{Eqn. 11-43}$$

An example of how the lineshapes of an AB system change with exchange rate is shown in Figure 11-4.

In some situations, the lineshapes of more complex systems than AB can be approximated quite satisfactorily by treating them as combinations of AB systems. Thus, systems such as ABX$_n$ can be handled by calculating separate AB spectra, where $J_{AX}$ and $J_{BX}$ act as perturbations on the chemical shifts, and then summing the separate

---

[5] J. Heidberg, J.A. Weil, G.A. Janusonis and J.K. Anderson, *J. Chem. Phys.* **41**, 1033 (1964)

lineshapes to obtain a comparison with experiment. Thus, the barrier to inversion of 1,1-difluorocyclohexane could be approximated by summing up AB spectra calculated

with appropriate increments of "effective shift differences" corresponding to the $J$ couplings with various possible spin states of the adjacent four protons.[6]

Similarly, the lineshapes for an AA'XX' $\rightleftarrows$ A$_2$X$_2$ exchanging system that involves inversion of configuration at C1 of 3,3-dimethylbutylmagnesium chloride, which exists almost exclusively as the *trans* configuration, could be very well simulated as a superposition of two independent AB systems.[7]

To clarify this exchange process, one hydrogen representative of the AA' group and one of the XX' group are marked with asterisks to show how inversion of the C-Mg bond at C1 followed by rotation about the C-C bond interchanges their positions and, if this process is fast enough, it averages the two coupling constants $^3J_{AX}$ and $^3J_{AX'}$.

## 11-6 Other Methods of Determining Exchange Rates by NMR

I can almost hear you telling me now that measuring exchange rates by comparison of experimental and calculated lineshapes, however serviceable, is ancient history and, then asking me what do I have that is more stylish in the present-day era of FT-NMR spectrometers? Indeed, there are other ways of measuring exchange rates and the most interesting ones are those that tie into our other experiences with NMR theory and practice.

---

[6]S.L. Spassov, D.L. Griffith, E.S. Glazer, K. Nagarajan and J. D. Roberts, *J. Am, Chem, Soc.*, **89**, 88 (1967)
[7]G.M. Whitesides, M. Witanowski and J.D. Roberts, *J. Am, Chem, Soc.*, **87**, 2854 (1965)

## 11-6 Other Methods of Determining Exchange Rates by NMR

**1. The Carr-Purcell Technique.** One of the simplest ideas relies on the Carr-Purcell or Carr-Purcell-Meiboom-Gill procedures for determination of $T_2$ described in Sections 2-3. If you look at Figure 11-1, you see in the fast-exchange regime that the lineshape is at first broad and then narrows as the system progresses beyond the coalescence point to faster exchange rates. This suggests a connection between $T_{2obsd}$, measured by the Carr-Purcell or Carr-Purcell-Meiboom-Gill pulse sequences, and the exchange rate. If such a connection exists, then it will be useful to determine exchange rates above the coalescence point where in general we expect the rates to be rather fast. The connection does exist and it is little more complicated than simply doing what we saw could be done to measure $T_2$ in Section 2-3. The reason is that we have to consider the length of the pulse interval in the sequence 90°-τ-180°-2τ-180°-2τ-180°- -, because, if τ is very short, the exchange process will not have time to influence $T_{2obsd}$. Here, again we encounter a sticky notation problem, because workers in the field use τ for lifetimes in exchange processes and also for pulse intervals without regard for possible confusion for the rest of us. So here, we use $\tau_p$ for the τ in the Carr-Purcell sequence. With this, $T_{2obsd}$ is related to τ of exchange by Equation 11-45.

$$1/T_{2obsd} = 1/T_2 + p_A p_B 4\pi^2 (v_{0A} - v_{0B})^2 \tau \left[ 1 - \frac{2\tau}{2\tau_p} \tanh \frac{2\tau_p}{2\tau} \right] \qquad \text{Eqn. 11-45}$$

You can see from this equation how the Carr-Purcell pulse interval $\tau_p$ influences $T_{2obsd}$. There will be a problem here in that diffusion arising from inhomogeneities in the magnetic field has an effect on $T_2$'s measured with long $\tau_p$ times and we do need long $\tau_p$ times when τ is relatively long to be able to observe the effects of exchange. The Carr-Purcell-Meiboom-Gill sequence was touted in Section 2-3, because it allowed very short $\tau_p$ times to minimize diffusion, but that is not what we want here. If diffusion is not a problem and long $\tau_p$ times are possible that are in fact are much greater than τ, then Equation 11-45 simplifies to Equation 11-46.

$$1/T_{2obsd} = 1/T_2 + p_A p_B 4\pi^2 (v_{0A} - v_{0B})^2 \tau \left[ 1 - \frac{2\tau}{2\tau_p} \right] \qquad \text{Eqn. 11-46}$$

**2. Saturation Transfer.** Another set of procedures related to what we have discussed before is **saturation transfer**, first proposed and investigated by Forsén and Hoffman.[8] The idea is simple, elegant and powerful. It is complementary to the $T_{2obsd}$

---
[8] S. Forsén and R.A. Hoffman, *J. Chem. Phys.* **39**, 2892 (1963); *J. Chem. Phys.* **40**, 1189 (1964).

procedure in that it is best suited to measure rates in the slow-exchange regimes.

This procedure has several forms. Let is consider a particularly simple example.. Suppose that we carry out a CW NMR experiment in the very slow exchange-rate regime on an uncoupled two-site exchanging system. We will use a low power $B_{1A}$ in the slow-passage mode (see Section 4-3), so slow that we will be able to sit indefinitely on the resonance line of one of the chemically shifted site species, call it A. If we do that, equilibrium will be reached between $B_{1A}$ power in and power out (by relaxation) in accord with Equation 4-23. We continue to observe the signal, but now we apply, at a strong power level $B_{1B}$, a continuous rf frequency at the chemical shift of the other nuclei B. If exchange is very slow, nothing will happen to the A nuclei being observed and the nuclei B irradiated at the second frequency will also reach equilibrium, but because the power level at the second frequency is high, the equilibrium $M_z$ of the second nucleus will be close to zero (Equation 4-22).

At this point, suppose we change the conditions in such a way as to make exchange faster. Now, the more magnetically saturated second nuclei will move to the site of the observed nuclei and, at the same time, the less saturated nuclei will move to the irradiation site. The result will be to cause the observed signal intensity of the A nuclei to diminish because the incoming B nuclei have a low $M_z$. Obviously, the A signal strength will depend on time, the exchange rate and the power levels. The start for mathematical treatment of this sort of situation is Equation 4-19 that relates the magnetization to the oscillator $B_1$ power. For non-exchanging A nuclei, we modify Equation 4-20 to get:

$$\frac{dM_{zA}}{dt} = \frac{M_{0zA}-M_{zA}}{T_{1A}} - v_A B_{1A} v_A \qquad \text{Eqn. 11-47}$$

With exchange, $M_{zA}$ will diminish by $-M_{zA}/\tau_A$ and increase by $M_{zB}/\tau_B$ so that if we correct Equation 11-47 to take this into account we get:

$$\frac{dM_{zA}}{dt} = \frac{M_{0zA}-M_{zA}}{T_{1A}} - \frac{M_{zA}}{\tau_A} + \frac{M_{zB}}{\tau_B} - v_A B_{1A} v_A \qquad \text{Eqn. 11-48}$$

An analogous equation can be written for $dM_{zB}/dt$:

$$\frac{dM_{zB}}{dt} = \frac{M_{0zB}-M_{zB}}{T_{1B}} - \frac{M_{zB}}{\tau_B} + \frac{M_{zA}}{\tau_A} - v_B B_{1B} v_B \qquad \text{Eqn. 11-49}$$

If we set up the experiment with a strong $B_{1B}$ power level, the $\gamma^2 B_{1B}^2 T_{1B} T_{2B}$ term in Equation 4-22 will dominate and the B nuclei will have essentially no Z magnetization. This will remove the $M_{zB}/\tau$ term from Equation 11-49 and the fact that the $B_{1A}$ power

## 11-6 Other Methods of Determining Exchange Rates by NMR

level will be low enough to allow detecting the A resonance in the slow-passage mode means that we can write Equation 11-50 for the change in $M_{zA}$ with time when the $B_{1B}$ power is turned on.

$$\frac{dM_{zA}}{dt} = \frac{M_{0zA} - M_{zA}}{T_{1A}} - \frac{M_{zA}}{\tau_A} \qquad \text{Eqn. 11-50}$$

It turns out that the A magnetization falls to an equilibrium value that is given by Equation 11-51 and, if we know $T_{1A}$, then we can calculate $\tau_A$ from the ratio of the signal strengths before and after the $B_{1B}$ power is turned on.

$$M_{zA} = M_{0zA} \frac{\tau_A}{\tau_A + T_{1A}} \qquad \text{Eqn. 11-51}$$

If the $B_{1B}$ power is now turned off, the A magnetization will return to its original value at a rate that will depend in a complex way on $\tau_A$, $\tau_B$, $T_{1A}$, $T_{1B}$ and $M_{0xA}$. Equations for the change in $M_{zA}$ as a function of time and the other parameters are available.[9]

**3. Pulse Methods.** Besides the procedures for obtaining exchange rates by the Carr-Purcell pulse sequences described in Section 11-6-1, it is also possible to use a technique for measuring exchange rates for uncoupled nuclei that is reminiscent of the inversion recovery sequence discussed in Section 7-1. Here, in the simple case of two sets of nuclei A and B, we apply a soft **selective** 180° pulse to the B nuclei (see Section 7-9) and, after a variable waiting period $t_1$, we apply a nonselective 90° analyzing pulse, acquire and transform, just as in the determination of $T_1$. If exchange transfers magnetic polarization from nuclei A to nuclei B, the magnetizations ($M_{zA}$ and $M_{zA}$) of the A and B nuclei will reflect that in the magnitudes of the signals resulting from the 90° pulse.

The results of such an experiment are shown in Figure 11-5 using as the example N,N-dimethylmethanamide. Here, the equations for changes in peak heights are given by Equations 11-52 and 11-53, where it is assumed that $T_1$ is the same in both sites.

$$M_{zA} + M_{zB} = M_{0zA}\left\{\exp\left(-\frac{t_2}{T_1}\right)\right\} \qquad \text{Eqn. 11-52}$$

$$M_{zA} - M_{zB} = M_{0zA}\left\{\exp\left[\left(-\frac{1}{T_1} - \frac{2}{\tau}\right)t_2\right]\right\} \qquad \text{Eqn. 11-53}$$

---

[9] Reference 3, pp. 53-56; F.A.L. Anet and A.J. Bourn, *J. Am. Chem. Soc.*, **89**, 760 (1967).

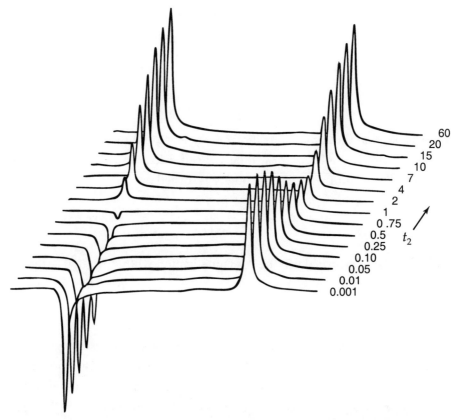

**Figure 11-5** Graph of a saturation transfer experiment carried out on the N-methyl groups of neat dimethylmethanamide at 85° with the $t_2$ in seconds. Spectra taken by C. Molodowitch

We use the variable $t_2$ as the delay time in Equations 11-52 and 11-53, as well as in Figure 11-5, to follow the usual convention for two-dimensional spectra.

It should be clear that Forsén-Hoffman methodology can be used with multi-site exchanges and with simple weakly coupled systems. But for more complicated systems, the density matrix formalism is the method of choice for determining the exchange rates from the experimental spectra.

There is an alternative method using 2D NMR called EXSY that has a pulse sequence is like that of NOESY (Section 8-4-2), except the mechanism of magnetization transfer is through exchange rather than cross relaxation. The pulse sequence starts off a $90°_x$ pulse and acquire. The $t_1$ period is varied to sort out the chemical shifts and the $t_2$ in the same way as COSY (Section 8-4) with $90°_x - t_1 - 90°_x - t_2$ then is followed by a period allows the mixing of the magnetizations at the two different sites by exchange. The resulting 2D spectrum has very strong cross peaks if significant exchange occurs in the $t_2$ period. The 2D plot for a two-site uncoupled system such as dimethylmethanamide will look like Figure 8-22, except there will be no fine structure in the peaks. The amount of exchange occurring during $t_2$ is determined by measuring the volume of the cross peaks as though the 2D plot were made as in the right-hand representation of Figure 8-1. The volume of the peaks will change with the length of the $t_2$ period and can be used to obtain the rate constant. The procedure is elegant, but not easily made accurate. Furthermore, a series of 2D spectra with different $t_2$ values has to be made to obtain the rate constant for any given set of experimental conditions. To determine an activation energy as a function of temperature will require quite a few sets of these 2D series. It is not surprising that the lineshape method still survives.

## Exercises

**Exercise 11-1 a.** Calculate the ratio of the coalescence $\tau$ values for a two-site uncoupled, equal-probability, exchange process with equal $T_2$ values that reaches coalescence at 115° C at 60 MHz and at 180° C at 500 MHz.
**b.** Calculate the activation energy ($E_a$) corresponding to this exchange process.

**Exercise 11-2 a.** Estimate the rate constant at coalescence for rotation about the C-N bonds of thiourea as shown in Figure 11-3 and compare your value with that you can deduce from the changes in lineshapes that were calculated for particular values of $\tau$.
**b.** Check to see how well Equation 11-1 applies to the lineshapes shown in Figure 11-4 where the shapes are complicated by spin-spin couplings.
**c.** In Section 1-2, there is a discussion of the uncertainty principle and how the lifetime of states is approximately related to the spectroscopic line width. Explain how could you apply the relationship suggested by the uncertainty principle that absorption line widths of 1 Hz correspond to a lifetime of states of about 1/6th of a second to estimate the lifetimes of the states at, or near, the coalescence point of Figures 11-3 and 11-4.

**Exercise 11-3 a.** Sketch the changes in lineshape and give the reasons for your expectations when considering the effect of temperature from the slow- to the fast-exchange regimes for a two-site system with $v_A - v_B = 25$ Hz, no spin-spin coupling and $p_A = p_B$, where the nuclei at one site have a very much faster relaxation rate than they do at the other site.
**b.** Calculate the relative position of the fast-exchange resonance line for the spectra shown in Figure 11-2.

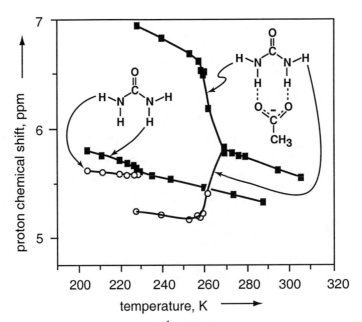

**Figure 11-6** Temperature dependences of the $^1$H NMR N-H chemical shift for urea and urea-acetate ion complex. Chemical shifts are relative to the dimethylmethanamide C-H resonance at 8.03 ppm. The points (■ and ○) correspond to *trans-* and *cis-*urea protons, respectively. Both sets of spectra were taken of 80% dimethylmethanamide/20% deuteriodimethyl sulfoxide solutions and had urea concentrations of 15 mM. One solution was also 15 mM in acetate ion. Spectra taken by K.H. Haushalter

**Exercise 11-4** Analyze Figure 11-6 with emphasis on explaining the slopes of the straight portions of the correlation lines and determining the rates of exchange at the coalescence points. You may wish to review Section 10-4 on proton shifts for information on proton shifts.

**Exercise 11-5** The proton spectrum of 1,2,2-trimethylaziridine at 35° has a single somewhat broadened doublet for its methylene protons separated by 0.71 ppm at 60 MHz and two ring methyl resonances separated by 0.11 ppm. The ring methyl resonances coalesce at 60° and the methylene proton resonances coalesce at 110°. The *N*-methyl resonance is a sharp singlet at all three temperatures. Use this data to calculate $E_a$, the energy of activation for the rate process and explain what caveats you may have about its reliability.

**Exercise 11-6** The 19F NMR spectrum of 1-aza-4,4-difluorocyclohexane dissolved in trichloro- methane or 2-propanone at room temperature gives a closely spaced pentet corresponding to a coupling of about 12 Hz (**a.** in Figure 11-7).

# Exercises

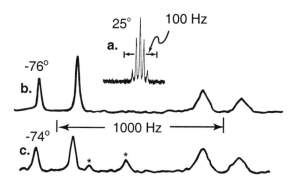

**Figure 11-7** Variable temperature $^{19}$F NMR spectra of 1-aza-4,4-difluorocyclohexane. Spectrum **a.** is independent of whether the solvent is trichloromethane or 2-propanone. Spectrum **b.** is at -76° C and is for trichloromethane as solvent. Spectrum **c.** was taken at -74° C with 2-propanone as solvent. Spectra taken by G.A. Yousif

At about -75°, the spectra are vastly different from room temperature as can be seen in Figure 11-7. Now, the spectrum in trichloromethane (**b.**) is basically an AB quartet with half of the AB quartet having a different line shape than the other half. The general motif of the spectrum in 2-propanone (**c.**) is similar, except that you will notice there are two additional weaker lines on the left-hand side (marked with asterisks) that have an integrated intensity of 0.24 relative to 0.76 for the other two lines on that side. Analyze these spectra touching on the differences between spectra at room and low temperatures, the magnitudes of the various couplings and chemical shifts, and what is happening in 2-propanone compared to trichloromethane solutions. Sketch out several of the intermediate $^{19}$F spectra that you would expect observe as a function of temperature between that of spectrum **a.** and **b.**

**Exercise 11-7 a.** Suppose you made a solution of 3,3-dimethylbutyl-1,2-$d_2$-magnesium chloride, a substance that surely exists, because of steric hindrance, almost completely in a conformation with the MgCl and *t*-butyl groups *trans* to one another. Neglecting any $J_{HD}$ couplings, explain what you would expect the proton spectrum of the *trans* conformation of this substance to look like in the fast- and slow-exchange extremes (review Section 11-5).

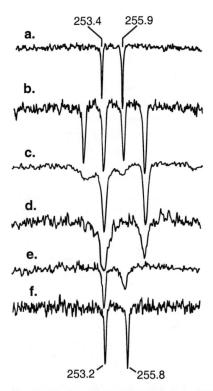

**Figure 11-8** Natural-abundance $^{15}$N NMR spectra of 0.25 M 1-aza-2-cyclononanone in dimethyl sulfoxide. Spectrum **a.** was taken with complete proton decoupling. Spectrum **b.** was the same as **a.** but with gated proton decoupling. Spectrum **c.** was like **b.**, but with 0.003 equivalents added of aqueous 1 M sodium hydroxide solution. Spectrum **d.** was the same as **c.**, except that 0.005 equivalents of the hydroxide solution were added. Spectrum **e.** was also the same as **c.**, except that 0.015 equivalents of the hydroxide solution were added. Spectrum **f.** is the same as **e.**, but now taken with proton decoupling as in **a**. Spectra taken by I. Yavari

**b.** Arguing from the information given for the patterns of proton spectra of $A_2X$ and $AA'X$ systems in Sections 9-4, 9-5 and 9-6, explain how you would expect the proton spectrum of 3,3-dimethylbutylmagnesium chloride would appear in the fast- and slow-exchange extremes.

**Exercise 11-8** Explain how you could achieve the same starting magnetization state for the *N*-methyl groups of *N,N*-dimethylmethanamide by using a combination of **hard** pulses rather than a 180° **selective** pulse at one methyl resonance as described in Section 11-6-**3**.

**Exercise 11-9** Figure 11-8 shows natural-abundance $^{15}$N spectra of 1-aza-2-cyclononanone in dimethyl sulfoxide with the addition of various amounts of aqueous sodium hydroxide solution, with and without complete decoupling of the protons attached to nitrogen. In this system, the sodium hydroxide catalyzes intermolecular proton exchange.

a. Explain why there are two different nitrogen resonances with different intensities in the proton-decoupled $^{15}N$ spectrum **a**.

b. Account for the four-line spectrum **b.** when there is gated decoupling of the protons.

c. Explain why adding increasing amounts of hydroxide changes the observed spectra through the sequence of **c.** to **e.** Be sure you account for the number and the relative positions of the peaks in **e**.

d. Explain why spectra **a.** and **f.** are essentially identical.

**Exercise 11-10 a.** Which curve in Figure 11-5 corresponds to the A nuclei and which to the B nuclei of Equations 11-52 and 11-53? How do the peak heights of these curves relate to $M_{zA}$ and $M_{zB}$?

**b.** Make plots of both $\ln[(M_{zA} + M_{zB})/M_{0zA}]$ and $\ln[(M_{zA} - M_{zB})/M_{0zA}]$ as a function of $t_2$ at 0.5 sec intervals over the range from 0 to 5 sec, taking $T_1 = 11.6$ sec, $\tau = 2.13$ sec and $M_{0zA} = 1$ (arbitrary units). What do the slopes of these plots represent in terms of the variables in Equations 11-52 and 11-53?

**c.** Take a ruler to Figure 11-5 and measure a few pairs of peak heights for the signals of the A and B nuclei as a function of $t_2$. Use the values you obtain and Equations 11-52 and 11-53 to estimate $T_1$ and $\tau$ for N,N-dimethylmethanamide under the particular conditions that the spectra were taken.

# Appendix 1
# NMR Reference Books

Freeman, R., *Spin Choreography. Basic Steps in High-Resolution NMR*, Oxford University Press, 1997. An extraordinary book of penetrating insights into the many difficult concepts of NMR by one of the great innovators of modern NMR techniques. Basically, this book is a collection of short, beautifully illustrated essays and it helps to know a lot about NMR and its jargon before wading in. The mathematics are relatively minimal, but product operators are discussed in knowing detail with respect to use of multiple-quantum coherences. Includes many very clear descriptions of complex NMR experiments.

Braun, S., Kalinowski, H.-O. and Berger, S., *100 and More Basic NMR Experiments. A Practical Course*, VCH Publishers, New York, 1996. An excellent book of NMR recipes that allows one to become acquainted with a wide variety of important modern FT NMR procedures in one, two and three dimensions. Literature references are given, along with short theoretical explanations that require knowledge of the product-operator symbolism (the basis for which is not explained, but available in Freeman's *Spin Choreography* see above). Unfortunately (but necessarily), the recipes are rather specific for Bruker spectrometers.

Berger, S., Braun, S. and Kalinowski, H.-O. and Berger, S., *NMR Spectroscopy of the Non-Metallic Elements.* John Wiley & Sons, New York, 1996. An excellent book that complements the authors' helpful volume on $^{13}C$ NMR. This one has three chapters on NMR principles and these are followed by six very fine chapters on $^{15}N$, $^{17}O$, $^{19}F$, $^{31}P$, $^{33}S$ and $^{129}Xe$ spectroscopy. A most useful book.

Günther, H., *NMR Spectroscopy.* John Wiley & Sons, New York, Second Edition, 1998. A superb, very throrough book by a most knowledgeable and experienced NMR spectroscopist with strong emphasis on organic chemistry.

Macomber, Roger S., *A Complete Introduction to Modern NMR Spectroscopy,* John Wiley & Sons, New York, 1998. A very good book for the serious intermediate NMR user that is quite up-to-date for the applications of most interest to organic chemists. Thoughtful exercises and a rather chatty style.

# NMR Reference Books

Canet, D., *Nuclear Magnetic Resonance. Concepts and Methods.* John Wiley & Sons, New York, 1996. Translation by the author of a 1991 book in French, rather mathematical, but directed largely to general principles with an unusually detailed coverage of relaxation.

Derome, A.E., *Modern NMR Techniques for Chemistry*, Pergamon Press, New York, 1987. An excellent book to own if you are going to do a lot of NMR. More oriented to spectrometer operations than analysis of spectra. Now, somewhat dated.

Sanders, J. K. M. and Hunter, B. K., *Modern NMR Spectroscopy, A Guide for Chemists*, Oxford University Press: New York, Second Edition, 1993. Similar to Derome, although not as detailed, but somewhat more up-to-date.

Sanderström, J., *Dynamic NMR Spectroscopy*, Academic Press, New York, 1982. Not a new book, but one that has excellent discussions of rate processes for a variety of types of exchanging systems.

Friebolin, H., B*asic One- and Two-Dimensional NMR Spectroscopy*, VCH Publishers, New York, Second Edition, 1993. A translation and update of a book previously published in German. Excellent discussions and it also contains useful data for spectral analysis in terms of molecular structures.

Nakanishi, K. (Editor), *One-dimensional and Two-dimensional NMR Spectra by Modern Pulse Techniques.* University Science Books, Sausalito, California, 1990. A superb collection of 92 examples of structural analysis by NMR that include almost all of the modern techniques. The examples are themselves chosen from a wide variety of natural products and each is presented along with a brief description of the basic principles involved. There is also a section of NMR principles in general. Highly recommended as a sort of encyclopedia of specific techniques.

Kalinowski, H.-O., Berger, S. and Braun, S., *Carbon-13 NMR Spectroscopy*, J. Wiley & Sons, New York, 1988. Expensive but very useful for $^{13}$C.

Wehrli, F. W. and Wirthlin, T., *Interpretation of Carbon-13 NMR Spectra*, Heyden & Son Ltd., London, 1976; 2nd ed., J. Wiley & Sons, Chichester, 1988. A quite good reference work for structural elucidation.

Croasman, W.R. and Carlson, R.M.K. (Editors), *Two-Dimensional NMR Spectroscopy*, VCH Publishers, 220 East 23rd Street, New York, 1994, 2nd ed. A useful multiple-author reference work that is reasonably up-to-date. The chapters vary in clarity and, unfortunately, the introductory chapter is not very comprehensive with respect to explanation of theoretical matters.

van de Ven, F., *Multidimensional NMR in Liquids*, VCH Publishers, Inc., 1995. Devoted much more to the basics than the Croasman-Carlson book, not a lot of discussion of different pulse sequences, fairly heavy on mathematics but not so much so as the Ernst-Bodenhausen-Wokaun volume.

Shaw, D., *Fourier Transform Spectroscopy*, Heyden & Son Ltd., London, 1978. Rather elementary, but useful.

Breitmaier, E. and Voelter, W., *Carbon-13 NMR Spectroscopy*, VCH Verlagsgesellschaft, FDR (Federal Republic of Germany), 1987. An update of a useful earlier book covering some of the newer techniques, but without addition of very much in the way of new spectroscopic data.

Ernst, R. R., Bodenhausen, G., and Wokaun, A., *Principles of Nuclear Magnetic Resonance in One and Two Dimensions*, Oxford Science Publications, New York, 1987. A classic, if you like rigorous mathematics. Ernst (Nobel Prize in Chemistry, 1991) has spawned a host of new experimental techniques and is now the best known theoretician in the field. Little is offered in the way of qualitative explanation.

Abragam, A., *The Principles of Nuclear Magnetism*, Oxford University Press, 1961. The classic book on NMR from the standpoint of physicists.

Pople, J.A., Schneider, W.G. and Bernstein, H.J., *High-resolution Nuclear Magnetic Resonance*, McGraw-Hill Book Company, Inc., New York, 1959. A very complete, very influential, but early book with emphasis primarily from the standpoint of physical chemists.

Freeman, R., *A Handbook of Nuclear Magnetic Resonance*, Addison Wesley Longman, Essex, England, Second Edition, 1998. A sort of small encyclopedia by an acknowledged leader in developing new observational methods. Many qualitative explanations and fascinating cartoons.

Farrar, T. C., *Introduction to Pulse NMR Spectroscopy*, Farragut Press, Chicago, 1989. A revision of an earlier book, not as complete as Derome, but excellent on many aspects of the theory.

Fukushima, E. and Roeder, S. B. W., *Experimental Pulse NMR, A Nuts and Bolts Approach*,

Addison-Wesley Publishing Company, Reading, Massachusetts, 1981. Many good explanations although little about newer techniques for structural analysis.

Grant, D.M. and Harris, R.M. (Editors-in-Chief), *Encyclopedia of Nuclear Magnetic Resonance*, J. Wiley & Sons, Ltd., Chichester, England, 1996. A massive multi-volume work that covers the waterfront of NMR. Several hundred technical articles written by experts in the field and a volume of personal histories by many of those who helped launch the technique in physics, chemistry, biology and medicine. Especially interesting is an overview of NMR history by E.D. Becker. The work is not an encyclopedia in the usual sense, but is more a collection of in-depth essays on particular topics. For many purposes, Freeman's *Hand book* (described above) will be found to be equally useful.

Neuhaus, D. and Williamson, M., *The Nuclear Overhauser Effect. In Structural and Conformational Analysis*, VCH Publishers, New York, 1989. A high-level book on this very important technique for structural analysis by NMR.

Duddeck, H. and Dietrich, W., *Structure Elucidation by Modern NMR*, Springer-Verlag, New York, 1989. A work book involving a number of problems using advanced and quite modern techniques for structure analysis.

Martin, M. L., Martin, G.J. and Delpuech, J.-J., *Practical NMR Spectroscopy*, Heyden, London, 1980. An excellent book for its stated purpose although now somewhat dated.

Traficante, D.D. (Editor), *Concepts in Magnetic Resonance. An Educational Quarterly*, John Wiley & Sons, Inc. New York, 1989-. A superb, relatively new journal devoted to educational discussions of fairly limited NMR topics in substantial, but not usually overwhelming, detail. Currently most articles are quite high level.

Young, S.W., *Magnetic Resonance Imaging. Basic Principles*, 2nd Edition, Raven Press, New York, 1988. A generally non-mathematical exposition of how NMR imaging works with clinical applications. Primarily for the medical world.

Callaghan, P.T., *Principles of Nuclear Magnetic Resonance Microscopy*, Oxford Science Publications, 1991. A rather rigorous and thorough mathematical approach to NMR imaging in general, with microscopy almost as an afterthought. Not a how-to-do-it book, but a how-to-do-understand the principles book. Substantial discussions of κ-space, if you need to know about it.

Roberts, J.D., *Nuclear Magnetic Resonance. Applications to Organic Chemistry*, McGraw-Hill Book Company, Inc., New York, 1959. The first book on NMR for organic chemists.

Roberts, J.D., *An Introduction to the Analysis of Spin-Spin Splitting in High Resolution Nuclear Magnetic Resonance Spectroscopy*, W.A Benjamin, Inc., New York, 1961. A textbook with problems covering basic theory of spin-spin splitting in NMR, primarily directed to organic chemists.

Lambert, J. B. and Rittner, R., *Recent Advances in Organic NMR Spectroscopy*, Norell Press, New Jersey, 1987. Treats some specialized topics.

In addition to the above books and current journals covering NMR, there are several multivolume works that can provide useful, but often more specialized information. Although many of these volumes are far from current, many contain compilations of very useful information that would be difficult to find elsewhere. Several examples follow:

*Nuclear Magnetic Resonance. A Specialist Periodical Report,* The Chemical Society, Burlington House, London.

*Progress in Nuclear Magnetic Resonance Spectroscopy,* Emsley, J.W., Feeney, J. and L.H. Sutcliffe, Editors, Pergamon Press, New York.

*Advances in Magnetic Resonance,* Waugh, J.S., Editor, Academic Press, New York.

*Topics in Carbon-13 NMR Spectroscopy,* Levy, G.C., Editor, Wiley-Interscience, New York.

*Annual Review of NMR Spectroscopy,* Mooney, E.F., Editor, Academic Press, New York.

# Appendix 2
# A Simple Program for Calculating the Fourier Transform

!This program is adaptation of a program from *The FFT fundamentals and Concepts* by
!R.W. Ramirez, Prentice-Hall, 1985, p. 69.  It is written in True BASIC, available from
!True BASIC, Inc., Hanover, N.H.  The source code can be used with True BASIC
!language systems on IBM PC, Macintosh, Amiga and SUN computers.

```
OPTION BASE 0
OPTION ANGLE RADIANS
DIM X(31), XR(31), XI(31)
MAT X=0
MAT XR=0
MAT XI=0
LET EM=1
!
!############### Generation of the waveform ##################
!
FOR I=0 TO 31
       LET X(I)=COS(2*PI*18.625*I/100)/(EXP(I/100/0.625))
       !The intervals are 1/100 second, the frequency is 18.625 hz
       !The term 1/(EXP(I/100/0.0625) corresponds to a T2 of 0.0625 sec.
NEXT I
!
!################## Plot of the waveform #####################
!
CLEAR
PRINT "32 POINT COS WAVE FORM "
SET WINDOW  -5,37,-2,2
FOR K=0 TO 31
       PLOT K,X(K);
NEXT K
```

```
PLOT
PLOT 0,0;31,0
PLOT 0,-1.2;0,1.2
PLOT
FOR J=0 TO 31 STEP 4
        PLOT J, +0.02;J,-0.02
        PLOT
NEXT J
FOR J=-1.0 TO 1.0 STEP 0.2
        PLOT  -0.15,J;0.15,J
        PLOT
NEXT J
PLOT
PRINT "Press any key to continue"
GET KEY Ky        ! Introduces a pause for inspection of the waveform
!
!################### Transform of datapoints ###################
!
FOR K=0 TO 31
        FOR N=0 TO 31
                LET AG=2*PI*K*N/32
                LET XR(K)=XR(K)+X(N)*COS(AG)/32
                LET XI(K)=XI(K)-X(N)*SIN(AG)/32
        NEXT N
NEXT K
!
!################ Print option for numerical values ################
!
CLEAR
PRINT "Touch P to print the digital FT numerical values, any other key to skip."
GET KEY Ky
IF Ky=80 OR Ky=112 THEN
        OPEN #1: PRINTER
        PRINT #1: "Digital FT, 32 points "
        PRINT #1: "INDEX     REAL PART    IMAG PART"
        LET A$=  " ###     ##.######   ##.######"
```

```
        FOR K=0 TO 31
            PRINT #1, USING A$:K,XR(K),XI(K)
        NEXT K
        CLOSE #1
        CLEAR
        PRINT "Touch any key to continue."
        GET KEY Ky
END IF
!
!############## Plot real and imaginary parts of transform ##############
!
CLEAR
PRINT "32 POINT/SECOND  DFT - REAL PART"
CALL REAL_PLOT
PRINT "Press any key to continue"
GET KEY Ky
CLEAR
PRINT "32 POINT/SECOND DFT - IMAGINARY PART"
CALL IMAG_PLOT
!
!#################### SUB REAL_PLOT #########################
!
SUB REAL_PLOT
        SET WINDOW  -5,37,-1,1
        FOR K=0 TO 31
            PLOT K,XR(K);
        NEXT K
        PLOT
        PLOT 0,0;32,0
        PLOT
        FOR J=0 TO 32 STEP 4
            PLOT J, 0.0;J,-0.03
            PLOT
        NEXT J
        PLOT TEXT, AT 0,-0.08:"0"
        PLOT TEXT, AT 8,-0.08:"25"
```

```
            PLOT TEXT, AT 16,-0.08:"50"
            PLOT TEXT, AT 23.5,-0.08:"-25"
            PLOT TEXT, AT 32,-0.08:"0"
END SUB
!
!####################### SUB IMAG_PLOT ###################
!
SUB IMAG_PLOT
      SET WINDOW  -5,37,-1,1
      FOR K=0 TO 31
            PLOT K,XI(K);
      NEXT K
      PLOT
      PLOT 0,0;32,0
      PLOT
      FOR J=0 TO 32 STEP 4
            PLOT J, 0.0;J,-0.03
            PLOT
      NEXT J
      PLOT TEXT, AT -0.1,-0.08:"0"
      PLOT TEXT, AT 7.8,-0.08:"25"
      PLOT TEXT, AT 15.8,-0.08:"50"
      PLOT TEXT, AT 23.7,-0.08:"-25"
      PLOT TEXT, AT 32,-0.08:"0"
END SUB
!
END
```

# Appendix 3
# A Program for Simple Numerical Integration of the Bloch Equations

NMR_PROGS This is a Macintosh version for doing a simple numerical integration of
! the Bloch equations in the CW or FT mode written in True BASIC. It should be easy
! to translate it into other languages and the source code will work with the True BASIC
! language system on IBM PC's or compatibles and SUN work stations. The program
! is here dimensioned for 1024 points, but this is easily changed. The program and its
! current initial choice of variables was used to produce the curves of Figure 4-10.

```
OPTION BASE 1           !Arrays dimensioned (1,N), not (0,N)
DIM X(1024),Y(1024),Z(1024),U0(1024)
LET Dt=0.001         ! The increment for the numerical integration
LET Omega=5          ! This is the resonance frequency in Hz
LET Freq0=10         ! The starting frequency for the sweep in Hz, here set to give the
                     ! conventional sweep with decreasing frequency to the right
LET T1=1             ! Longitudinal relaxation time
LET T2=1             ! Transverse relaxation time
LET Mz0=5E-6         ! Equilibrium magnetization along the z-axis
LET B1=1.0           ! Actually gamma*B1, a useful value for the conditions
LET Sweep_time=10                ! Time in sec. for the sweep from end to end
LET Sweep_width=10               ! Sweep width in Hz
LET R=Sweep_time/(Dt*1024)       ! R determines how many Dt intervals in each
                                 ! data point
IF R-INT(R)<>0 THEN              ! An integral number of Dts for each data point?
    LET R=INT(R+1)
    LET Dt=Sweep_time/(R*1024)   ! Dt reset to give the proper time intervals/data
                     ! point
END IF
!
! Start of magnetization calculations
DO
```

```
        CLEAR
        PRINT "Press the appropriate key:"
        PRINT
        PRINT "To have the sample at equilibrium with the applied"
        PRINT "      field and B1 at the start of the sweep  =  1"
        PRINT
        PRINT "To have the sample at equilibrium with the applied"
        PRINT "      field, but B1 off until the start of the sweep  =  5"
        PRINT
        PRINT "The sample is placed in the applied field and the"
        PRINT "      the sweep and B1 are started at once    =  0"
        GET KEY Ky
        SELECT CASE Ky
        CASE 48
                CALL No_equil2
                EXIT DO
        CASE 53
                CALL No_equil1
                EXIT DO
        CASE 49
                CALL Yes_equil
                EXIT DO
        CASE ELSE
                SOUND 500, 0.2
                PRINT "Pick an allowed key! "
                PAUSE 3
        END SELECT
LOOP
CLEAR          !Clear screen
PRINT TAB (3,20);"System busy making magnetization calculations"
LET Tyme=0                          ! Starts at time zero
LET Omega_rad=Omega*2*Pi            ! Converts Hz to rads/sec
LET Freq=Freq0*2*Pi                 ! Starting sweep frequency ,rads/sec
LET Dfreq=Sweep_width*Dt*2*Pi/Sweep_time      ! Frequency increment as sweep
                                              ! moves along in time
FOR F=1 TO 1024
```

```
        FOR C=1 TO R
            LET V=V+(-(Omega_rad-Freq)*U+B1*Mz-V/T2)*Dt
            LET U=U+(-U/T2+(Omega_rad-Freq)*V)*Dt
            LET Mz=Mz+((Mz0-Mz)/T1-B1*V)*Dt
            IF Freq0<Omega THEN    !Checks to see whether desired sweep is
                                   !increasing frequency to the right
                LET Freq=Freq+Dfreq
            ELSE
                LET Freq=Freq-Dfreq    !Decreasing frequency to the right
            END IF
!           LET Tyme=Tyme+Dt
            IF Tyme>(Sweep_time+Dt) THEN EXIT FOR
        NEXT C
        LET X(F)=Tyme*Sweep_width/Sweep_time
        LET Y(F)=V
        LET U0(F)=U
        LET Z(F)=Mz
NEXT F
!
! End of numerical integration
!
LET X_max=Tyme*Sweep_width/Sweep_time  ! Plot abcissa maximum
LET X_min=X(1)                         ! Plot abcissa min
CLEAR                                  !Clear CRT screen
SET WINDOW -X_max*.1,1.1*X_max,-Mz0*1.2,Mz0*1.2    ! Window made siightly
                                                   ! larger than plot
FOR J=1 TO 1024
   PLOT X(J),Y(J);        !Plotting V
NEXT J
PLOT
FOR J=1 TO 1024
   PLOT X(J),U0(J);       !Plotting U
NEXT J
PLOT
FOR J=1 TO 1024
   PLOT X(J),Z(J);        !Plotting Mz
```

```
NEXT J
PLOT
! End of plotting routine
!
STOP
!
!################ SUB No_equil2 ####################
!
! Start with no magnetization of the sample at all.
!
SUB No_equil2
      LET U=0
      LET V=0
      LET Mz=0
END SUB
!
!################ SUB No_equil1 ####################
!
! Start with equilibrium Mz, but no U or V magnetization.
!
SUB No_equil1
      LET U=0
      LET V=0
      LET Mz=Mz0
END SUB
!
!################ SUB Yes_equil ####################
!
! Start with sample in equilibrium with both applied field and B1.
! Calculates equilibrium U, V and Mz for the selected "slow-passage" condition.
!
SUB Yes_equil
      LET U=Mz0*B1*T2^2*(Omega-Freq0)*2*Pi
      LET U=U/(1+T2^2*((Omega-Freq0)*2*Pi)^2+B1^2*T1*T2)
      LET V=Mz0*B1*T2/(1+T2^2*((Omega-Freq0)*2*Pi)^2+B1^2*T1*T2)
      LET Mz=Mz0*(1+T2^2*((Omega-Freq0)*2*Pi)^2)
```

```
        LET Mz=Mz/(1+T2^2*((Omega-Freq0)*2*Pi)^2+B1^2*T1*T2)
END SUB
!
!##########################################
!
END
```

# Appendix 4
# A Quantitative Approach to Spin-Spin Splitting

Chapter 9 presents the fundamental quantum-mechanical equations for calculating the energy changes in the transitions that characterize spin-spin splitting, but does not show how they are used in actual calculations. Also, while a qualitative argument is made for expecting that the transition probability for $\alpha\alpha \to 1/\sqrt{2}(\alpha\beta+\beta\alpha)$ for an $A_2$ system is more favorable than $\alpha\alpha \to \alpha\beta$ for an AX system (see Section 9-2), nothing is said about how this is arrived at, although Equations 9-7 to 9-9 provide a means for making the calculations when the shifts and coupling constants are known. The purpose of this Appendix is to provide a brief introduction to the quantum-mechanical treatment of simple spin 1/2 systems in hope of promoting understanding and also to show the kind of steps that are involved in the LAOCOON program when it is used to calculate a spectrum given particular shifts and coupling constants. If you have not made quantum-mechanical calculations before, the treatment of nuclear-spin systems is a wonderfully simple place to start, free of many of the difficulties and approximations encountered in treating the electronic structures of atoms and molecules. Furthermore, the results are experimentally verifiable to a high degree of precision.

As we have earlier, we will use $\alpha$ and $\beta$ to designate the quantum-mechanical spin functions of individual nuclei. These functions have some special properties that we need to define but will not try to justify. First, they are **normalized**. This means that, if we have a spin variable $\phi$, that is a measure of the angle of rotation (in degrees) of the spin about its axis, then Equation A4-1 holds.

$$\int_0^{360°} \alpha^2 d\phi = \int_0^{360°} \beta^2 d\phi = 1 \qquad \text{Eqn. 4A-1}$$

Equation A4-1 rewritten in the simpler Dirac notation becomes Equation A4-2.

$$\langle \alpha|\alpha \rangle = \langle \beta|\beta \rangle = 1 \qquad \text{Eqn. 4A-2}$$

The other property $\alpha$ and $\beta$ have is that they are **orthogonal**, which means that Equation A4-3 holds.

$$\langle\alpha|\beta\rangle = \langle\beta|\alpha\rangle = 0 \qquad \text{Eqn. 4A-3}$$

Normalized and orthogonal wave functions are widely used in other quantum calculations, although in many it is necessary to use complex conjugates of the wave functions in normalization. However, that need not concern us here. The energy of a nuclear spin state $\alpha$ is given by Equation A4-4 in Dirac notation where $H$ is the Hamiltonian energy operator.

$$E = \langle\alpha|\mathcal{H}|\alpha\rangle / \langle\alpha|\alpha\rangle \qquad \text{Eqn. 4A-4}$$

In a magnetic field $B$, there are magnetic quantum members along the Z axis, $I_z = \pm 1/2$. We know from Chapter 1 that $E = -\gamma h B_0/4\pi$, $I_z = +1/2$ ($\alpha$) and $E = \gamma h B_0/4\pi$ when $I_z = -1/2$ ($\beta$), so that $\Delta E = \gamma h B_0/2\pi$. On this basis, we can take H to be $-\gamma h B_0 I_z/2\pi$ and hence that

$$E = \langle\alpha|\gamma h B_0 I_z/2\pi|\alpha\rangle / \langle\alpha|\alpha\rangle \qquad \text{Eqn. 4A-5}$$

$$= -\gamma h B_0/4\pi$$

We can reduce the labor of writing the constants by defining:

$$\mathcal{Y} = \gamma h/2\pi$$

$$E_\alpha = -\mathcal{Y} B_0/2 \qquad \text{Eqn. 4A-6}$$

$$E_\beta = \mathcal{Y} B_0/2 \qquad \text{Eqn. 4A-7}$$

If we now take into account nuclear shielding by the electrons and define $B$ as the applied field we have:

$$E_\alpha = -\mathcal{Y}(1-\sigma_1)B/2 \qquad \text{Eqn. 4A-8}$$

and we can redefine the operator H to be $-\mathcal{Y}(1-\sigma_1)B$. For two independent nuclei (no coupling) in the state $\alpha\beta$ with different shielding constants

$$E = \langle\alpha\beta|\mathcal{H}_1 + \mathcal{H}_2|\alpha\beta\rangle / \langle\alpha\beta|\alpha\beta\rangle \qquad \text{Eqn. 4A-8}$$

where $H_1$ operates only on nucleus (1) and $H_2$ operates only on nucleus (2). By definition, $\langle\alpha\beta|\alpha\beta\rangle = 1$, so we have:

$$E = \langle\alpha\beta|\mathcal{H}_1|\alpha\beta\rangle + \langle\alpha\beta|\mathcal{H}_2|\alpha\beta\rangle \qquad \text{Eqn. 4A-9}$$

$$= -\gamma(1-\sigma_1)B_0/2 + \gamma(1-\sigma_2)B/2$$

$$= (\sigma_1 - \sigma_2)\gamma B_0/2$$

When two nuclei have a spin-spin coupling a new operator has to be introduced.

$$E = \langle\alpha\beta|\mathcal{H}_1 + \mathcal{H}_2 + \mathcal{H}_{12}|\alpha\beta\rangle / \langle\alpha\beta|\alpha\beta\rangle \qquad \text{Eqn. 4A-10}$$

and this can be broken down as before to:

$$E = \langle\alpha\beta|\mathcal{H}_1|\alpha\beta\rangle + \langle\alpha\beta|\mathcal{H}_2|\alpha\beta\rangle + \langle\alpha\beta|\mathcal{H}_{12}|\alpha\beta\rangle \qquad \text{Eqn. 4A-11}$$

$H_{12}$ has the form $J_{12} \cdot I(1) \cdot I(2)$, which is the same as $J_{12}[I_x(1) \cdot I_x(2) + I_y(1) \cdot I_y(2) + I_z(1) \cdot I_z(2)]$. Dirac has shown that for spin 1/2 nuclei $H_{12}$ is equivalent to $(J_{12}/4)(2p_{12} - 1)$ where $p_{12}$ represents a **permutation** operator that interchanges the numbering of nucleus (1) and nucleus (2). The results of using this operator are shown in Equations 12-15.

$$p_{12}|\alpha\beta\rangle = |\beta\alpha\rangle \qquad \text{Eqn. 4A-12}$$

$$p_{12}|\beta\alpha\rangle = |\alpha\beta\rangle \qquad \text{Eqn. 4A-13}$$

$$p_{12}|\beta\beta\rangle = |\beta\beta\rangle \qquad \text{Eqn. 4A-14}$$

$$p_{12}|\alpha\alpha\rangle = |\alpha\alpha\rangle \qquad \text{Eqn. 4A-15}$$

From the examples, we can see that $p_{12}|\alpha\beta\alpha\rangle = |\beta\alpha\alpha\rangle$ where $p_{12}$ has no effect on $\alpha(3)$ because it only interchanges $\alpha(1)\beta(2)$. Using Dirac's procedure, we can calculate:

$$\langle\alpha\beta|\mathcal{H}_{12}|\alpha\beta\rangle = \langle\alpha\beta|(J/4)(2p_{12}-1)|\alpha\beta\rangle$$

$$= (J_{12}/4)(\langle\alpha\beta|2p_{12}|\alpha\beta\rangle - \langle\alpha\beta|\alpha\beta\rangle)$$

$$= (J_{12}/4)(2\langle\alpha\beta|\beta\alpha\rangle - \langle\alpha\beta|\alpha\beta\rangle)$$

$$= (J_{12}/4)(2\cdot 0 - 1) = -J/4$$

Note that this is the same result we have assumed before (Section 6-4). For the state ββ, we have:

$$\langle\beta\beta|\mathcal{H}_{12}|\beta\beta\rangle = (J_{12}/4)(\langle\beta\beta|2P_{12}|\beta\beta\rangle - \langle\beta\beta|\beta\beta\rangle)$$

$$= (J_{12}/4)(2\langle\beta\beta|\beta\beta\rangle - \langle\beta\beta|\beta\beta\rangle) = +J/4$$

and again this is the result assumed earlier.

Another thing that we want to be able to calculate is the relative transition probabilities, because these determine the integrated line intensities. This is not difficult for the AX case, where the states are all symmetric and, for convenience, we designate $\alpha\alpha$ as $s_1$, $\beta\alpha$ as $s_2$, $\alpha\beta$ as $s_3$ and $\beta\beta$ as $s_4$.

Assuming that we want an upward (energy-absorbing) transition, as from $\alpha\alpha \to \alpha\beta$, we write Equation A4-16, where $F^+$ is a raising operator that operates on $\alpha\alpha$.

$$P_{s_1 \to s_2} = (\langle\alpha\beta|F^+|\alpha\alpha\rangle)^2 \qquad \text{Eqn. 4A-16}$$

The nature of the $F^+$ operator is revealed by Equations 17 to 20:

$$F^+|\alpha\alpha\rangle = |\alpha\beta\rangle + |\beta\alpha\rangle \qquad \text{Eqn. 4A-17}$$

$$F^+|\alpha\beta\rangle = |\beta\beta\rangle \qquad \text{Eqn. 4A-18}$$

$$F^+|\beta\alpha\rangle = |\beta\beta\rangle \qquad \text{Eqn. 4A-19}$$

$$F^+|\beta\beta\rangle = 0 \qquad \text{Eqn. 4A-20}$$

With this knowledge, Equation A4-16 becomes:

$$P_{s_1 \to s_2} = [\langle\alpha\beta|\alpha\beta\rangle + \langle\alpha\beta|\beta\alpha\rangle]^2$$

$$= (1+0)^2 = 1$$

It is straightforward to show that the transition probabilities for $\alpha\alpha \to \beta\alpha$, $\beta\alpha \to \beta\beta$ and $\alpha\beta \to \beta\beta$ will also be unity; also that $\alpha\alpha \to \beta\beta$ will have zero probability.

Let us now turn to the AB and $A_2$ cases where we already know matters are more complex. The key here is the mixing of states and mixing can only be important when the states have the same total magnetic quantum number. Thus, $\alpha\alpha$ or $\beta\beta$ will not mix, but $\alpha\beta$ and $\beta\alpha$ can. What we need to determine is the degree of mixing, the energy levels that result and the transition probabilities to, and from, the mixed states.

For the AB case, the equation for the mixed wave function $\psi$ is

$$\psi = c_1\alpha\beta + c_2\beta\alpha \qquad \text{Eqn. 4A-21}$$

where the mixing coefficients $c_1$ and $c_2$ are to be determined. A necessary condition for normalization is that $c_1^2 + c_2^2 = 1$. Proceeding as for Equation A4-10, we can show that the energy of the system is given by Equation A4-22.

$$E = \frac{\langle (c_1\alpha\beta + c_2\beta\alpha)|\mathcal{H}|(c_1\alpha\beta + c_2\beta\alpha)\rangle}{\langle c_1\alpha\beta + c_2\beta\alpha | c_1\alpha\beta + c_2\beta\alpha \rangle} \qquad \text{Eqn. 4A-22}$$

The denominator is $c_1^2\langle\alpha\beta|\alpha\beta\rangle + c_1c_2\langle\alpha\beta|\beta\alpha\rangle + c_1c_2\langle\beta\alpha|\alpha\beta\rangle + c_2^2\langle\beta\alpha|\beta\alpha\rangle = c_1^2 + c_2^2$, so it is simpler to write Equation A4-23, knowing that $\langle\alpha\beta|\mathcal{H}|\beta\alpha\rangle = \langle\beta\alpha|\mathcal{H}|\alpha\beta\rangle$.

$$E = \frac{c_1^2\langle\alpha\beta|\mathcal{H}|\alpha\beta\rangle + 2c_1c_2\langle\alpha\beta|\mathcal{H}|\beta\alpha\rangle + c_2^2\langle\beta\alpha|\mathcal{H}|\beta\alpha\rangle}{c_1^2 + c_2^2} \qquad \text{Eqn. 4A-23}$$

All of these integrals are of types we have encountered earlier for the AX case. What we want now are the values of $c_1$ and $c_2$ that give the lowest mixed energy state and, to determine these, we take partial derivatives and set them equal to zero.

$$\frac{\partial E}{\partial c_1} = 0 \text{ and } \frac{\partial E}{\partial c_2} = 0 \qquad \text{Eqn. 4A-24}$$

The results are simple as is shown by the simultaneous Equations 25 and 26.

$$c_1\langle\alpha\beta|\mathcal{H}|\alpha\beta\rangle - E + c_2\langle\alpha\beta|\mathcal{H}|\beta\alpha\rangle = 0 \qquad \text{Eqn. 4A-25}$$

$$c_1\langle\alpha\beta|\mathcal{H}|\beta\alpha\rangle + c_2\langle\beta\alpha|\mathcal{H}|\beta\alpha\rangle - E = 0 \qquad \text{Eqn. 4A-26}$$

There are different ways to solve these equations, the most straightforward is to find the value of $E$ that makes them consistent and we can do this by solving the determinant of Equation A4-27.

$$\begin{vmatrix} \langle\alpha\beta|\mathcal{H}|\alpha\beta\rangle - E & \langle\alpha\beta|\mathcal{H}|\beta\alpha\rangle \\ \langle\alpha\beta|\mathcal{H}|\beta\alpha\rangle & \langle\beta\alpha|\mathcal{H}|\beta\alpha\rangle - E \end{vmatrix} = 0 \qquad \text{Eqn. 4A-27}$$

Equation A4-27 is the simplest example of a very general procedure that can be used find the energies possible for any number of mixed states with the same total magnetic

# A Quantitative Approach to Spin-Spin Splitting

quantum number. In such determinants there will be diagonal elements of the type $\langle\alpha\beta\alpha|\mathcal{H}|\alpha\beta\alpha\rangle - E$ and off-diagonal elements of the type $\langle\alpha\beta\alpha|\mathcal{H}|\beta\alpha\alpha\rangle$.

Now, we evaluate the specific integrals using the procedures described earlier:

$$\langle\alpha\beta|\mathcal{H}|\alpha\beta\rangle = \langle\alpha\beta|\mathcal{H}_1 + \mathcal{H}_2 + \mathcal{H}_{12}|\alpha\beta\rangle$$

$$= (\sigma_1 - \sigma_2)\mathcal{Y}B_0/2 - J_{12}/4$$

$$\langle\beta\alpha|\mathcal{H}|\beta\alpha\rangle = \langle\beta\alpha|\mathcal{H}_1 + \mathcal{H}_2 + \mathcal{H}_{12}|\beta\alpha\rangle$$

$$= (\sigma_2 - \sigma_1)\mathcal{Y}B_0/2 - J_{12}/4$$

$$\langle\alpha\beta|\mathcal{H}|\beta\alpha\rangle = \langle\alpha\beta|\mathcal{H}_{12}|\beta\alpha\rangle$$

$$= J_{12}/4\langle\alpha\beta|2P_{12} - 1|\beta\alpha\rangle$$

$$= J_{12}/4\langle\alpha\beta|2|\alpha\beta\rangle - \langle\alpha\beta|\beta\alpha\rangle = J_{12}/2$$

The determinant now becomes:

$$\begin{vmatrix} (\sigma_1 - \sigma_2)\mathcal{Y}B_0/2 - J_{12}/4 - E & J_{12}/2 \\ J_{12}/2 & (\sigma_2 - \sigma_1)\mathcal{Y}B_0/2 - J/4 - E \end{vmatrix} = 0 \qquad \text{Eqn. 4A-29}$$

Solution of this determinant gives:

$$E = -J_{12}/4 \pm (1/2)\sqrt{J^2 + \mathcal{Y}^2 B_0^2(\sigma_1 - \sigma_2)^2} \qquad \text{Eqn. 4A-30}$$

Equation A4-30 can be used to derive Equation 9-7 because $\mathcal{Y}B_0(\sigma_1 - \sigma_2)$ is the chemical-shift difference, $\Delta\nu$, between nuclei A and B. Now, with the energies of the states, we can calculate the transition probabilities and, for this, we need to compute $c_1$ and $c_2$. The ratio, $c_1/c_2$, is obtained by substituting the energy values into the cofactors (with proper algebraic signs) of Equation A4-29 and making sure the final values are normalized.

$$\frac{c_1}{c_2} = \frac{\text{cofactor}_1}{\text{cofactor}_2} = \frac{-(\langle\beta\alpha|\mathcal{H}|\beta\alpha\rangle - E)}{\langle\alpha\beta|\mathcal{H}|\beta\alpha\rangle}$$

$$= \frac{-\left[(\sigma_2 - \sigma_1)\mathcal{Y}B_0/2 \pm (1/2)\sqrt{J_{12}^2 + \mathcal{Y}^2 B_0^2(\sigma_1 - \sigma_2)^2}\right]}{J_{12}/2} \qquad \text{Eqn. 4A-31}$$

If the shift difference is very large compared to $J$ then you can see that $c_1/c_2$ will either be close to zero or very large and this means that mixing is very small so that $\alpha\beta$ and $\beta\alpha$ excellent approximations to the wave functions, in effect then giving an AX system.

However, when the shift difference is zero and $J \neq 0$ then the values of $c_1/c_2$ are $\pm 1$. If we normalize, *i.e.*, require that $c_1^2 + c_2^2 = 1$, then $c_1 = 1/\sqrt{2}$ and $c_2 = \pm 1/\sqrt{2}$ so that the mixed functions become the same as assumed earlier (Section 9-2):

$$\psi_1 = (\alpha\beta + \beta\alpha)/\sqrt{2} \qquad \text{Eqn. 4A-32}$$

$$\psi_2 = (\alpha\beta - \beta\alpha)/\sqrt{2} \qquad \text{Eqn. 4A-33}$$

Next, let us take a look at the relative transition probabilities for the changes $\alpha\alpha \to (\alpha\beta + \beta\alpha)/\sqrt{2}$ and $\alpha\alpha \to (\alpha\beta - \beta\alpha)/\sqrt{2}$ that are $s_1 \to s_2$ and $s_1 \to a$, respectively.

$$P_{s_1 \to s_2} = (\langle(\alpha\beta + \beta\alpha)/\sqrt{2}|F^+|\alpha\alpha\rangle)^2$$

$$= (\langle(\alpha\beta + \beta\alpha)/\sqrt{2}|\alpha\beta + \beta\alpha\rangle)^2$$

$$= (1/2)(\langle\alpha\beta|\alpha\beta\rangle + \langle\alpha\beta|\beta\alpha\rangle)^2 = 2$$

and

$$P_{s_1 \to a} = (\langle(\alpha\beta - \beta\alpha)/\sqrt{2}|F^+|\alpha\alpha\rangle)^2$$

$$= (1/2)(1-1) = 0$$

Again, these are the results that I gave you to take on faith earlier (Section 9-2).

To drive home the procedure in a more general case, let me outline what you would do to solve for the energy levels, and so on, of an ABC system where we would have mixing of $\alpha\alpha\beta$, $\alpha\beta\alpha$ and $\beta\alpha\alpha$, as well as (separately) of $\alpha\beta\beta$, $\beta\alpha\beta$ and $\beta\beta\alpha$.

The determinant for the first case would be:

$$\begin{vmatrix} \langle\alpha\alpha\beta|\mathcal{H}|\alpha\alpha\beta\rangle - E & \langle\alpha\alpha\beta|\mathcal{H}|\alpha\beta\alpha\rangle & \langle\alpha\alpha\beta|\mathcal{H}|\beta\alpha\alpha\rangle \\ \langle\alpha\alpha\beta|\mathcal{H}|\alpha\beta\alpha\rangle & \langle\alpha\beta\alpha|\mathcal{H}|\alpha\beta\alpha\rangle - E & \langle\alpha\beta\alpha|\mathcal{H}|\beta\alpha\alpha\rangle \\ \langle\alpha\alpha\beta|\mathcal{H}|\beta\alpha\alpha\rangle & \langle\alpha\beta\alpha|\mathcal{H}|\beta\alpha\alpha\rangle & \langle\beta\alpha\alpha|\mathcal{H}|\beta\alpha\alpha\rangle - E \end{vmatrix} = 0$$

The integrals in the diagonal terms are straightforward to evaluate, thus for $\alpha\alpha\beta$:

$$\langle\alpha\alpha\beta|\mathcal{H}_1 + \mathcal{H}_2 + \mathcal{H}_3 + \mathcal{H}_{12} + \mathcal{H}_{13} + \mathcal{H}_{23}|\alpha\alpha\beta\rangle$$

$$= \tfrac{1}{2}B_0[-1+\sigma_1+\sigma_2+\sigma_3]+\langle\alpha\alpha\beta|\mathcal{H}_{12}|\alpha\alpha\beta\rangle+\langle\alpha\alpha\beta|\mathcal{H}_{13}|\alpha\alpha\beta\rangle+\langle\alpha\alpha\beta|\mathcal{H}_{23}|\alpha\alpha\beta\rangle$$

$$= \tfrac{1}{2}B_0[-1+\sigma_1+\sigma_2+\sigma_3]+(J_{12}-J_{13}-J_{23})/4$$

The off-diagonal terms are evaluated as before using the example here of ααβ, αβα:

$$\langle\alpha\alpha\beta|\mathcal{H}|\alpha\beta\alpha\rangle=\langle\alpha\alpha\beta|\mathcal{H}_{12}+\mathcal{H}_{13}+\mathcal{H}_{23}|\alpha\beta\alpha\rangle$$

$$=\langle\alpha\alpha\beta|\mathcal{H}_{12}|\alpha\beta\alpha\rangle+\langle\alpha\alpha\beta|\mathcal{H}_{13}|\alpha\beta\alpha\rangle+\langle\alpha\alpha\beta|\mathcal{H}_{23}|\alpha\beta\alpha\rangle$$

$$=(J_{12}/4)\langle\alpha\alpha\beta|2P_{12}-1|\alpha\beta\alpha\rangle+(J_{13}/4)\langle\alpha\alpha\beta|2P_{13}-1|\alpha\beta\alpha\rangle+(J_{23}/4)\langle\alpha\alpha\beta|2P_{23}-1|\alpha\beta\alpha\rangle$$

$$=(J_{12}/4)\langle\alpha\alpha\beta|2|\beta\alpha\alpha\rangle+(J_{13}/4)\langle\alpha\alpha\beta|2|\alpha\beta\alpha\rangle+(J_{23}/4)\langle\alpha\alpha\beta|2|\alpha\alpha\beta\rangle$$

$$0+0+2J_{23}/4=J_{23}/2$$

Proceeding in this way, we can calculate each of the elements of the determinant. Because the determinant is third-order, analytical solutions for the various $E$ values are rather messy. However, with numerical values of the various $\sigma_n$ and $J_{ij}$, it is straightforward to extract the corresponding $E$ values from the resulting third-order equation by standard procedures.

To get the coefficients, the ratio of cofactors gives:

$$\frac{c_1}{c_2}=\frac{\begin{vmatrix}\langle\alpha\beta\alpha|\mathcal{H}|\alpha\beta\alpha\rangle-E & \langle\alpha\beta\alpha|\mathcal{H}|\beta\alpha\alpha\rangle \\ \langle\alpha\beta\alpha|\mathcal{H}|\beta\alpha\alpha\rangle & \langle\beta\alpha\alpha|\mathcal{H}|\beta\alpha\alpha\rangle-E\end{vmatrix}}{\begin{vmatrix}\langle\alpha\alpha\beta|\mathcal{H}|\alpha\beta\alpha\rangle & \langle\alpha\beta\alpha|\mathcal{H}|\beta\alpha\alpha\rangle \\ \langle\alpha\alpha\beta|\mathcal{H}|\beta\alpha\alpha\rangle & \langle\beta\alpha\alpha|\mathcal{H}|\beta\alpha\alpha\rangle-E\end{vmatrix}}$$

$$\frac{c_2}{c_3}=\frac{-\begin{vmatrix}\langle\alpha\alpha\beta|\mathcal{H}|\alpha\beta\alpha\rangle & \langle\alpha\beta\alpha|\mathcal{H}|\beta\alpha\alpha\rangle \\ \langle\alpha\alpha\beta|\mathcal{H}|\beta\alpha\alpha\rangle & \langle\beta\alpha\alpha|\mathcal{H}|\beta\alpha\alpha\rangle-E\end{vmatrix}}{\begin{vmatrix}\langle\alpha\alpha\beta|\mathcal{H}|\alpha\beta\alpha\rangle & \langle\alpha\beta\alpha|\mathcal{H}|\alpha\beta\alpha\rangle-E \\ \langle\alpha\alpha\beta|\mathcal{H}|\beta\alpha\alpha\rangle & \langle\alpha\beta\alpha|\mathcal{H}|\beta\alpha\alpha\rangle\end{vmatrix}}$$

and here again it is easiest to substitute numbers and solve the resulting determinants.

None of the above should be taken to mean that there are not complications. Thus, even the simple-looking $A_3$ system will be found to mix at the $I_z$ levels of $\pm 1/2$ to yield for each of these levels two degenerate (equal-energy) antisymmetric states and a single symmetric state (see Exercise 12). Determining the coefficients for the antisymmetric states is not straightforward. You can be thankful that the LAOCOON program does all of this for you automatically when you give it shifts and coupling constants for up to seven nuclei of spin 1/2 and, as described earlier (Section 9-8), also helps you make an iterative fit to a given spectrum for up to seven nuclei of spin 1/2.

**Exercises**

**Exercise 1** Verify that $[\alpha(1)\beta(2) - \beta(1)\alpha(2)]/\sqrt{2}$ is a normalized function.

**Exercise 2** Compute a normalization factor for $[2\alpha\alpha\beta - \alpha\beta\alpha - \beta\alpha\alpha]$.

**Exercise 3** Show that $\alpha\alpha$ and $(\alpha\beta - \beta\alpha)/\sqrt{2}$ are orthogonal functions.

**Exercise 4** Evaluate each of the following integrals for nuclei with the same $\gamma$ and $\sigma$ values.
   a. $\langle \alpha | \mathcal{H}_1 | \beta \rangle$
   b. $\langle \alpha\alpha | \alpha\alpha \rangle$
   c. $\langle \alpha\alpha | \beta\alpha \rangle$
   d. $\langle (\alpha\beta + \beta\alpha)/\sqrt{2} | \mathcal{H}_1 | \alpha\beta \rangle$
   e. $\langle \beta\beta\alpha | \alpha\beta\beta \rangle$
   f. $\frac{1}{2} \langle (\alpha\beta - \beta\alpha) | \mathcal{H}_1 | (\beta\alpha + \alpha\beta) \rangle$
   g. $\langle \alpha\alpha\beta | \mathcal{H}_1 + \mathcal{H}_2 + \mathcal{H}_3 | \alpha\alpha\beta \rangle$
   h. $\frac{1}{4} \langle (\alpha\beta - \beta\alpha)(\alpha\beta + \beta\alpha) | (\alpha\beta - \beta\alpha)(\alpha\beta + \beta\alpha) \rangle$ (four different nuclei)

**Exercise 5** Evaluate each of the following terms:
   a. $F_+ | \alpha\beta\alpha \rangle$
   b. $F_- | (\alpha\beta - \beta\alpha)/\sqrt{2}$ ($F_-$ is a lowering operator)
   c. $F_+ | (\alpha\beta + \beta\alpha)/\sqrt{2}$
   d. $F_+ | (\alpha\beta + \beta\alpha)(\alpha\beta - \beta\alpha)/\sqrt{2}$ (four different nuclei)

**Exercise 6** Evaluate each of the following:
   a. $[\langle \beta\beta | F_+ | \alpha\alpha \rangle]^2$
   b. $[\langle \beta\beta | F_+ | (\alpha\beta + \beta\alpha)/\sqrt{2} \rangle]^2$
   c. $[\langle \beta\beta | F_+ | (\alpha\beta - \beta\alpha)/\sqrt{2} \rangle]^2$

d. $\left[\langle\beta\beta\beta|F_+|\alpha\beta\alpha\rangle\right]^2$

**Exercise 7** Evaluate:
  a. $P_{12}|\beta\beta\rangle$
  b. $P_{13}|\alpha\beta\beta\rangle$
  c. $P_{14}|\alpha\alpha\beta\beta\rangle$
  d. $P_{13}|\alpha\beta(\alpha\beta-\beta\alpha)\rangle$ (four different nuclei)

**Exercise 8** Show that $\langle(\alpha\beta-\beta\alpha)/\sqrt{2}|(J_{12}/4)(2P_{12}-1)|(\alpha\beta-\beta\alpha)/\sqrt{2}\rangle = 3J/4$

**Exercise 9** Evaluate for $i = 1$ to 3 and $j > i$.
  a. $\langle\alpha\alpha\alpha|(J_{ij}/4)(2P_{ij}-1)|\alpha\alpha\alpha\rangle$
  b. $\langle\alpha\beta\alpha|(J_{ij}/4)(2P_{ij}-1)|\alpha\beta\alpha\rangle$
  c. $\langle(2\alpha\beta\alpha-\alpha\beta\alpha-\beta\alpha\alpha)|(J_{ij}/4)(2P_{ij}-1)|(2\alpha\beta\alpha-\alpha\beta\alpha-\beta\alpha\alpha)\rangle$

**Exercise 10** Show that for orthogonal functions $\psi_1$ and $\psi_2$ that $\psi_2 = (\alpha\beta-\beta\alpha)/\sqrt{2}$ will give the same energy as $\psi_2 = (\beta\alpha-\alpha\beta)/\sqrt{2}$ and that $\langle(\alpha\beta-\beta\alpha)/\sqrt{2}|\mathcal{H}|(\alpha\beta+\beta\alpha)/\sqrt{2}\rangle = \langle(\alpha\beta+\beta\alpha)/\sqrt{2}|\mathcal{H}|(\alpha\beta-\beta\alpha)/\sqrt{2}\rangle$.

**Exercise 11** Plot $E$ of Equation A4-23 against the ratio of $c_1/c_2$ with the constraint that $c_1^2 + c_2^2 = 1$ and taking $\langle\alpha\beta|\mathcal{H}|\alpha\beta\rangle = 1$ Hz, $\langle\beta\alpha|\mathcal{H}|\beta\alpha\rangle = -9$ Hz and $J_{12} = 16$ Hz. Determine, at least approximately, the maximum and minimum values of $E$ and the values of $c_1$ and $c_2$ to which they correspond. These maximum and minimum $E$ values correspond to the energies of the mixed states. Now, assume that the shift of nucleus 1 is 90 Hz and that of nucleus 2 is 100 Hz and calculate the energies of the states $\alpha\alpha$ and $\beta\beta$ (don't forget the coupling). From these values and the values of $E$ for the mixed states, calculate the expected line positions. Use the $c_1$ and $c_2$ values obtained from your plot for the mixed states to derive spin functions for those states. Now, modify Equation A4-16 to calculate the relative transition probabilities for $\alpha\alpha \rightarrow$ mixed states and from this data and the line positions sketch the expected appearance of the spectrum.

**Exercise 12** Analysis of the $A_3$ system involves the solution of two third-order determinants when $I_z = \pm 1/2$. A satisfactory set of functions for these levels are $(\alpha\alpha\beta+\alpha\beta\alpha+\beta\alpha\alpha)/\sqrt{3}$, $(\alpha\alpha\beta+\alpha\beta\alpha-2\beta\alpha\alpha)/\sqrt{6}$ and $(\alpha\alpha\beta-\alpha\beta\alpha)/\sqrt{2}$ for the case where $I_z = +1/2$, along with $(\beta\beta\alpha+\beta\alpha\beta+\alpha\beta\beta)/\sqrt{3}$, $(\beta\beta\alpha+\beta\alpha\beta-2\alpha\beta\beta)/\sqrt{6}$ and $(\beta\beta\alpha-\beta\alpha\beta)/\sqrt{2}$ fwhen $I_z = -1/2$. The $J$ couplings between the pairs are by definition equal so a single $J$

value suffices. Use the wave functions given above, along with $\alpha\alpha\alpha$ and $\beta\beta\beta$, to construct the energy levels taking $v_A = 0$ Hz and $J = 10$ Hz. Calculate the transition energies and probabilities for the transitions for which $\Delta I_z = +1$.

**Exercise 13** Set up and solve the two-row determinants needed to calculate the permitted energy levels of a general $A_2B$ spin system. Use your results to calculate the transition energies for $v_A - v_B = 20$ Hz and $J_{AB} = 10$ Hz.

**Exercise 14** Set up the two-row determinant for mixing of $\alpha\beta\beta$ and $\beta\alpha\beta$ for an ABX system and evaluate each of the elements and show that

$$E = v_x/2 - J_{AB}/4 \pm (1/2)\left[v_A - v_B + (J_{AX} - J_{BX})/2\right]^2 + J_{AB}^2\right]^{1/2}.$$

**Exercise 15** Show that the appearance of an ABX system is independent of the sign of $J_{AB}$.

**Exercise 16** Two of the spin functions of an ABX system can be written with normalized coefficients as:

$$\psi_3 = (c_1\alpha\beta + c_2\beta\alpha)\beta$$
$$\psi_5 = (c_3\alpha\beta - c_4\beta\alpha)\alpha$$

Use the following equation to derive the probabilities of a transition between a state described by $\psi_3$ and one described by $\psi_5$ in terms of $c_1$, $c_2$, $c_3$ and $c_4$.

$$P_{3\to 5} = \left[\langle(c_1\alpha\beta + c_2\beta\alpha)\beta|F^+|(c_3\alpha\beta + c_4\beta\alpha)\rangle\right]^2$$

# INDEX

Absorption mode (v) 56-58
    slow passage 60
Absorption-mode spectrum
    Fourier transform and 48-52
    separation from dispersion mode 66
Acetonitrile nitrogen-15 shift 253
Acetylacetone proton shifts 255
Acetylene (see Ethyne)
Acquisition time
    resolution and 83-87
    signal-to-noise and 83-85
    time averaging and 46
Acronyms 148
Adiabatic rapid-passage spectra 74-76
    advantages of 76
    oscillator power and 76
Alcohols (see also Ethanol)
    hydrogen bonding and proton shifts of 254
Aliasing and Fourier transform 51
Aminobenzene nitrogen-15 shielding 248
Analog-to-digital conversion and integrals 105
Aniline nitrogen-15 shielding 248
Antiphase nuclear moments 26
Antisymmetric nuclear states 217-221
Apodization and truncation 89
Azabenzene
    nitrogen-15 shielding in 248
    nitrogen-15 shifts and hydrogen bonding to 253
    NOE of 133
Aziridine inversion rates 272

$B_1$ power (see Oscillator power)
Benzene and proton relaxation by oxygen 122
Bloch equations
    derivation of 54-58
    fast-passage spectra and 62-68
    field sweep and 66
    frequency sweep and 66
    limitations of 58, 143
    magnetization components in 54
    numerical integration errors in 64-65
    optimum power calculation 60
    precession in 54
    program source code for 295-299
    reaction-rate derivation with 263
    relaxation in 54-55
    saturation in 60
    slow passage and 58-62
Bloch, F. 54, 204
    early NMR spectra and 76
Bloch-Siegert effect 22
    chemical shifts and 204
    decoupling and 205
Boltzmann distribution and magnetic field 7-8
Boltzmann's constant 7
Boron-11 nuclear properties 4
Bothner-By, A.A. 235
Brownian motion and relaxation 121

CAMELSPIN 208
Carbon-12 nuclear properties 4
Carbon-13
    adiabatic rapid-passage spectra for 76
    DEPT use with 176
    differential NOE's 134
    digitized CW spectra 79-80
    dipole-dipole relaxation of 124
    energy transfer and cross polarization and 194
    ethane shifts 256
    ethyne shifts 256
    formate spectrum 130
    gated decoupling and 135
    hydrogen-bonding effects on shifts of 254
    INADEQUATE structure determination with 209
    INEPT with 148
    insulin spectra 112
    integrals in spectra of 103-105
    J-modulated 185-189
    labeled-formate energy diagram 124
    magnetization transfer to 144
    maximum entropy and 110
    monosodium glutamate spectrum 81
    natural-abundance spectra 79-81
    nuclear properties 4
    phenyllithium or sodium shielding and 248
    proton broad-band decoupled spectra and 79-80
    proton NOE and 104
    proton spin-spin coupling to 144
    relaxation effects on 97
    shift range of 42, 247
    shift ranges 197
    vinylpyridine spectrum 111
    WALTZ decoupling for 197
Carbon spectra (see Carbon-13)
Carbon-hydrogen bond proton shifts 255
Carr-Purcell echo train 34
Carr-Purcell sequence

decoupling in 150
determination of $T_2$ by 34
diffusion and 34-35
drawbacks of 34-35
pulses in 33
Carr-Purcell-Meiboom-Gill sequence
   advantages of 35
   pulse in 35
   reaction-rate measurement with 277
   spin locking and 191-193
Characteristic time definition 7
Chemical exchange (see also Reaction rates)
   scalar coupling relaxation and 139
Chemical shifts
   acetylacetone protons 255
   acetylenic protons 255
   Bloch-Siegert effect and 204
   carbon-13
      ethane 256
      ethyne 256
   carbon-13 range of 247-256
   color and 248
   diamagnetic effects on 249
   electronegativity effects on 247-248
   electronic transitions and 248
   elements of 1
   ethylenic protons 255
   fluorine-range of 247
   INEPT and 152
   molecular tumbling and 15
   nitrogen-15 (see also Nitrogen-15)
      aniline shielding in 248
      hydrogen bonding and 249
      protonation effects on 248-249, 253
      pyridine and hydrogen bonding and 253
      range of 247-256
      second-order paramagnetic effect and 252-253
   orbitals used in bonding and 247
   orientation in magnetic field and 15-16, 137
   paramagnetic effects on 249-253
   phenyllithium or sodium and shielding 248
   phosphorus range of 247
   proton
      alcohols and hydrogen-bonding effects on 254
      diamagnetic effects on 255
      paramagnetic effects on 255
      range of 247-256
      second-order paramagnetic effect and 253-256
   rate measurement effects of 273
   second-order paramagnetic effect and 249-253
   shielding constant for 15
   shift tensors and 137

spin-spin splitting and 2
standards for 2
units for 2
Chemical-shift anisotropy
   dipole-dipole relaxation competition with 138
   electrical dissymmetry and 138
   field dependence of 138
   longitudinal relaxation by 137
   principles of 137
Chemical-shift tensors 137
Chlorine-35 and quadrupolar relaxation 124
Chromium acetylacetonate and nitrogen-15 relaxation 134
Coalesence point
   equation for 262
   NMR rate measurements and 261
Cofactors in spin-spin splitting 306, 308
Coherence transfer (see Magnetization transfer)
Coherences (see Multiple-quantum coherences)
Combination transitions 125, 224-227
Compass analogy for nuclear magnets 6
Complex conjugate spin functions 302
Continuous wave spectra 58, 66-72
   adiabatic rapid-passage spectra 74-76
   efficiency of 80
   rapid-scan 72-74
   reaction rates and 263
   relaxation wiggles in 67-68
   resolution in 68-72
Correlation spectra 72-74
Correlation time (see Molecular correlation time)
COSY
   ethanol and 183
   homonuclear 205
   magnitude spectra and 207
   phase cycling in 189, 190
   phase relations in 207
   product operators for 206
   pulse sequence for 206
   two-D NMR and 182-184
Coupling constant signs 227, 231
Cross polarization and Hartmann-Hahn condition 193-195
Cross relaxation and two-dimensional spectra 190-191
CW spectra (see Continuous wave spectra)
Cyclohexane inversion 276

DANTE for soft pulses 178
Data registers overflow 87
Dead time
   free induction decay and 89
   integrals and 105
Deceptively simple spectra 233-235

Decoupling
  Bloch-Siegert effect in 205
  broad-band WALTZ procedure for 197
  carbon-13 and NOE 80
  Carr-Purcell sequence and 150
  gated 135
  INEPT and 150, 164-170
  inverse gated 136
  MLEV procedure for 198
  nitrogen-15 and 132
  NOE in 80
  noise 137
    WALTZ alternative to 197
  nuclear Overhauser effect and 126
  off-resonance 136
  on-resonance 137
  signal intensities and 130
  WALTZ procedure for 197-199
Density matrix
  multiple-quantum coherences and 173
  reaction-rate determination and 275
  reaction rates and 280
  relaxation in 173
DEPT
  differentiation of groups with 176
  INEPT comparison 175
  multiple-quantum coherences in 176
  pulse sequence of 175
  spin-spin couplings in 176
Derome, A.E. 206
Detection sensitivity table 4
Deuterium
  longitudinal relaxation and 122
  nuclear properties 4
Deuterium oxide and water suppression 199
Diamagnetic shielding
  elements of 15
  equation for 15
  shielding constant for 15
Diffusion
  Carr-Purcell sequence and 34-35
  magnetic field gradients and 196
  NMR measurement of 36
Difluorocyclohexane inversion rates 276
Digital filters and water suppression 199
Dimethylformamide
  rotation barrier of 260
  rotation-rate determination of 279
Dipole-dipole relaxation 124
  chemical-shift anisotropy competition with 138
  efficiency of 126
  extramolecular deuterium and 122
  nitrogen-15 and 132
  NOE intensity and 131

  paramagnetic ions and 126
  pyridine 133
  pyridinium salts and 138
  urea 133
Dipole-dipole relaxation transitions 125
Dirac notation 302
Dirac permutation operator 303
Dispersion mode ($u$) 56-58
  slow passage 60
Dispersion-mode spectra 74-76
  Fourier transform and 48-52
Double-quantum transitions 125

Echo (see Spin echo)
Electronegativity and chemical shifts and 247-248
Electrons
  magnetic moment of 3
  spin of 3
Emission signals 130
  nitrogen-15 and 132
Ernst equation 95
Ethanol
  COSY of 183
  hydrogen-bonding effects on proton and carbon shifts of 254
  NMR spectra of 15
  proton spectrum 68
Ethanonitrile nitrogen-15 shifts 253
Ethyne shielding 16, 255
Euler method 64-65
Exchanging systems (see Reaction rates)
Exponential multiplication
  Lorentz-to-Gaussian transformation 99-102
  resolution and 99-102
  sensitivity and 99-102
EXSY pulse sequence for exchange 280

Fast-passage spectra and numerical integration 62-68
FID (see Free induction decay)
Field Homogeneity (see also Resolution)
Field sweep spectra and Bloch equations 66
Fluorine-19 nuclear properties 4
Folding over
  Fourier transform and 89-90
  quadrature detection and 90-91
Forbidden transitions 125
Formate ion 213
  dipole-dipole relaxation in 124
Forsen, A. 277, 280
Fourier integral 47
Fourier transform
  aliasing in 51
  alternatives to 105-112

linear prediction for 110-112
maximum entropy procedure for 110
noise reduction and 106-110
procedures for 106-110
analog-to-digital conversion in 105
carbon-13 spectra and 79-81
computer program for 48
dead time and 89
digital 46-48, 82
efficiency of 80
examples 48-52
exponential multiplication and 99-102
FIDs and 45
folding over in 89-90
frequency limits in 51
imaginary and real curves from 48
imaginary part of 46
inverse and rapid-scan spectra 72-74
magnitude spectra and 94
matched filter for 101-102
mathematical filter and 25
mathematics of 45-48
NMR spectroscopy and 79
non-uniqueness of 45
number of points in 85
Nyquist frequency in 51
parameter choices in 82-105
phase correction in 92-94
program source code for 291-294
pulse angles in 94-99
pulse width and 82
real part of 46
repetition rate and 94, 99
sampling rate and 51
time averaging and 46
time-domain data and 25
transverse relaxation effects on 98-99
truncation and 88-89
zero filling in 85
Fourier, J. 79
Free-induction decay (FID)
apodization and 88-89
Carr-Purcell-Meiboom-Gill sequence and 37
components of 106-110
dead time and 89
definition 24
exponential multiplication of 99-102
Fourier transform and 45
Fourier transform of 25
INEPT and 151
line width and 25
mathematical simulation of 106-110
maximum entropy simulation of 110
quadrature and 91
quadrature detection and 90

resolution and 83-87
truncation of 88-89
zero filling and 85
Freeman, R. 206, 208, 209
Frequency domain and NMR spectra 24
Frequency sweep spectra and Bloch equations 66
FT (see Fourier transform)
*FT-NMR Problems* 1

Gated decoupling
principles of 136
sequence for 135
Grignard reagent inversion processes 276
Gyromagnetic ratio (see Magnetogyric ratio)
Gyroscope analogy to nuclear precession 14

Hamiltonian energy operator 216, 302
Hartmann-Hahn condition
cross polarization by 193-195
TOCSY and HOHAHA use of 208
two-dimensional NMR and 193-195
Heisenberg uncertainty principle 9
resolution and 68
HETCOR uses 208
Hoffman, R.A. 277, 280
HOHAHA uses 208
Homospoil 195-196
HQMC uses 209
Hydrogen bonding
nitrogen-15 shifts and 249
proton shifts of alcohols and 254
shift effects of 255
Hydrogen nuclear properties 4
Hydrogen *ortho* and *para* states 219

Imine inversion rates 272
INADEQUATE uses 209
INEPT 148-171
chemical shifts in 152
coupling constant and τ 159
decoupled refocused 170
decoupling and 164-170
decoupling in 150
definition of 148
intensity ratios in 175
magnetogyric ratio and 171
methine groups and 148-166
methylene groups and 160
phase cycling in 161-163, 175
phase problems in 153
pulse 180° in 150, 152
pulse sequence 175
quaternary carbons and 170
refocused 164

# Index

selection of τ 159
selective magnetism transfer and 148
signal dependence on τ 155
signal enhancement and magnetogyric ratio 171
signal strengths in 170
structural relations 159
summary 170-171
$T_1$ relaxation and 150
waiting period (τ') in 170
Insulin carbon spectrum 112
Integrals
    baseline smoothing and 104
    dead time and 105
    differential NOE's and 134
    independence of FID decay rate 51
    influences on 102
    number of points per Hz and 104
    paramagnetic ions and 134
    phasing and 104
    pulse frequency and 104
    pulse time and 104
    quadrature detection and 104
    repetition rate and 103
    signal-to-noise and 105
Integrals and analog-to-digital conversion 105
Inverse Fourier transform
    mathematics of 72-74
    rapid-scan spectra and 72-74
Inverse-gated decoupling
    integrals and 104
    NOE and 136
    proton-carbon NOE and 104
    sequence for 136
Inversion recovery measurement of $T_1$ 37-38

Kupce, E. 208

Laboratory reference frame 23
LAOCOON
    analysis of spectra with 235-243
    program 301, 309
Lattice definition 6, 120
Leakage and adiabatic rapid-passage spectra 75-76
Leakage current 67
Left-hand rule for precession 20
Line broadening and quadrupolar relaxation 124
Line width 60
    free induction decay rate and 25
    lifetime of states and 9-10
    magnetic field homogeneity and 25
    signal decay rates and 45
    transverse relaxation and 60, 68
    uncertainty principle and 9

Linear prediction for FID simulation 110-112
Lineshapes
    AA'XX' and $A_2X_2$ systems 276
    AB system exchange equations 275
    $ABX_n$ systems 275
    activation parameters from 271
    errors in measuring rates with 273
    intramolecular processes and 271-275
    multisite processes and 269-271
    parameters for analysis of 273
    reaction rates and 260-276
    spin-spin splitting effects on 275-276
Lithium-7 nuclear properties 4
Longitudinal relaxation 124
    chemical-shift anisotropy and 137
    chromium acetylacetonate and 134
    concept of 118
    definition of 37
    equation for components of 122
    Ernst equation and 95
    integrals and 103-105
    intramolecular mechanisms for 122-139
    inversion recovery rate equation for 38
    measurement of 37
    molecular motions and 121
    N-methylpyrrolidine nitrogen-15 spectrum 134
    nitrogen-15 and 134
    nitrogen-15 and paramagnetic ions 134
    nitrogen-15 and viscosity in 135
    operation of 26-27
    oxygen-induced 122
    paramagnetic ions and 120-122
    pulse sequence of 38
    quadrupolar mechanism for 123-124
    radiation damping and 120
    relaxation agents for 94-95
    repetition time and 94-99
    signal averaging and 95-99
    transverse relaxation equal to 118
Lorentzian line shapes 99-102

Magnetic field
    homogeneity of 25
    precession frequency and 14
    relaxation when inhomogeneous 27
    resonance frequency and 2
    selective population transfer and 145
    units for 2
Magnetic field gradients
    diffusion and 196
    diffusion measurement with 36, 195
    eddy currents and 196
    homospoil and 195-196
    multiple-quantum coherences and 201-204

resolution improvement and 196
reverse 196, 204
two-dimensional spectra and 195
Magnetic states
   distribution in magnetic field 5-8
   equilibration of 7
   multiple-quantum coherences and 172
Magnetization transfer 144
   carbon to proton 147
   cross polarization and 193
   energy diagram for 144
   magnetic field and 145
   magnetogyric ratio and 131, 145, 147
   NOE advantage over 146
   selective 143-147
   signal enhancement compared to NOE 146
   signal enhancement in 145
   spin-spin coupling and 147, 171
   symmetric states and 148
   time averaging with 146
   vector model and 147
Magnetogyric ratio
   INEPT and 171
   magnetization transfer and 147
   NOE and 131
   selective population transfer and 145
Magnitude spectra 94
   COSY and 206
Matched filter 101-102
      Maximum entropy and free-induction decay simulation 110
Methanoate ion
   dipole-dipole relaxation in 124
Methanoate ion NMR spectrum 213
Methine groups
   INEPT for 148-166
   J-modulated spectra of 184-187
Methyl groups
   J-modulated spectra of 188
   spectral editing 169
   spin-rotation relaxation of 139
Methylene groups
   J-modulated spectra of 187-188
   spectral editing 168
MLEV decoupling sequence 198
Moffitt,W. 249
Molecular correlation time
   dipole-dipole relaxation and 126
   nitrogen-15 relaxation and 134
   quadrupole relaxation and 123
Molecular motions and Fourier components 121
Molecular tumbling
   nitrogen-15 relaxation and 134
   NOE and 135

spin-spin interactions and 128
Molecular weight by NMR 105
Monosodium glutamate
   carbon-13 spectrum 81
   carbon 13 relaxation effects on 97
Multiple-quantum transitions
   dipole-dipole relaxation by 126
Multiple-quantum coherences 172
   density matrix and 173
   DEPT and 176
   field gradient effects on 201-204
   magnetic field inhomogeneities and 174
   multiple-quantum filters and 174
   relaxation and 172, 173
   relay of 208
   single quantum coherences from 174
   spectral editing and 174
   two-D NMR and 183
Multiple-quantum filters 174
Multiple-quantum states (see Multiple-quantum coherences)
Multiple-quantum transitions and the density matrix 173
Multisite exchange equations 269-271

Nakanishi, K. 207
Natural abundance table 4
Nist, B.I. 236
Nitrogen-14 nuclear properties 4
Nitrogen-15
   chemical shifts (see also Chemical shifts)
   chemical-shift anisotropy and 138-139
   color in relation to 249
   differential NOE's 134
   electron-pair shielding and 249
   longitudinal relaxation of 134
   NOE of 132
   nuclear properties 4
   protonation effects on shifts of 249, 253
   pyridine NOE of 133
   pyridine shielding in 248
   relaxation and paramagnetic ions 134
   relaxation and viscosity 135
   second-order paramagnetic effect and 249-253
   shift range of 42
   sign of nuclear moment 132
   urea NOE 133
NMR
   comparision with other forms of spectroscopy 8-10
   nuclear excitation 20-21
   oscillator coil for 19
   receiver coil for 16
   signal detection system 16
NMR acronyms 148

# Index

NMR magnitude spectra 94
NMR reference books 286-290
NMR signal frequency stepdown 24
NMR spectra
    adiabatic rapid-passage spectra 74-76
    ethanol 68
    field sweep 66
    insulin 112
    linear prediction and 110-112
    maximum entropy and 110
    phase correction in 92-94
    porcine insulin 112
    rapid-scan technique 72-74
    relaxation wiggles in 67-68
    resolution in 68-72
    signal-to-noise improvement of 106-112
    Torrey oscillations in 62
NMR transition intensities 130
NOE (see Nuclear Overhauser effect)
NOESY
    principles of 208
    uses of 208
Noise (see Signal-to-noise)
Noise decoupling 137
    WALTZ alternative to 197
Normalization of nuclear wave functions 217
Normalized spin functions 301
Nuclear magnetism
    nuclear spin and 3
    vector components of 17
Nuclear magnetization excitation 19
Nuclear moments
    NOE and 131
    table of 4
Nuclear Overhauser effect 134
    carbon-13 spectra and 80
    chemical-shift anisotropy and 138
    equation for carbon-13 131
    equation for nitrogen-15 133
    integrals and 104
    intensity equations for 131
    inverse-gated decoupling and 104, 136
    magnetization transfer advantage over 146
    magnetogyric ratio and 131
    maximum intensity of 131
    molecular correlation times and 134
    nitrogen-15 and 132
    NOESY and 208
    nuclear moments and 131
    origin of 126-136
    paramagnetic ions and 134
    proton-carbon and 104
    ROSEY (CAMELSPIN) and 208
    signal enhancement vs. magnetization transfer 146
    two-dimensional spectra and 190-191
    viscosity and 133-135, 138-139
Nuclear properties table 4
Nuclear quadrupole and scalar coupling relaxation 139
Nuclear spin ($I$) 2
    concept of 123
    magnetic states and 3
    magnetic states table 5
    nuclear angular momentum and 3
    nuclear magnetism and 3
    quadrupolar relaxation and 123-124
    spin-spin splitting and 212
    table 4
    values of 3
Nuclear spin states
    energy diagram for 124
    notations for 124
    populations of 126
        temperature effect on 126
Numerical integration
    Bloch equations 62-68
    errors in 64-65
    field sweep spectra 66
    initial conditions for 62
Nyquist frequency 82
    Fourier transform and 82
    sampling rate and 51
    spurious peaks and 82

Off-resonance decoupling 136
    noise procedure for 137
    structural analysis with 137
On-resonance decoupling 137
*Ortho*-hydrogen 219
Orthogonal spin functions 301
Oscillator
    Bloch-Siegert effect from 22
    nuclear excitation by 20-21
Oscillator coil 19
Oscillator power ($B1$)
    adiabatic rapid-passage spectra and 76
    Bloch equations and 55-58
    frequency stability of 69
    off-resonance decoupling and 137
    optimum value 60
    vector resolution of 19
Oxygen-induced relaxation 122
Oxygen-16 nuclear properties 4
Oxygen-17 nuclear properties 4
Oxygen-18 nuclear properties 4

*Para*-hydrogen 219
Paramagnetic ions
    differential NOE's and 134

dipole-dipole relaxation and 126
longitudinal relaxation and 120-122
nitrogen-15 relaxation and 134
2,4-Pentanedione proton shifts 255
Phase
    angles for 20
    coherence 18
    COSY and 207
    INEPT and 153
    random 18
    resonance signal and 17-19
Phase coherence
    spin echos and 33
    spin exchange and 119
    transverse relaxation and 118
Phase correction
    CW spectra 66
    equations for 92-94
    Fourier transform and 92-94
    FT NMR spectra 67
    integrals and 104
    introduction to 51
    leakage current for 67
    magnitude spectra and 94
Phase cycling
    COSY and 189-190
    INEPT and 161-163, 175
    two-dimensional spectra and 189-190
Phasing problems and repetition rate 98-99
Phosphorus-31 nuclear properties 4
Planck's constant 7
Polymer molecular weight by NMR 105
Polystyrene molecular weight by NMR 105
Population inversion 130
Porcine insulin carbon NMR apectrum 112
Precession 14
    left-hand rule for 20
Precession frequency
    Bloch equations and 54
    independence of nuclear orientation 31
    longitudinal relaxation and 121
    magnetic field and 14
Precession frequency table 4
Predictor-corrector method 63-65
Product operators
    COSY and 206
    vector model and 174
Proton nuclear properties 4
Proton shifts
    C-H bonds of 255
    range of 247
    second-order paramagnetic effect and 253
Proton spectra with splitting removed 209
Pulse (see also Soft pulses)
    180°
        changes of nuclear vectors in 31
        definition of 31
    90° 22
        carbon-13 and 42
        nitrogen-15 and 42
        time and frequency effects on 40-42
        variation with precession frequency 30
    adiabatic rapid-passage and 76
    DANTE sequence for 178
    Ernst equation and 95
    INEPT 180° 150, 152
    INEPT 90° 150
    less than 90° 95
    multiple-quantum coherences and 90° 174
    selective 177
    selective 180° 145
    shapes for hard pulses 82-83
    soft 180° 177
Pulse angle
    choices for 94-99
    integrals and 104
    precession frequency and 41
    pulse time and 40
Pulse frequency
    folding over and 89-90
    integrals and 104
    quadrature detection and 90-91
Pulse sequence
    Carr-Purcell spin echo 33
    Carr-Purcell-Meiboom-Gill 35
    COSY 206
    DANTE 178
    DEPT and 175
    EXSY 280
    gated decoupling and 135
    HQMC 209
    INEPT 148, 175
    inverse gated decoupling 136
    inversion recovery 38
    MLEV 198
    multiple-quantum coherence excitation with 201-204
    reaction-rate determination with 279-281
    reaction rates by EXSY 280
    refocused INEPT 164
    spin locking 191-193
    two-dimensional NMR 184
    WALTZ 197-199
    water suppression and 200-201
Pulse shape 90° pulses and 82-83
Pulse time
    90° 22
    integrals and 104
Pulse width 90° and 82
Pyridine

Index

chemical-shift anisotropy and 138
nitrogen-15 shielding in 248
nitrogen-15 shifts and hydrogen bonding to 253
NOE of 133
Pyridinium trifluoroacetate chemical-shift anisotropy 138

Quadrature detection
   folding over and 90-91
   Fourier transform and 90-91
   integrals and 104
   operation of 90-91
Quadrupolar relaxation 124
   mechanism of 123-124
   molecular correlation time and 123
   structural use of 124
Quadrupole moment origin of 123
Quadrupole moment table 4
Quantum-mechanical interpretation of NMR 3-10
Quaternary carbons INEPT and 170

Radiation damping 120
Raising operator and spin-spin splitting 304
Ramirez, R.W. 291
Rapid-passage spectra 74-76
Rapid-scan spectra 72-74
   advantages of 74
   instrument for 74
Reaction kinetics (see Reaction rates)
Reaction rates
   AB system lineshape equations for 275
   activation parameters 271
   chemical shift and lineshapes for 273
   coalesence equation for 262
   cyclohexane inversion and 276
   density matrix and 280
   equation derivation for 263-269
   EXSY for determination of 280
   Grignard inversion 276
   imine inversion 272
   integral measurement of 260
   lineshape analyses to measure 271-275
   lineshape errors in measuring 273
   lineshape measurements for 260-276
   lineshape parameters for 273
   multisite exchanges 269-271
   NMR measurements of 260-281
   pulse procedure for measurement of 279-281
   saturation-transfer measurement of 277-279
   spin-echo sequence for measurement of 277
   spin-spin splittings and lineshapes for 275-276
   thiourea bond rotations 271
Receiver bandwidth 16

Receiver coil 16
Reference books on NMR 286-290
Reference frame
   laboratory 23
   rotating 22
Refocused INEPT (see INEPT)
Relaxation 124
   benzene and oxygen 122
   Bloch formulation of 54-55
   Brownian motion and 121
   chemical-shift anisotropy and 137
   cross relaxation as 190-191
   density matrix and 173
   differential and water suppression 200
   dipolar 190-191
   dipole-dipole efficiency 126
   electric-field gradient and 123
   elements of 6
   energy transfer in 121
   extranuclear magnetic nuclei and 122
   field dependence of 138
   INEPT $T_1$ 150
   inhomogeneous magnetic field and 27
   intensity in NOE and 131
   intermolecular longitudinal mechanisms of 120
   intramolecular longitudinal 120
   laboratory frame 23
   longitudinal
      equation for components of 122
      intermolecular mechanisms for 120-122
      intramolecular mechanisms for 122-139
      paramagnetic ions and 120-122
      spin-lock field and 192-193
   multiple-quantum coherences and 172
   nitrogen-15 and 132, 134
   oxygen-induced 122
   paramagnetic ions and nitrogen 134
   quadrupolar mechanism of 123-124
   radiation damping and 120
   rate constants for 6
   rates of 6
   ROSEY (CAMELSPIN) and 209
   rotating frame 23
   scalar coupling and 139
   spin exchange and 119
   spin lattice 120
   spin rotation and 139
   spin-lock field and 191-193
   $T_1$ constant for 7
   time constants of 6
   transverse by homospoil and 195-196
   viscosity and 118-119
Relaxation agents 94-95
Relaxation wiggles

CW spectra and 67-68
FT spectra and 67-68
resolution and 71
Repetition time
   choices for 94-99
   integrals and 103
   longitudinal relaxation and 94, 99
Resolution
   acquisition period and 83-87
   exponential multiplication and 99-102
   gradient pulses to improve 196
   Heisenberg uncertainty principle and 68
   integrals and 104
   magnetic field homogeneity and 68
   measurement of 71-72
   number of data points and 83, 85-87
   pulling together of lines by 69
   relaxation wiggles and 71
   sampling period and 83
   sensitivity tradeoff 99
   transverse relaxation and 83-87
Resolution in NMR spectra 68-72
Resonance frequency (see also Precession frequency)
   field strength and 2
   nuclear constant ($\gamma$) and 2
   units for 2
Resonance frequency table 4
ROSEY uses 208
Rotating reference frame 22
Rotation barriers
   dimethylformamide and 260
Runge-Kutta method 63-65
Russell-Chapman computer program 1

Sample spinning
   inhomogeneity removal by 28
   side bands from 28
Sampling rate
   Fourier transform and 51
   Nyquist frequency and 51, 82
Saturation transfer and reaction rates 277-279
Scalar coupling relaxation 139
   chemical exchange and 139
Second-order paramagnetic effect 249-253
   proton shifts and 253-256
Selective magnetization transfer (see Magnetization transfer)
Selective polarization transfer (see Magnetization transfer)
Selective population transfer (see Magnetization transfer)
Sensitivity resolution tradeoff 99-102
Sensitivity of detection table 4
Shielding
   constants for 15
   diamagnetic 15
   paramagnetic 15
Shielding in spin-spin splitting equations 302
Siegert, A. 204
Signal averaging
   longitudinal relaxation and 95-99
   pulse angle and 95-99
   transverse relaxation and 95-99
Signal integrals (see Integrals)
Signal intensities elements of 1
Signal-to-noise
   exponential multiplication and 100-102
   integrals and 105
   matched filter and 101-102
   proton decoupling and 103, 105
   time averaging and 46, 96
Single-quantum transitions 125
Singlet states of magnetic nuclei 220
Slow-passage spectra 58-62
   comparison with FT spectra 62
   reaction rates and 263
Soft pulses
   DANTE sequence for 178
   excitation with 176
   pulse 180° with 177
   reaction-rate determination with 279
   shapes of 176-177
Solvent suppression procedures for 199
Spectral editing
   methyl groups 169
   methylene groups 168
   multiple-quantum coherences and 174
Spin (see Nuclear spin)
Spin echo
   decay envelope of Carr-Purcell sequence 34
   field inhomogeneities and 33
   formation of 32
Spin exchange and transverse relaxation 119
Spin functions (see Spin-spin splitting)
Spin locking
   Carr-Purcell-Meiboom-Gill sequence and 191-193
   pulse sequence for 191-193
   relaxation in 191-193
   two-dimensional spectra and 191-193
Spin temperature and Hartmann-Hahn condition 195
Spin-echo sequence and reaction rate measurement 277
Spin-lattice relaxation (see Relaxation)
Spin-rotation relaxation
   principles of 139
   temperature effects on 139
Spin-spin splitting

# Index

$A_2B$ systems 223-227
    signs of couplings and 227
$A_2X$ systems 227-231
$A_3$ systems in 309
AA'X systems 232-233
AA'XX' systems 232-233
AB and AX systems coupling constant signs and 227
AB and AX systems of 213-221
AB determinant for 305
AB line position and intensity equations 221
AB system equations for 302-307
AB system integral evauations in 306
ABC system analysis of 235-243
ABC system determinant in 307
ABC system integral evaluation for 307
ABX systems 233-235
    signs of couplings in 233-235
allowed transitions in 218
analysis of 235-243, 301, 309
antisymmetric states 217-221
carbon-13 by carbon-13 209
chemical shift and 213-221, 302
cofactor evaluation for 308
cofactor evaluation for AB in 306
combination transitions in 224-227
complex-conjugate spin functions and 302
coupling constant in equations for 303
deceptively simple spectra 233-235
decoupling and 126
DEPT and 176
dipole-dipole interactions and 128
Dirac notation for spin functions in 301
Dirac permutation operator 303-304
electron-mediated 128
elements of 2
equivalent nuclei 214-221
equivalent-appearing nuclei and 232-233
forbidden transitions in 218
Hamiltonian energy operator in 216, 302
INEPT and 159
LAOCOON for analysis of 235-243
magnetic field strength and 212
magnetization transfer and 144, 147, 171
methanoate and 214
mixing of spin states in 217-221, 304-307
molecular tumbling and 128
more lines than expected in 223-227, 232-233, 235
negative couplings in 228
    rationalization of 231
normalization condition 217
normalized functions for 301
nuclear spin and 2, 212
nuclear spin state classifications 221-222
nuclear wave functions and 216
off-resonance decoupling and 136
operator results in 302
orthogonal spin functions for 301
patterns of 212
permutation operator for 303
positive and negative couplings and 227-231
principles of 212-243
proton spectra removal of 209
raising operator in 304
reaction-rate measurement and 263
signs of 128-130
simple energy picture of 127
symmetric states 217-221
transition probabilities in 304-307
transition probablities in 218
units for 2
virtual coupling 233-235
Spinning side bands
    identification of 28
    sample spinning and 28
Spurious peaks and Nyquist frequency 82
Sulfur-32 nuclear properties 4
Sulfur-33 nuclear properties 4
Symmetric nuclear states 217-221
Symmetric states and magnetization transfer 148

$T_1$ relaxation (see Longitudinal relaxation)
$T_2$ relaxation (see Transverse relaxation)
Temperature and Boltzmann distribution 7-8
Tetramethylsilane (TMS) as shift standard 16
Thiourea and bond rotation 271
Time averaging
    acquisition time and 46
    carbon-13 spectra and 80
    efficiency of 80
    Fourier transform and 46
    magnetization transfer and 146
    optimal values of 96
    phase problems in 98-99
    repetition rate and 46
Time domain
    FID and 24
    Fourier transform and 25
TMS 16
TOCSY uses 208
Torrey oscillations 62
Transition probabilities 218
    spin-spin splitting and 301, 304-307
Transitions
    single quantum 5
    energy of 8
    energy units for 8
Transverse relaxation
    Carr-Purcell sequence for 34-35

Carr-Purcell-Meiboom-Gill sequence for 35
definition of 37
determination of rate of 29
diffusion effects on 34, 37
equation for 27
field inhomogeneity and 118
homospoil and 195, 196
inhomogeneous magnetic field and 27
intrinsic to sample 28
line-width estimation of 60
longitudinal relaxation equal to 118
phasing and 98-99
rate constant $T_2'$ 27
rate equations for Carr-Purcell sequence 34
resolution and 83-87
signal averaging and 95-99
spin echos and 29, 33-35
spin exchange and 119
viscosity and 118-119
Trifluoroethanol hydrogen bonding and nitrogen-15 shifts 249
Triplet states of magnetic nuclei 220
True BASIC
    Bloch equations code for 295-299
    Fourier transform code for 291-294
Truncation
    dead time and 89
    resolution loss in 88-89
    signal-to-noise ratio and 88-89
Tumbling (see Molecular tumbling)
Two-D NMR (see Two-dimensional NMR)
Two-dimensional NMR 182-210
    CAMELSPIN 208
    COSY as 182-184
    cross polarization and 193-195
    cross relaxation in 190-191
    ethanol and 183
    frequencies and 182
    Hartmann-Hahn effect in 193, 208
    HETCOR 208
    HMQC 209
    homonuclear COSY 207
    INADEQUATE 209
    J-modulated 184-189
    methine groups in J-modulated spectra of 184
    methyl groups in J-modulated spectra of 188
    methylene groups in J-modulated spectra of 187-188
    multiple-quantum coherences in 183
    NOE and 190-191
    NOESY 208
    on-resonance decoupling with 137
    phase cycling in 189-190
    proton spin-spin coupling removal from 209
    pulse sequence for 184
    ROSEY 208
    spin locking and 191-193
    TOCSY 208
    WALTZ decoupling and 197-199
    water suppression and 199

Uncertainty principle 9
    resolution and 68
Urea nitrogen-15 NOE 133

Vector model
    limitations of 143, 172
    magnetization transfer and 147
    precession, excitation and relaxation in 14-29
Vinylpyridine carbon-13 spectrum 111
Virtual coupling 213, 233-235
Viscosity
    longitudinal relaxation and 121
    NOE and 133, 135
    paramagnetic ion relaxation and 121
    transverse relaxation and 118-119

WALTZ decoupling 197-199
Water suppression
    digital filters for 199
    longitudinal relaxation in 200
    procedures for 199
    pulse sequence for 200-201
Waugh equation graph 96
Waugh, J.S. 96
Wiberg, K.B. 236
Wiggles (see Relaxation wiggles)

Zero filling and spectral smoothing 85